Advances and Applications in Unmanned Aerial Vehicles

Advances and Applications in Unmanned Aerial Vehicles

Guest Editors

Octavio Garcia-Salazar
Anand Sanchez-Orta
Aldo Jonathan Muñoz-Vazquez

Basel • Beijing • Wuhan • Barcelona • Belgrade • Novi Sad • Cluj • Manchester

Guest Editors

Octavio Garcia-Salazar
Aerospace Engineering
Research and
Innovation Center
Autonomous University
of Nuevo Leon
Apodaca
Mexico

Anand Sanchez-Orta
Robotics and Advanced
Manufacturing Division
CINVESTAV
Saltillo
Mexico

Aldo Jonathan Muñoz-Vazquez
Department of
Multidisciplinary Engineering
Texas A&M University
McAllen
United States

Editorial Office
MDPI AG
Grosspeteranlage 5
4052 Basel, Switzerland

This is a reprint of the Special Issue, published open access by the journal *Machines* (ISSN 2075-1702), freely accessible at: www.mdpi.com/journal/machines/special_issues/ZW0ZK0ZIEN.

For citation purposes, cite each article independently as indicated on the article page online and using the guide below:

Lastname, A.A.; Lastname, B.B. Article Title. *Journal Name* **Year**, *Volume Number*, Page Range.

ISBN 978-3-7258-3126-5 (Hbk)
ISBN 978-3-7258-3125-8 (PDF)
https://doi.org/10.3390/books978-3-7258-3125-8

Contents

About the Editors

Octavio Garcia-Salazar

Octavio Garcia-Salazar obtained a B.Sc. in electronic engineering and an M.Sc. in electrical engineering with specialization in robotics at the Technological Institute of La Laguna, Torreon Coahuila, Mexico, in 2000 and 2003, respectively. He obtained a Ph.D. in control systems at the University of Technology of Compiègne, France, in 2009. From January 2010 to December 2011, he held a post as a CNRS postdoctoral researcher in autonomous systems at the laboratory LAFMIA UMI 3175 CNRS-CINVESTAV Mexico. From January 2012 to December 2012, he was a visiting researcher at CINVESTAV Monterrey. Since January 2013, he has been a Research Professor at the Aerospace Engineering Research and Innovation Center in the Faculty of Mechanical and Electrical Engineering of the Autonomous University of Nuevo Leon, Apodaca, Nuevo Leon, Mexico. His research interests are guidance, navigation, and control of UAVs, UAS, flight dynamics, avionics, and robotics.

Anand Sanchez-Orta

Anand Sanchez-Orta received his M.Sc. degree in Automatic Control from the Autonomous University of Nuevo León, Mexico, and his Ph.D. in Information and Systems Technologies from the University of Technology of Compiègne (UTC), France, in 2001 and 2007, respectively. After two years as a research assistant at the UTC, he held a CNRS research engineer position at the same university. He joined the Robotics and Advanced Manufacturing Division of the Research Center for Advanced Studies (CINVESTAV) in 2009, where he is currently a Research Professor. His research interests include control theory, estimation, and visual serving with applications in robotics.

Aldo Jonathan Muñoz-Vazquez

Aldo Jonathan Muñoz-Vazquez received his master's and doctoral degrees in Robotics and Advanced Manufacturing from the Center for Research and Advanced Studies of the National Polytechnic Institute of Mexico (CINVESTAV-IPN), at the Saltillo Campus in Saltillo, Coahuila, Mexico, in 2013 and 2017, respectively. Dr. Muñoz-Vazquez completed his postdoctoral stay at the Autonomous University of Tamaulipas, at the Reynosa-Rodhe Campus, in Reynosa, Tamaulipas, Mexico, in 2017; during this stay, he developed several different control strategies for robust power tracking in wind energy conversion systems. In the spring of 2018, he served as a Full-Time Professor at the Polytechnic University of Victoria, in Ciudad Victoria, Tamaulipas, Mexico. From July 2018 to September 2019, he was part of the "CONACYT-Lectureships" program of the Mexican National Council of Science and Technology, commissioned to the Autonomous University of Chihuahua, in Chihuahua, Chihuahua, Mexico, participating as a member of the research project: "Smart Environments". Since the fall of 2019, he has been at the Higher Education Center at McAllen of Texas A&M University, now as an Instructional Associate Professor, teaching engineering courses in different areas of engineering. He is an Associate and Academic Editor of *Fractal and Fractional*, *Transactions of the Institute of Measurement and Control*, *Measurement and Control*, and *Frontiers in Control Engineering: Nonlinear Systems* and has also served as an Associate and Guest Editor for other reputed international journals. He has reviewed more than four hundred scientific manuscripts for several different journals and conferences. His research and teaching interests include applied mathematics, control theory, robotics, and artificial intelligence applications.

Preface

In the last decade, unmanned aerial vehicles (UAVs), commonly known as drones, have been used for different purposes, including applications such as search and rescue, highway patrol, and infrastructure inspections for power lines, bridges, and factories. These applications require UAVs that can operate under specific conditions or complex environments. As a result, new configurations are being developed, including improved sensors, longer flight times, and enhanced autonomy.

This Special Issue provides the recent advances and applications in unmanned aerial vehicles, considering both theories and experiments for multirotor, fixed-wing, and non-conventional (convertible) vehicles. After a stringent peer review process, twelve papers were finally included in this Special Issue, which covers the following aspects: modeling, guidance, navigation, control, simulation, and applications. The order of the accepted papers is below:

1. Improved PVTOL Test Bench for the Study of Over-Actuated Tilt-Rotor Propulsion Systems.

2. Intelligent PIV Fuzzy Navigation and Attitude Controller for an Octorotor Mini-UAV.

3. The Coupled Wing Morphing of Ornithopters Improves Attitude Control and Agile Flight.

4. Q-Learning with the Variable Box Method: A Case Study to Land a Solid Rocket.

5. Design and Determination of Aerodynamic Coefficients of a Tail-Sitter Aircraft by Means of CFD Numerical Simulation.

6. Performance Evaluation of an H-VTOL Aircraft with Distributed Electric Propulsion and Ducted-Fans Using MIL Simulation.

7. Design and Performance of a Novel Tapered Wing Tiltrotor UAV for Hover and Cruise Missions.

8. Variations in Finite-Time Multi-Surface Sliding Mode Control for Multirotor Unmanned Aerial Vehicle Payload Delivery with Pendulum Swinging Effects.

9. An Inverse Kinematics Approach for the Analysis and Active Control of a Four-UPR Motion-Compensated Platform for UAV–ASV Cooperation.

10. The Viability of a Grid of Autonomous Ground-Tethered UAV Platforms in Agricultural Pest Bird Control.

11. Use of Unmanned Aerial Vehicles for Building a House Risk Index of Mosquito-Borne Viral Diseases.

12. A Conceptual Framework for Economic Analysis of Different Law Enforcement Drones.

As Guest Editors of this Special Issue, we would like to extend our gratitude to all the authors of the published articles and those who expressed interest in contributing to this Special Issue. We especially appreciate the reviewers who thoroughly examined each submitted article. Their scientific expertise provided valuable insights and constructive feedback that helped enhance the authors' research.

Octavio Garcia-Salazar, Anand Sanchez-Orta, and Aldo Jonathan Muñoz-Vazquez

Guest Editors

Article

Improved PVTOL Test Bench for the Study of Over-Actuated Tilt-Rotor Propulsion Systems

Luis Amezquita-Brooks [1,†] , Eber Maciel-Martínez [1,†] and Diana Hernandez-Alcantara [2,*]

1   Facultad de Ingenieria Mecanica y Electrica, Universidad Autonoma de Nuevo Leon,
    San Nicolás de los Garza 66455, NL, Mexico; luis.amezquitabrk@uanl.edu.mx (L.A.-B.);
    eber.macielmr@uanl.edu.mx (E.M.-M.)
2   School of Engineering and Technologies, Universidad de Monterrey,
    San Pedro Garza García 66238, NL, Mexico
*   Correspondence: diana.hernandeza@udem.edu
†   These authors contributed equally to this work.

Abstract: In recent years, applications exploiting the advantages of tilt-rotors and other vectored thrust propulsion systems have become widespread, particularly in many novel Vertical Takeoff and Landing (VTOL) configurations. These propulsion systems can provide additional control authority, enabling more complex flight modes, but the resulting control systems can be challenging to design due to the mismatch between the vehicle degrees of freedom and physical input variables. These propulsion systems present both advantages and difficulties because they can exert the same overall forces and moments in many different propulsive configurations. This leads to the traditional non-uniqueness problem when using the inverse dynamics control allocation approach, which is the basis of many popular VTOL control algorithms. In this article, a modified Planar VTOL (PVTOL) test bench configuration, which considers an arbitrary number of co-linear tilting rotors, is introduced as a benchmark for the study of the control allocation problem. The resulting propulsion system is then modeled and linearized in a closed and compact form. This allows a simple and systematic derivation of many of the currently used control allocation approaches. According to the proposed PVTOL configuration, a two-rotor test bench is implemented experimentally and a decoupling control allocation strategy based on Singular Value Decomposition (SVD) analysis is developed. The proposed approach is compared with a traditional input mixer algorithm based on physical intuition. The results show that the SVD-based solution achieves better cross-coupling reduction and preserves the main properties of the physically derived approach. Finally, it is shown that the proposed PVTOL configuration is effective for studying the control allocation problem experimentally in a controlled environment and could serve as a benchmark for comparing different approaches.

Keywords: PVTOL; test bench; VTOL; control allocation; cross-coupling

Citation: Amezquita-Brooks, L.;
Maciel-Martínez, E.;
Hernandez-Alcantara, D. Improved
PVTOL Test Bench for the Study of
Over-Actuated Tilt-Rotor Propulsion
Systems. *Machines* **2024**, *12*, 46.
https://doi.org/10.3390/
machines12010046

Academic Editor: Tao Li

Received: 21 November 2023
Revised: 18 December 2023
Accepted: 18 December 2023
Published: 10 January 2024

1. Introduction

Recently, *Vertical Takeoff and Landing (VTOL)* aircraft configurations have attracted a high degree of interest from the research community and the aeronautics industry. The novelty of these vehicles introduces a set of challenging issues because many of the well-established design methodologies, used for conventional aircraft configurations, do not apply directly. This has also attracted the attention of the authorities, which have proposed several novel specifications to address many of these challenges (see, for example, [1,2]).

The majority of the reports found in the literature deal with a specific VTOL configuration. For instance, in [3,4], the authors study a commonly used arrangement, in which the tilt-rotors are used to generate either lift force in hover or thrust force in horizontal flight (i.e., as the one used in the V-22 Osprey). In this case, the authors propose a nonlinear observer and control approach to stabilize the vehicle with good results. Many other aspects of this configuration have been researched, for instance, in [5], the dynamic properties of the vehicle are studied by calculating the linear stability and control derivatives of the vehicle. This study shows that it is possible to evaluate the well-known flight modes of a traditional airplane such as phugoid, short period, Dutch roll, and spiral subsidence. Using a similar approach. In [6], it is shown that a linear control approach can be used to stabilize the full nonlinear dynamics of the vehicle. In this case, the authors noticed a serious degree of cross-coupling when the vehicle operated in helicopter mode. A similar configuration, which considers that the aircraft remains in hover with a vertical fuselage orientation, is studied in [7]. In this report, the authors used a scheduled linear control approach to stabilize the vehicle. These reports show that, although this VTOL configuration results in a highly coupled and nonlinear system, the use of linear control approaches is preferred in some instances due to its simplicity and the possibility of evaluating the closed loop dynamics in terms of classical flight dynamics modes.

Although the configuration mentioned in the previous paragraph is common, it is far from being the only one used in VTOL applications. In particular, tilt-rotors can be found in several configurations, ranging from all the rotors being tilt-rotors to a single rotor used as a tilt-rotor. For instance, in [8], a fixed-wing aircraft with one fixed and two tilting rotors is presented. In this case, a set of nested control loops with linear controllers is used to stabilize the vehicle. Several reviews dealing with VTOL configurations can be found in the literature [9–11]. From these reports, the following insights can be obtained:

- There is a high degree of heterogeneity in the propulsion system configurations, including number, distribution, and operating conditions.
- There is an increasing use of distributed tilting rotors (or similar thrust vectoring elements).
- The multiplicity of propulsors and control surfaces results in a high number of control inputs, introducing an over-actuation problem, which is normally denoted as the *control allocation problem* in the literature (although under-actuation problems have also been treated as allocation problems by some authors).
- The control allocation problem is commonly dealt with by separating the vehicle into two parts: (1) a rigid body, which is driven by virtual control inputs (i.e., force and moment vectors), and (2) the propulsion subsystem, which is driven in such a way that the desired force and moment vectors are obtained. In practice, it is difficult to pair specific physical inputs to a particular force or moment component without introducing cross-coupling in the rigid body control loop.
- Although many control approaches have been proposed, linear *Proportional Integral Derivative (PID)* controllers are still used in many practical setups due to their simplicity and low computational overhead.

The control allocation problem is one of the main issues found in novel vehicle configurations with many tilting propulsors. The simplest approach for the allocation problem consists of pairing specific physical inputs with the desired outputs in a fully decentralized control manner. In other cases, physical insight allows proposing a simple allocation strategy [8]. Although these approaches have been used for traditional configurations (i.e., classical pairings, such as elevator surface and pitch angle), many novel VTOL configurations cannot be operated using this scheme due to excessive cross-coupling.

In cases where the cross-coupling introduced by a simple control allocation algorithm is excessive, a preferred approach consists of establishing virtual (decoupled) inputs and attempting to solve the inverse actuator dynamics to determine the necessary physical inputs [9]. This approach is easy and effective if the propulsion subsystem inverse dynamics exist and are unique. A well-known example of this is the typical quad-rotor control allocation algorithm, where a vector of four virtual inputs is related to the angular velocities of the four propellers through a fixed and invertible 4x4 matrix [12]. It is also interesting to note that this approach has proven to be highly successful even though the propulsion model used to derive it is arguably over-simplified, as it neglects several important aerodynamic factors [13]. This proves that it is possible to obtain effective allocation solutions using simplified propulsion models.

In more recent years, VTOL configurations have become more complex so that inverse propulsion dynamics either do not exist or are not unique. In particular, over-actuated systems are more common due to the increased number of actuators. In these cases, a simple solution consists of using physical insights to add additional constraints to reduce the dimension of the control input vector. For example, in [14], the control allocation for a spacecraft using two kinds of actuators was solved by a proportional distribution of the desired control torque considering the characteristics of each actuator. Physical insight has also been used to refine *Cascade-Generalized Inversion* (*CGI*) approaches, such as in [15], where a test bench for a vehicle with multiple tilting rotors was controlled by grouping several physical actuators in a hierarchical manner. There are also several mathematical tools which can aid in solving the allocation problem, such as the Moore–Penrose pseudo-inverse matrix and the *Karush–Kuhn–Tucker* (*KKT*) optimality conditions [16]. Many of these approaches rely on a linear approximation approach, whose range can be expanded by the use of scheduling techniques [16]. Non-linear approaches have been also proposed; however, the resulting solutions can be too taxing for the computational systems on board small vehicles. Therefore, fast linear or scheduling approaches are still preferred in many applications [16,17].

More recently, in [17], a review article dealing with the taxonomy of vehicles with multiple rotors showed that there are still many challenges in the study of these propulsion systems. Among these, the heterogeneity of the propulsion configurations, the aerodynamic interaction of the various actuators (also noted in [13]), and the actuator limitations (including actuator saturation and response times) are the most relevant. One of the tools observed in [17] which has been used to overcome some of these problems is the use of co-linear or co-planar propulsor arrangements.

The previous paragraphs illustrate how the popularization of VTOL vehicles with a wide range of configurations, particularly those that contain an arbitrary number of tilting rotors, has motivated the study of the control allocation problem, from both theoretical and practical points of view. In this context, it is appropriate to recall that during the emergence of the modern quad-copter drone, in the early 2000s and later, several novel technical and theoretical developments were made. One of the tools that proved to be very valuable for researchers was the so called *Planar Vertical Takeoff and Landing* (*PVTOL*) configuration, which was widely adopted by researchers in a simplified two *degrees of freedom* (*DoF*) test bench version. This simplified model allowed both researchers and prospective control engineers to study novel control strategies and to perform preliminary experimental tests in a safe environment before performing actual flight tests. The sheer number of contributions to control, sensor, and modelling theory that still rely on the PVTOL configuration is a testament to its usefulness. From neural network backstepping control [18] to the study of robust feedback linearization [19] and noise-rejecting active disturbance controllers [20], the PVTOL model is still widely used by researchers. In many ways, the PVTOL has become a de facto benchmark platform.

While the PVTOL is a good approximation of simple vehicles such as the quad-copter, this simplification fails to capture the main difficulties of many current VTOL vehicles. In particular, many of these vehicles include some form of tilt-rotor or other similar thrust vectoring propulsion system. There have been some attempts to propose a more general PVTOL configuration, which also include the complications of tilting rotors. For example, in [21], a tilting rotor PVTOL was introduced with good results; however, in this report, only simulations are presented and the allocation problem is reduced by neglecting the reactive torque of the propellers.

Finally, a summary of the main elements found in the literature review is as follows:

- There is a high level of heterogeneity in the propulsion configuration of recent VTOL vehicles with a movement towards a higher number of actuators.
- The resulting control allocation problems are highly dependent on the propulsion configuration. Therefore, the study of this issue can yield vehicle-specific results.

- The use of simplified test benches (such as the PVTOL) can aid in the study of prospective control strategies while preserving many of the dynamic and experimental complexities, allowing a more rapid development of novel/better solutions.
- The typical PVTOL test bench configuration is unable to represent many of the difficulties found in recent VTOL vehicles.
- Although non-linear control and allocation approaches can deliver improved performance for wider operating ranges, in many *Unmanned Aerial Vehicle* (*UAV*) applications, linear controllers and allocation approaches are still being used and researched due to their simplicity and efficiency.

Considering the previous context, in this article, the control allocation problem is studied both theoretically and experimentally. In particular, the following elements are presented:

- A novel PVTOL test bench configuration containing an arbitrary number of co-linear tilting rotors is proposed as a progression of the traditional PVTOL test bench configuration. This configuration can reproduce several of the interesting issues found in novel VTOL vehicles, mainly the control allocation and cross-coupling problems. In addition, this configuration can be easily extended to include an arbitrary number of co-planar tilting propulsors, so that a wider range of vehicles can be mimicked.
- A general method for obtaining a closed form of the linearization of the propulsion model for the modified PVTOL configuration is presented. This can be useful because, as mentioned before, many control allocation strategies depend on it.
- A simple test bench, based on the modified PVTOL configuration, is implemented experimentally. This simple test bench allows testing control allocation and cross-coupling problems.
- A simple decoupling control allocation scheme for the test bench, based on a linear approximation derived through *Singular Value Decomposition* (*SVD*), is presented. This approach allows defining an optimal solution for the allocation problem of the modified PVTOL configuration. The resulting optimization can solve the over-actuation problem by introducing a wide range of considerations. In this case, low-error and practical physical considerations are used.
- The proposed SVD control allocation approach is compared with a more traditional input mixer algorithm, derived from physical insight, both through simulations and using the experimental test bench. The results show that the proposed control allocation scheme allows decreasing the cross-coupling with a simple static decoupling matrix.

The authors also want to emphasize that the test bench presented here (as well as the proposed allocation algorithm) should be considered an initial step towards a generalized study of the control allocation problem through the use of a prospective standardized test bench configuration based on the modified PVTOL configuration. The authors believe that the proposed test bench could be used by the scientific community for this purpose; however, those seeking additional complexity may consider adding more co-linear propulsors (or even co-planar propulsors in an arbitrary matrix configuration). This would result in an even more complex over-actuation problem. The authors hope that the use of such a family of test benches could potentially become a widely used benchmark as an aid in the study and development of novel propulsion systems for VTOL vehicles.

2. Materials and Methods

2.1. A PVTOL with an Arbitrary Number of Co-Linear Tilting Rotors

This section introduces a PVTOL test bench with an arbitrary number of co-linear tilting rotors and derives the linearization of the resulting propulsion system model in a compact and closed form. This propulsion configuration was chosen because it can replicate many of the control allocation problems found in a wide range of vehicles and can be easily extended to a more complex arbitrary co-planar propulsion system by simple aggregation. In comparison with the typical PVTOL experimental test bench configuration, which is restricted to a single DoF (pitch angle), the proposed configuration adds yaw angle as an additional DoF. This enables the introduction of a multivariable control problem (i.e., two outputs instead of one), which facilitates a better study of the control allocation problem.

It must be noted that adding further DoFs to the test bench (for example, roll angle) could increase its research potential at the cost of increased mechanical complexity. In this regard, it is important to recognize that a successful test bench benchmark must balance the trade-off between its mechanical/manufacturing complexity and its research value potential. To support this claim, consider the classical PVTOL test bench, which is normally restricted to pitch angle movement. Although this restriction is clearly not completely representative of vehicles such as the quad-copter, as mentioned in the introduction, this has not precluded the classical PVTOL test bench from being used for a wide range of studies related to quad-rotor applications. In this case, the proposed configuration remains simple enough to be cost- and mechanically accessible, but complex enough for the control allocation and multivariable cross-coupling problems to appear. Therefore, we believe that the proposed configuration (in particular, the two-rotor version presented in an upcoming section) represents a similar sweet spot for these problems, as the classical PVTOL test bench is for the typical quad-rotor.

Figure 1 shows the propulsion system with an arbitrary number (n) of tilting rotors, where the distance of each rotor from the center of gravity is denoted by l_i and the tilt angle of each rotor is denoted by δ_i, with $i = 1, 2, 3, ...n$. The figure also depicts the angular positions and rates of the PVTOL in a typical inertial *North-East-Down* (*NED*) reference frame, where θ and ψ denote the pitch and yaw angles, respectively, and q and r denote the pitch and yaw rates, respectively.

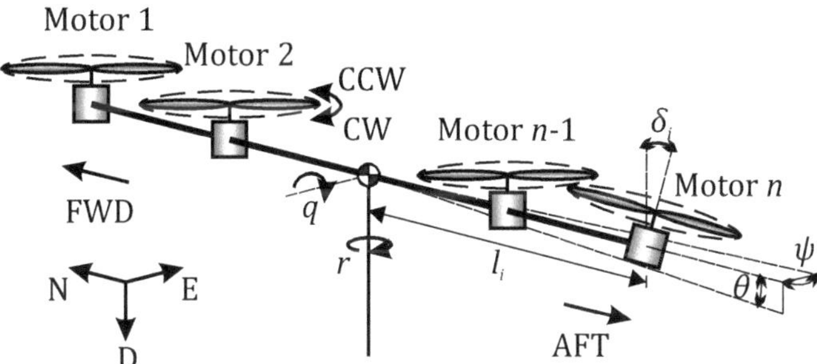

Figure 1. Modified PVTOL configuration with n tilting propulsors.

Each motor/propeller induces a thrust force T_i and a moment R_i aligned with its tilt angle, as shown in Figure 2. Because R_i is produced by rotational drag, its direction is opposite to the propeller rotation. According to simplified propeller modelling, typically used for multi-rotor vehicles, force T_i and moment R_i can be modelled as [13]:

$$\begin{aligned} T_i &= \tfrac{1}{2}\rho S C_T \omega_i{}^2 \\ R_i &= \tfrac{1}{2}\rho S C_Q \omega_i{}^2 \end{aligned} \tag{1}$$

where S is the disc area of the propeller, C_T is the thrust coefficient, C_Q is the reactive moment coefficient, ρ is the air density, and ω_i is the angular speed of propeller i. Accordingly, R_i can be rewritten as $R_i = k_i T_i$ with $k_i = C_Q/C_T$, which is convenient to reduce the number of independent input variables.

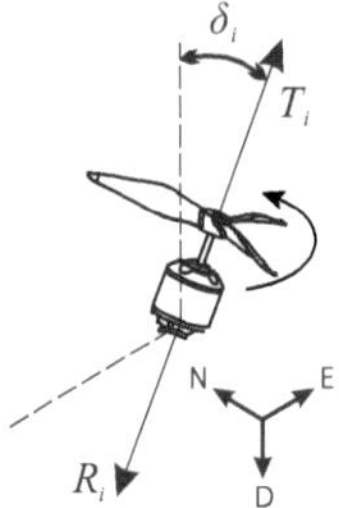

Figure 2. Thrust force (T_i) and reactive moment (R_i) produced by a single propeller.

This article studies the control allocation problem for the pitching and yawing moments, denoted as M and N, respectively, since these are the two natural degrees of freedom for a test bench with the modified PVTOL configuration of Figure 1. Following this configuration, the resulting pitching and yawing moments are:

$$M = \sum_{i=1}^{n} T_i l_i \cos \delta_i (-1)^{\alpha_i} - \sum_{i=1}^{n} T_i k_i \sin \delta_i (-1)^{\beta_i}$$
$$N = \sum_{i=1}^{n} T_i k_i \cos \delta_i (-1)^{\beta_i} + \sum_{i=1}^{n} T_i l_i \sin \delta_i (-1)^{\alpha_i} \tag{2}$$

with:

$$\alpha_i = \begin{cases} 0, & \text{if position FWD} \\ 1, & \text{if position aft} \end{cases} \tag{3}$$

$$\beta_i = \begin{cases} 0, & \text{if rotation CCW} \\ 1, & \text{if rotation CW} \end{cases} \tag{4}$$

where CW, CCW, and FWD denote *clockwise*, *counterclockwise*, and *forward* respectively.

For instance, considering a simple two-rotor arrangement, with a FWD propeller ($i = 1$) rotating CCW and an aft propeller ($i = 2$) rotating CW, the resulting moment equations yield:

$$M = T_1 l_1 \cos \delta_1 - T_2 l_2 \cos \delta_2 - T_1 k_1 \sin \delta_1 + T_2 k_2 \sin \delta_2$$
$$N = T_1 k_1 \cos \delta_1 - T_2 k_2 \cos \delta_2 + T_1 l_1 \sin \delta_1 - T_2 l_2 \sin \delta_2 \tag{5}$$

Although the non-linear propulsion model given by Equations (2) and (5), is useful for many applications, such as simulation, evaluation, and non-linear control design, many control allocation methods are based on a linear representation of the propulsion subsystem. As follows, the linear approximation of the two-propeller case will be derived and then extended for the general case.

The linear approximation of (5) is given by the Jacobian matrix P of the moment vector $[M \, N]^T$, that is:

$$\begin{bmatrix} \Delta M \\ \Delta N \end{bmatrix} = P \begin{bmatrix} \Delta T_1 \\ \Delta \delta_1 \\ \Delta T_2 \\ \Delta \delta_2 \end{bmatrix} \tag{6}$$

with

$$P = \begin{bmatrix} \frac{\partial M}{\partial T_1} & \frac{\partial M}{\partial \delta_1} & \frac{\partial M}{\partial T_2} & \frac{\partial M}{\partial \delta_2} \\ \frac{\partial N}{\partial T_1} & \frac{\partial N}{\partial \delta_1} & \frac{\partial N}{\partial T_2} & \frac{\partial N}{\partial \delta_2} \end{bmatrix}_0 \tag{7}$$

where $[\Delta M \, \Delta N]^T$ and $[\Delta T_1 \, \Delta \delta_1 \, \Delta T_2 \, \Delta \delta_2]^T$ are the resulting linear output and input vectors, and the sub-index 0 denotes evaluation in the equilibrium point.

Matrix P is also called the *propulsion matrix* in the literature and its properties (e.g., range) are a determining factor when calculating the physical control inputs (i.e., $[T_1 \, \delta_1 \, T_2 \, \delta_2]^T$)

necessary to reach a desired combination of input moments (i.e., $[M\ N]^T$). In this case, since the control inputs vector $\in \mathbb{R}^4$, while the moment vector $\in \mathbb{R}^2$, this PVTOL configuration yields an over-actuation problem, which will be studied in more detail later in the article.

From (6), it follows that:

$$P = \left[\begin{array}{cccc} l_1\cos\delta_1 - k_1\sin\delta_1 & -T_1 l_1\sin\delta_1 - k_1 T_1\cos\delta_1 & -l_2\cos\delta_2 + k_2\sin\delta_2 & T_2 l_2\sin\delta_2 + k_2 T_2\cos\delta_2 \\ l_1\sin\delta_1 + k_1\cos\delta_1 & T_1 l_1\cos\delta_1 - k_1 T_1\sin\delta_1 & -l_2\sin\delta_2 - k_2\cos\delta_2 & -T_2 l_2\cos\delta_2 + k_2 T_2\sin\delta_2 \end{array}\right]_0 \tag{8}$$

Matrix P can be further decomposed as:

$$P = AP_1 = \left[\begin{array}{cccc} l_1 & -k_1 & -l_2 & k_2 \\ k_1 & l_1 & -k_2 & -l_2 \end{array}\right] \left[\begin{array}{cccc} \cos\delta_1 & -T_1\sin\delta_1 & 0 & 0 \\ \sin\delta_1 & T_1\cos\delta_1 & 0 & 0 \\ 0 & 0 & \cos\delta_2 & -T_2\sin\delta_2 \\ 0 & 0 & \sin\delta_2 & T_2\cos\delta_2 \end{array}\right]_0 \tag{9}$$

Furthermore , matrix P_1 can also be decomposed as:

$$P_1 = BC = \left[\begin{array}{cccc} \cos\delta_1 & -\sin\delta_1 & 0 & 0 \\ \sin\delta_1 & \cos\delta_1 & 0 & 0 \\ 0 & 0 & \cos\delta_2 & -\sin\delta_2 \\ 0 & 0 & \sin\delta_2 & \cos\delta_2 \end{array}\right]_0 \left[\begin{array}{cccc} 1 & 0 & 0 & 0 \\ 0 & T_1 & 0 & 0 \\ 0 & 0 & 1 & 0 \\ 0 & 0 & 0 & T_2 \end{array}\right]_0 \tag{10}$$

Therefore, matrix P can be rewritten as:

$$P = ABC = \left[\begin{array}{cccc} l_1 & -k_1 & -l_2 & k_2 \\ k_1 & l_1 & -k_2 & -l_2 \end{array}\right] \left[\begin{array}{cccc} \cos\delta_1 & -\sin\delta_1 & 0 & 0 \\ \sin\delta_1 & \cos\delta_1 & 0 & 0 \\ 0 & 0 & \cos\delta_2 & -\sin\delta_2 \\ 0 & 0 & \sin\delta_2 & \cos\delta_2 \end{array}\right]_0 \left[\begin{array}{cccc} 1 & 0 & 0 & 0 \\ 0 & T_1 & 0 & 0 \\ 0 & 0 & 1 & 0 \\ 0 & 0 & 0 & T_2 \end{array}\right]_0 \tag{11}$$

Following the structure of (11), a generalization for the system with n propulsors can be derived. In particular, following a similar analysis, the linearization of the full non-linear propulsion subsystem (2) is given by:

$$\left[\begin{array}{c} \Delta M \\ \Delta N \end{array}\right] = P \left[\begin{array}{c} \Delta T_1 \\ \Delta \delta_1 \\ \vdots \\ \Delta T_n \\ \Delta \delta_n \end{array}\right] \tag{12}$$

where $P = ABC$ and $A \in \mathbb{R}^{2 \times 2n}$, $B \in \mathbb{R}^{2n \times 2n}$, and $C \in \mathbb{R}^{2n \times 2n}$.

As follows, each of the elements which comprise (12) will be derived in a closed and compact form. First, matrix A, which contains information regarding the location and orientation of each rotor, can be written as the concatenation of sub-matrices A_i:

$$A = \left[\begin{array}{cccc} A_1 & A_2 & \cdots & A_n \end{array}\right] \tag{13}$$

with

$$A_i = \left[\begin{array}{cc} l_i(-1)_i^\alpha & k_i(-1)^{\beta_i+1} \\ k_i(-1)_i^\beta & l_i(-1)_i^\alpha \end{array}\right] \tag{14}$$

In addition, B comprises a set of 2D rotation sub-matrices as:

$$B = diag\{B_1, B_2, \ldots, B_n\} \tag{15}$$

with

$$B_i = \begin{bmatrix} \cos \delta_i & -\sin \delta_i \\ \sin \delta_i & \cos \delta_i \end{bmatrix}_0 \tag{16}$$

Finally, matrix C contains information regarding the equilibrium thrust of each propeller:

$$C = diag\{C_1, C_2, \dots, C_n\} \tag{17}$$

with

$$C_i = \begin{bmatrix} 1 & 0 \\ 0 & T_i \end{bmatrix}_0 \tag{18}$$

This completes the derivation of the linear approximation of the propulsion subsystem of the PVTOL of Figure 1. The structure obtained here is useful because it allows separating the three main physical properties that determine the resulting propulsion matrix P. In particular:

- Matrix A contains information regarding the physical position and rotation direction of each propeller. Since these properties are normally constant, updating matrix A is not necessary when calculating a linear approximation on a different operating point. However, analyzing its structure could be useful to determine which particular propulsive configuration is better for particular applications.
- Matrix B contains information regarding the tilt angle of each of the propellers. Depending on the operating range and behavior of the propulsion system, this matrix could be the most sensible for operating point modifications, and should be updated accordingly in scheduling approaches.
- Matrix C contains information regarding the thrust force of each propeller. This matrix could also require regular updates in a scheduling approach depending on the operating behavior of the vehicle.

In many cases, matrices B and C may not require a full update if only a sub-set of rotors change their operating point. Instead, updating sub-matrices B_i and C_i may be sufficient.

Although, in this case, the derivation and decomposition of the propulsion matrix was performed for the modified PVTOL, it may be possible that a similar decomposition could be useful for other complex propulsion systems, which is the motivation for this analysis (i.e., as a proof of concept).

2.2. Experimental Test Bench

In this section, an experimental test bench considering the modified PVTOL configuration proposed in the previous section is introduced. The schematic of the test bench is shown in Figure 3a, while Figure 3b shows the actual prototype. Although this is a simple two-rotor setup, this configuration yields an over-actuated system (as noted in the previous section); thus, it is sufficient to study the control allocation problem.

The test bench consists of a base with a transverse rod connecting a servo-motor at each end. Two bearings allow the system to move freely on two degrees of freedom. Brushless motors with propellers are attached to each of the servo-motors. A summary of the technical specifications of the test bench is as follows:

- Two 2304 Racestar brushless motors (81 W max power).
- Two Gemfan 51466 three-blade propellers, one CW and one CCW.
- Two MG995 servo-motors.
- One BNO055 absolute orientation sensor.

The orientation angles are measured using the BNO055 absolute orientation sensor array, which consists of a 3-axis accelerometer and gyroscope combined with a Kalman filter for the pitch and yaw angle reconstruction. These kinds of sensors are normally able to deliver measurements with a mean error of under 1.5 deg/s and 0.5 m/s^2 for the gyroscope and accelerometer, respectively [22–24]. This configuration can yield a low level of angle measurement error in applications where no axial acceleration is present, so that the main source of measured acceleration is the gravity, as in the present application [25].

The test bench was operated with an external power source with a fixed voltage of 12 V, which provided enough current for all the operating conditions tested. The control algorithms were implemented in an ATmega2560 microcontroller, which is a low-power 8-bit micro controller capable of 6 MIPS at 16 MHz. The possibility to implement the proposed control allocation algorithm in this low-computational-power microcontroller is a salient characteristic, since most low-cost autopilot UAV solutions have limited computational power.

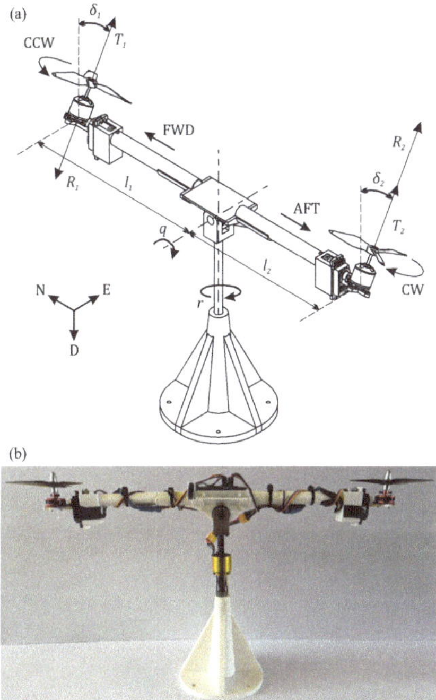

Figure 3. (**a**) Test bench schematic. (**b**) Physical prototype.

The microcontroller code was directly generated with Simulink/Matlab without further optimization operating at a sampling rate of 5 ms. Figure 4 shows the overall Simulink program used for the implementation of the control algorithms presented in this article.

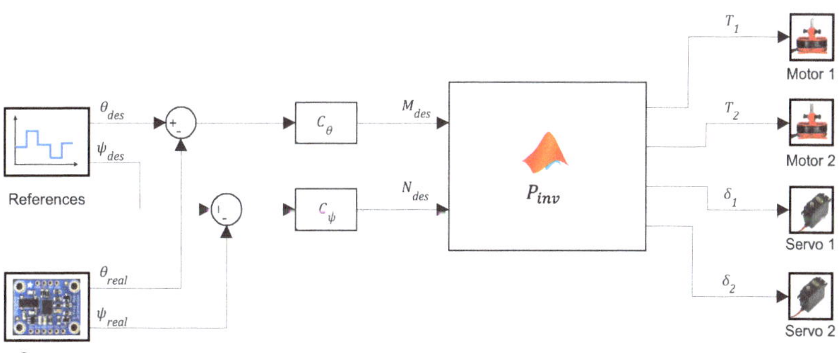

Figure 4. Simulink program for test bench.

An Atmel-based microcontroller board was programmed for the test bench. A summary of the resulting input and output variables for the test bench is shown next.

- Input variables:
    - Right motor thrust T_1;
    - Left motor thrust T_2;
    - Tilt angle of right motor δ_1;
    - Tilt angle of left motor δ_2.
- Output variables:
    - Pitch rate $\dot{\theta} = q$;
    - Yaw rate $\dot{\psi} = r$;
    - Pitch angle θ;
    - Yaw angle ψ.

Further details regarding the experimental and simulation setups are shown in Section 3, together with the corresponding experiments and simulations.

Recalling Equation (5), the resulting motion equations are given by:

$$I_{yy}\dot{q} = T_1 l_1 \cos\delta_1 - T_2 l_2 \cos\delta_2 - T_1 k_1 \sin\delta_1 + T_2 k_2 \sin\delta_2$$
$$I_{zz}\dot{r} = T_1 k_1 \cos\delta_1 - T_2 k_2 \cos\delta_2 + T_1 l_1 \sin\delta_1 - T_2 l_2 \sin\delta_2$$

$$(19)$$

where I_{yy} and I_{zz} are the pitch and yaw inertial masses, respectively.

Accordingly, the resulting model considering the linearized propulsion model yields:

$$\begin{bmatrix} I_{yy} & 0 \\ 0 & I_{zz} \end{bmatrix} \begin{bmatrix} \Delta\dot{q} \\ \Delta\dot{r} \end{bmatrix} = P \begin{bmatrix} \Delta T_1 \\ \Delta\delta_1 \\ \Delta T_2 \\ \Delta\delta_2 \end{bmatrix} \tag{20}$$

with $P \in \mathbb{R}^{2\times 4}$ as in Equation (11).

2.3. Control Allocation Problem

The objective is to design a linear controller for the pitch and yaw angles of the test bench using the input vector $[T_1 \, \delta_1 \, T_2 \, \delta_2]^T$. According to (20), this results in an over-actuated and a multivariable control problem. Although there are several control approaches which can deal with multivariable cross-coupling, an ideal approach would be to have an input-decoupling scheme which incorporates two virtual inputs, that is:

$$\begin{bmatrix} I_{yy} & 0 \\ 0 & I_{zz} \end{bmatrix} \begin{bmatrix} \Delta\dot{q} \\ \Delta\dot{r} \end{bmatrix} = \begin{bmatrix} M_{des} \\ N_{des} \end{bmatrix} \tag{21}$$

where M_{des} and N_{des} are the two arbitrary virtual inputs.

Therefore, in this case, the control allocation problem consists in finding a matrix P_{inv} such that $I = P P_{inv}$ so that the required physical input variables can be calculated for a particular value of the desired virtual inputs, that is:

$$\begin{bmatrix} T_1 \\ \delta_1 \\ T_2 \\ \delta_2 \end{bmatrix} = P_{inv} \begin{bmatrix} M_{des} \\ N_{des} \end{bmatrix} \tag{22}$$

In this article, two approaches to solve the control allocation problem will be compared: (1) a decentralized control approach based on physical intuition and (2) a mathematical approach based on SVD and optimization.

2.3.1. Physical Intuition-Based Decentralized Control Allocation

In this section, a proposal for the input signal mixer will be derived based on physical intuition and the following assumptions:

- A decentralized control approach for the pitch and yaw angles is sufficient to stabilize the test bench. This implies that pitch and yaw moments are derived separately without considering their interaction. This normally implies, for instance, that when calculating the pitch angle input mixer (directly related to M_{des}), it is assumed that $N_{des} = 0$. In addition, if the pitch angle mixer also affects the yawing moment N, then these effects are neglected, or at best considered as input perturbations for the yaw angle controller. That is, any residual cross-coupling is neglected.
- A small range of operation for the tilt-rotors is enough to stabilize the test bench. This assumption allows maintaining the focus of this study on the linear elements of the control allocation problem, which is the basis of most scheduling approaches. Later, it will be confirmed experimentally that indeed only small tilt angles are required for the stabilization of this PVTOL configuration. Nonetheless, extension of these results for wider operating ranges will be forthcoming in future studies.

For small angles, the system tends to behave as a classical PVTOL configuration with a single degree of freedom in the pitch angle and propeller thrust as inputs. In this case, the resulting pitching moment yields:

$$M = l_1 T_1 - l_2 T_2 \tag{23}$$

Thus, it is reasonable to use the differential thrust of propellers to induce pitching moment. In addition, when no pitching moment is required, it is also desirable for both propellers to maintain a specific equilibrium thrust T_0. This yields the well-known PVTOL pitching moment input mixer:

$$T_1 = T_0 + \frac{M_{des}}{2l_1} \quad T_2 = T_0 - \frac{M_{des}}{2l_2} \tag{24}$$

On the other hand, in the case of the yaw angle, if $T_1 \approx T_2$ (i.e., $M_{des} \approx 0$), then the yawing moment yields:

$$N = T_1(k_1 \cos \delta_1 - k_2 \cos \delta_2) + T_1(l_1 \sin \delta_1 - l_2 \sin \delta_2) \tag{25}$$

Accordingly, noticing that $T_1 \approx T_2$ in (24) implies $T_1 \approx T_0$, the small angle approximation for the yawing moments is:

$$N = T_0 l_1 \delta_1 - T_0 l_2 \delta_2 \tag{26}$$

Therefore, the input mixer for the yawing moment is proposed as:

$$\delta_1 = \frac{N_{des}}{2T_0 l_1} \quad \delta_2 = -\frac{N_{des}}{2T_0 l_2} \tag{27}$$

Equations (24) and (27) will be used for the decentralized control allocation approach. It is clear that in this case, the pitching moment is achieved by modifying the propeller thrust, whereas the yawing moment is achieved by modifying the rotor tilt angle. Therefore, this solution is also easy to understand and to implement.

2.3.2. Singular Value Decomposition Control Allocation

In this section, a simple control allocation solution will be derived using the SVD analysis of the linearization of the propulsion system. Recalling Equation (6), the SVD of matrix P yields:

$$P_{2 \times 2n} = U_{2 \times 2} S_{2 \times 2n} V_{2n \times 2n}^T \tag{28}$$

where the sub-indexes show the size of each matrix.

Matrix S is a rectangular diagonal matrix called the *singular values matrix*, whose diagonal elements s_i are always positive, typically written in descending order, and are called the *singular values* of matrix P. In addition, matrices S and V are orthogonal. The rank of matrix P is equal to the number of non zero singular values. Thus, matrix P can be considered as a linear transformation of R^{2n} into R^{r_s}, where $r_s = rank(S)$ is the number of non-zero s_i values, which in this case, since $S \in R^{2 \times 2n}$, can only be at most 2. Finally, this analysis shows that for the modified PVTOL configuration with arbitrary number of tilting rotors, a fully decoupling control allocation solution exists iff $r_s = 2$.

The SVD of matrix P also allows viewing the linear transformation between an input vector $u \in R^{2n}$ and an output vector $y \in R^2$ as a series of intermediate linear transformations, as shown in Figure 5, where the intermediate transformed vectors are $y_2 \in R^{2n}$ and $y_1 \in R^2$, that is:

$$y_2 = V^T u \quad y_1 = S y_2 \quad y = U y_1 \tag{29}$$

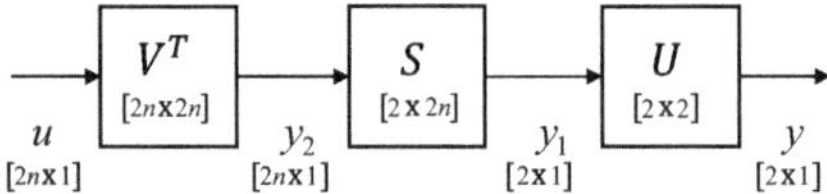

Figure 5. Decomposition of the linear transformation matrix P into intermediate transformations.

Figure 5 shows that the loss of dimension introduced by matrix P occurs in the intermediate transformation $y_1 = S y_2$ and is due to the rank of matrix S. In particular, matrix S has the following structure:

$$S = \begin{bmatrix} S_1 & 0_{2 \times 2n-2} \end{bmatrix} \tag{30}$$

where $0_{2 \times n-2}$ denotes a zero matrix of the indicated size and S_1 is given by:

$$S_1 = diag\{s_1, s_2\} \tag{31}$$

If the rank of P is two (i.e., S_1 is invertible), then all the possible input combinations u which produce a particular desired output y can be calculated with:

$$\begin{aligned}
y_1 &= U^T y \\
y_{2S} &= S_1^{-1} y_1 \\
y_2 &= \begin{bmatrix} y_{2S} \\ y_{2N} \end{bmatrix} \\
u &= V y_2
\end{aligned} \tag{32}$$

where $y_{2N} \in R^{2n-2}$ is an arbitrary vector which does not have any effect over the resulting output vector y.

If $y_{2N} = 0$, then Equation (32) yields the Moore–Penrose pseudo-inverse. In particular, y_{2N} introduces additional degrees of freedom which account for the null subspace of S. The process of using this pseudo-inverse transformation to calculate an input u which produces a particular desired output y is represented in Figure 6. This allows injecting an arbitrary vector y_{2N} which can be selected according to additional practical considerations. Note that in this case, the SVD analysis was preferred over the direct pseudo-inverse approach because of the additional flexibility that the SVD decomposition yields, such as more precise control of the null space injection and information contained in matrices S and V.

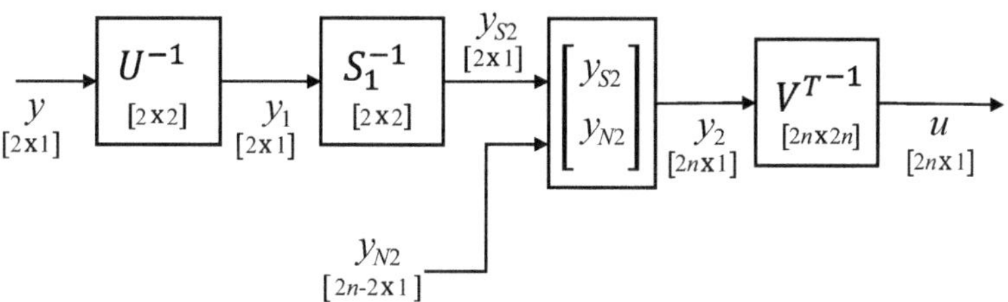

Figure 6. Graphical representation of using the pseudo-inverse to calculate input u from the desired output y.

2.3.3. SVD-Based Control Allocation Configuration

In the previous section, a control allocation approach based on SVD was presented for the general case of n co-linear tilting rotors. This method introduced additional degrees of freedom which could be potentially used to better comply with additional specifications. In this section, the SVD control allocation method will be applied to the two-rotor modified PVTOL test bench and the additional degrees of freedom will be configured according to practical considerations.

Considering the physical parameters of the test bench, and the equilibrium points reported in Table 1, the propulsion matrix P (i.e., the linear approximation of the propulsion system) yields:

$$P = \begin{bmatrix} 0.2050 & -0.0424 & -0.2050 & 0.0424 \\ 0.0103 & 0.8446 & -0.0103 & -0.8446 \end{bmatrix} \tag{33}$$

Table 1. Test bench parameters and equilibrium points.

Variable	Value
T_{1_0}	4.12 N
T_{2_0}	4.12 N
δ_{1_0}	0°
δ_{2_0}	0°
$l_{1,2}$	0.205 m
$k_{1,2}$	0.0103 m

Using Equations (28) and (32) with $u = [T_1\ \delta_1\ T_2\ \delta_2]^T$ and $y = [M_{des}\ N_{des}]^T$ yields:

$$\begin{bmatrix} T_1 \\ \delta_1 \\ T_2 \\ \delta_2 \end{bmatrix} = \begin{bmatrix} 2.433 & 0.122 & 0.707 & -0.025 \\ -0.029 & 0.592 & 0.025 & 0.707 \\ -2.433 & -0.122 & 0.707 & -0.025 \\ 0.029 & -0.592 & 0.025 & 0.707 \end{bmatrix} \begin{bmatrix} M_{des} \\ N_{des} \\ a \\ b \end{bmatrix} \tag{34}$$

where the vector containing the additional degrees of freedom is given by $y_{2N} = [a\ b]^T$.

Parameters a and b can be selected freely according to the designer considerations. In this case, the first consideration is that it is desirable that $\delta_1 = \delta_2 = 0$ when $M_{des} = N_{des} = 0$. Physically, this implies that, in equilibrium, the rotors should point upwards. Therefore, from (34), it is required that

$$0.025a + 0.707b = 0 \tag{35}$$

Thus, b is set as:

$$b = -0.035a \tag{36}$$

Next, parameter a is selected so that the quadratic error considering the non-linear propulsion model is minimized. This yields the following minimization problem:

$$\min_a J - (e_M^2 + e_N^2) \tag{37}$$

with:

$$\begin{bmatrix} e_M \\ e_N \end{bmatrix} = \begin{bmatrix} M \\ N \end{bmatrix} - P\,P_{inv}(a)\begin{bmatrix} M_{des} \\ N_{des} \end{bmatrix} \tag{38}$$

where M and N are as in Equation (5), P is from (33), and $P_{inv}(a)$ is matrix (34) with an arbitrary a value.

A simple numerical exercise, shown graphically in Figure 7, reveals that the optimal value for this parameter is $a = 5.82$.

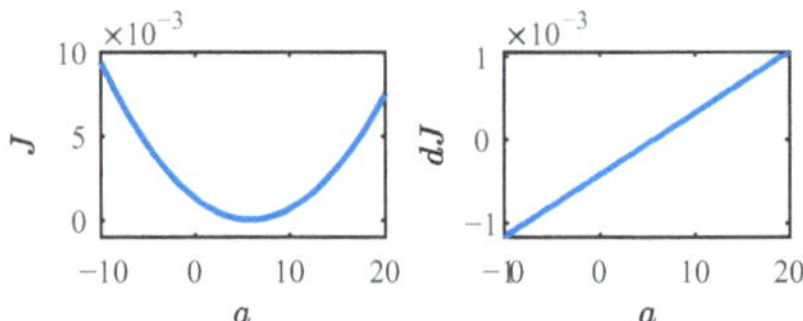

Figure 7. Cost function J and its derivative (dJ) as a function of a.

Finally, the resulting input mixer yields:

$$\begin{bmatrix} T_1 \\ \delta_1 \\ T_2 \\ \delta_2 \end{bmatrix} = \begin{bmatrix} 2.433 M_{des} + 0.122 N_{des} + 4.12 \\ -0.029 M_{des} + 0.592 N_{des} \\ -2.433 M_{des} - 0.122 N_{des} + 4.12 \\ 0.029 M_{des} - 0.592 N_{des} \end{bmatrix} \tag{39}$$

2.3.4. Final Mixer Algorithm Comparison

In this section, the resulting mixer algorithms using both approaches will be compared for T_1 and δ_1 (note that T_2 and δ_2 are similar). In particular, the decentralized control approach using the parameters of Table 1 and Equations (24) and (27) yields:

$$T_{1,dec} = 4.12 + 2.43 M_{des} \quad \delta_{,dec} = 0.592 N_{des} \tag{40}$$

On the other hand, from Equation (39), using the SVD approach yields:

$$T_{1,dec} = 4.12 + 2.43 M_{des} + 0.122 N_{des} \quad \delta_{,dec} = 0.592 N_{des} - 0.029 M_{des} \tag{41}$$

Examination of Equations (40) and (41) reveals that both approaches arrive at similar results, with the main difference being that the SVD approach introduces additional decoupling factors. In addition, this also shows that the selection of the free parameters during the SVD approach design is an important element which allowed for the incorporation of similar physical insights as the decentralized approach.

3. Results

3.1. Simulation Results

In this section, the effectiveness of both control allocation approaches is tested through simulations using the control scheme shown in Figure 8, where the inverse propulsion system dynamics (i.e., P_{inv}) is obtained according to the control allocation schemes of the previous section.

A pair of PID controllers was designed for the pitch and yaw angles, C_θ and C_ψ, respectively, using Equation (21) as a design model. That is, it was assumed that the propulsion system is fully decoupled and compensated. This assumption yields a pair of simple double-integrator dynamics, which are controlled through direct pole placement and the specifications of Table 2. These specifications aim at introducing a certain degree of bandwidth separation between the orientation angles and consider a slower response time for the propeller tilting servos than the propeller thrust. This aspect will become an important issue for the experimental implementation and is commented on later in the article. The resulting PID gains are reported in Table 3.

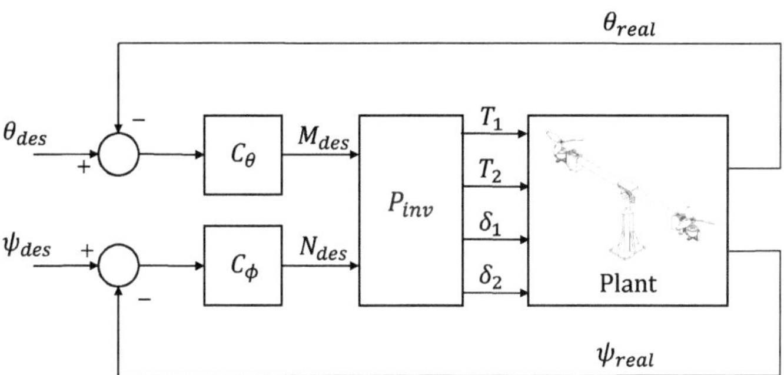

Figure 8. Test bench control system scheme. In this case, P_{inv} represents the control allocation algorithm.

Table 2. Specifications.

Requirement	Pitch θ	Yaw ψ
Overshoot M_p	<35%	<25%
Settling Time T_s	<11 s	<15 s

Table 3. Simulation PID controller gains.

Constant	Pitch	Yaw
Proportional	2	1.5
Integral	0.1	0.1
Derivative	1.5	1

The simulations were performed with both control allocation approaches (i.e., *decentralized* and *SVD-decoupled*), the previously mentioned PID controllers and the non-linear model of the modified PVTOL test bench from Equation (19). In order to assess the resulting cross-coupling due to the input mixer algorithms, the following conditions were simulated:

1. Pitch-Varying Case: The pitch θ reference changes in steps, while the yaw ψ reference remains constant at zero.
2. Yaw-Varying Case: The yaw ψ reference changes in steps, while the pitch θ reference remains constant at zero.

3.1.1. Pitch-Varying Case

Figure 9a shows the simulated responses of the pitch and yaw angles for the decentralized approach, while Figure 9b presents the same variables using the SVD-decoupling algorithm. This figure shows that while the pitch angle response of both approaches is similar and within the control specifications, the cross-coupling introduced in the yaw angle is significantly greater in the case of the decentralized input mixer.

In order to better assess the effectiveness of the SVD-decoupling control allocation approach, Figure 10 presents a comparison between the reference moments (i.e., M, N and M_{des}, N_{des}) and the actual moments exerted by the propulsion system for the simulation of Figure 9 with the SVD-decoupling approach. This figure shows that the resulting SVD-decoupling mixer has a very low error level (N and M deltas) and very good tracking capabilities. The greatest error level is observable in the case of the yawing moment at the time of the pitch angle step movement, which is when the propulsion system cross-coupling is at its highest.

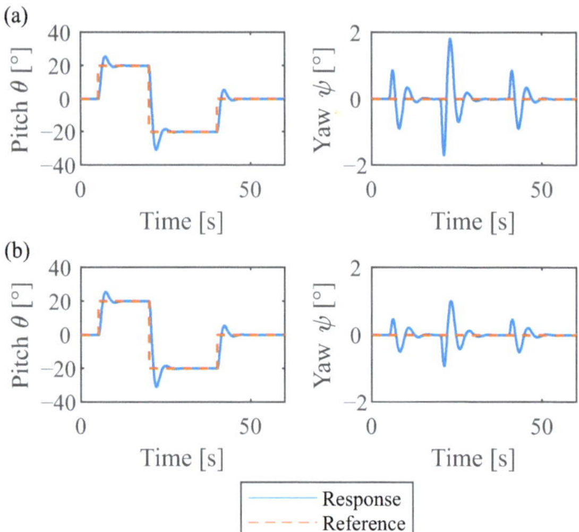

Figure 9. Pitch θ varying simulated responses. (a) Decentralized control allocation approach. (b) SVD-decoupling control allocation approach.

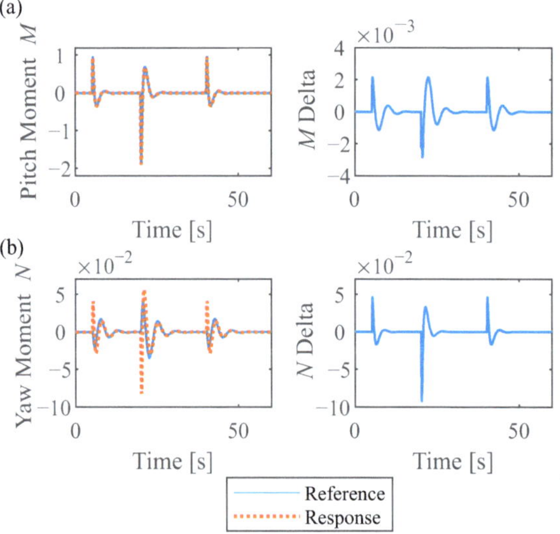

Figure 10. Comparison between reference moments and the actual moments exerted by the propulsion system for the SVD-decoupling control allocation approach (a) Pitching moment M. (b) Yawing moment N.

3.1.2. Yaw-Varying Case

For the yaw variation simulations, Figure 11a shows the simulated responses of the pitch and yaw angles for the decentralized approach, while Figure 11b presents the same variables using the SVD-decoupling algorithm. Similarly to the pitch-varying case, the cross-coupling in the pitch angle due to variations in the yaw angle reference is reduced by using the SVD-decoupling mixer.

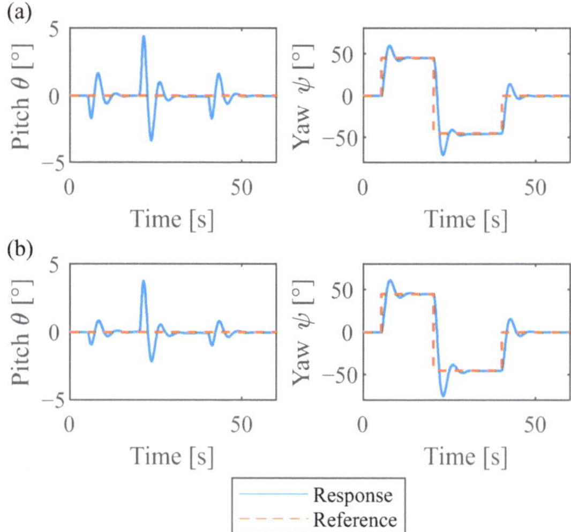

Figure 11. Yaw ψ varying simulated responses. (**a**) Decentralized control allocation approach. (**b**) SVD-decoupling control allocation approach.

Finally, Figure 12 presents a comparison between the reference moments (i.e., M, N and M_{des}, N_{des}) and the actual moments exerted by the propulsion system for the simulation of Figure 11 with the SVD-decoupling approach. The greatest error level is observable in the case of the pitching moment at the time of the yaw angle step movement, which is when the propulsion system cross-coupling is at its highest.

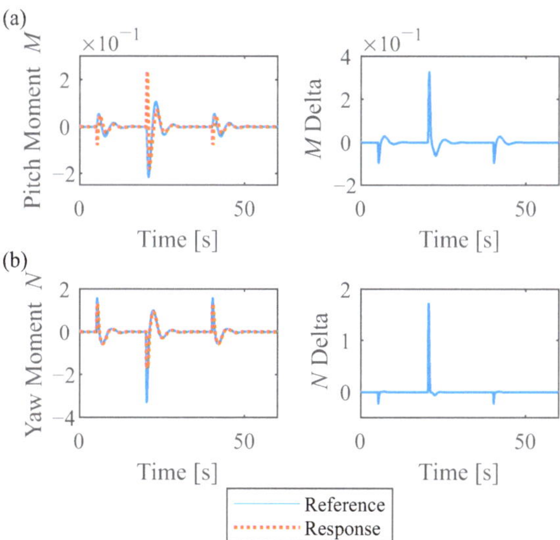

Figure 12. (**a**) Pitch moment M for varying yaw ψ input from decoupling control. (**b**) Yaw moment N for varying yaw ψ input from decoupling control.

3.1.3. Discussion

The previous results show that, in a simulated environment, the SVD-decoupling approach is effective in reducing the resulting cross-coupling, even when considering non-linear propulsion dynamics. In addition, the proper setting of the free variables when calculating the propulsion system inverse dynamics allowed obtaining similar responses to the physical insight approach. That is, the SVD-decoupling approach allows integrating physical insight easily. This is an important feature, because in most real applications, practical considerations have to be taken into account. For example, Figure 13 presents a comparison between the tilt-rotor angles for the simulations of the previous section considering both control allocation approaches. This figure shows that both approaches essentially produce the same overall tilt-rotor angles; however, the SVD-decoupling approach introduces slight adjustments, which resulted in a reduced cross-coupling.

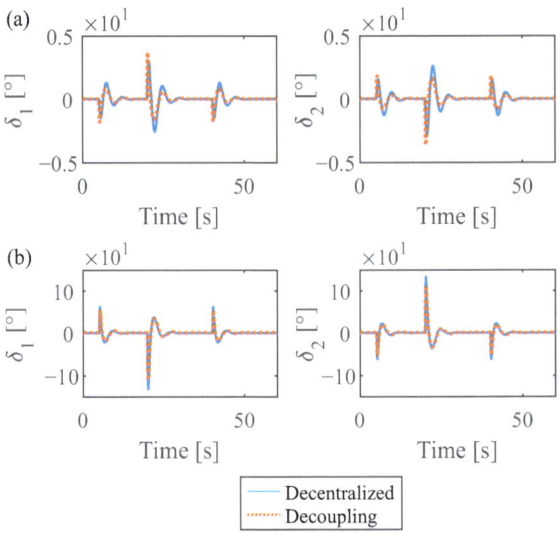

Figure 13. (a) Rotor angles δ_1, δ_2 from decentralized and decoupling controllers for a Pitch θ varying input. (b) Rotor angles δ_1, δ_2 from decentralized and decoupling controllers for a Yaw ψ varying input.

3.2. Experimental Results

The control system of Figure 8 was implemented experimentally using the test bench of Figure 3 and a similar set of experiments to those presented in the simulations was performed.

The PID gains of the previous section (i.e., simulation) were taken as a baseline in the tuning of the prototype's controller. Since the simulated test bench equations (Equation (19)) did not account for friction damping, a slight adjustment of the PID gains was necessary to maintain the control specifications. A comparison of the frequency-domain response of the original (i.e., simulated) and the experimentally adjusted PID controllers is presented in Figure 14. This figure confirms that the main differences between these controllers are (1) decreased phase (damping) in the experiment and (2) increased gain for the experimental PID controller, which is in line with the effects of a higher level of friction in the experimental setup. The resulting PID gains used for the experiments are presented in Table 4. Both control allocation schemes were implemented using the same PID controllers.

Table 4. Experimental PID controller gains.

Constant	Pitch	Yaw
Proportional	1.5	1.5
Integral	1	0.5
Derivative	0.5	1

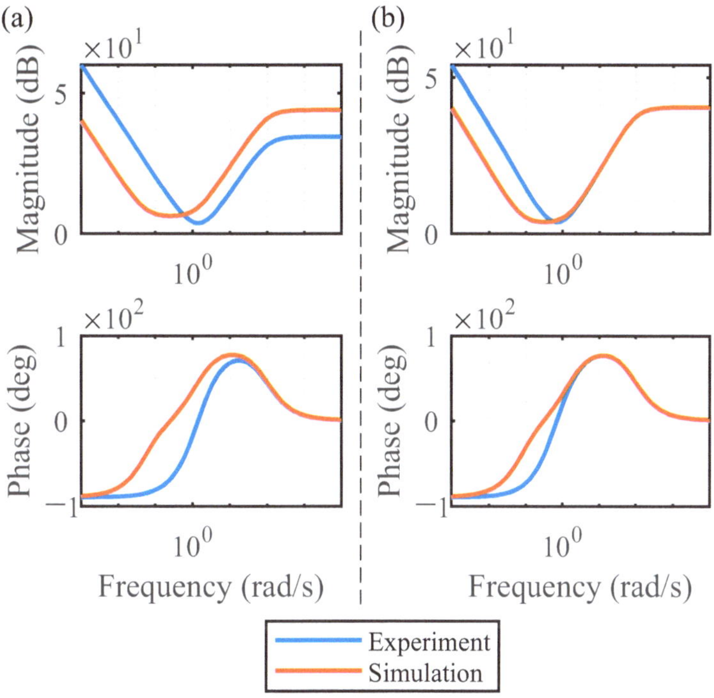

Figure 14. (**a**) Pitch θ moment PID controller bode diagrams for simulation and experiment. (**b**) Yaw ψ moment PID controller bode diagrams for simulation and experiment.

3.2.1. Pitch-Varying Case

The experimental responses of the test bench when the pitch reference is modified for both control allocation approaches are shown in Figure 15. In comparison with the simulated responses from Figure 9, this figure shows that the pitch angle responses are more affected by the amplitude of the pitch angle variation in both approaches. For instance, the overshoot when moving the pitch angle reference from $20°$ to $-20°$ is larger than in the simulated response. This could be mostly due to actuator saturation (absolute or rate) or other non-modelled dynamics. The effect is that the cross-coupling reduction in the yaw angle of the SVD-decoupling approach is reduced in the largest pitch reference modification ($40°$ magnitude), but it is maintained in the other cases ($20°$ magnitude). In addition, a greater reduction was observed for the positive pitch angle variation than the corresponding negative of the same magnitude. This suggests that there could be imbalances in the test bench, either mechanical or due to actuator mismatch.

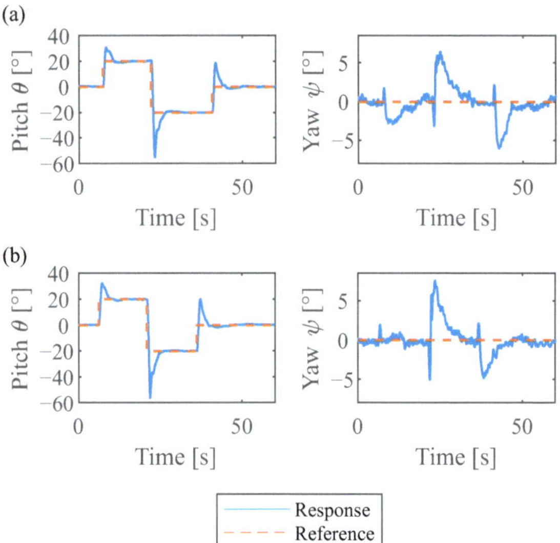

Figure 15. Pitch θ varying experimental responses. (**a**) Decentralized control allocation approach. (**b**) SVD-decoupling control allocation approach.

3.2.2. Yaw-Varying Case

Figure 16 shows the experimental results when varying the yaw angle reference. In this case, the cross-coupling reduction in the pitch angle introduced by the SVD-decoupling approach is maintained in all cases, while the overall yaw angle response performance is similar for both approaches. The contrast with the pitch-varying case, where the cross-coupling was greater in the largest magnitude reference change, suggests an asymmetry in the response times of the main pitch and yaw actuators (i.e., electronic speed controllers for pitch and tilt servos for yaw in this case).

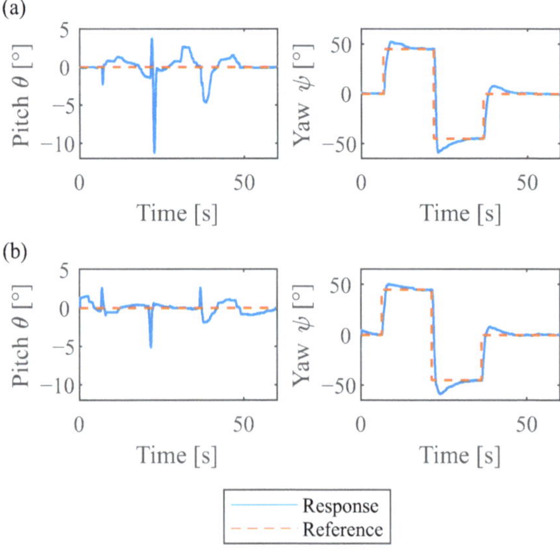

Figure 16. Yaw ψ varying experimental responses. (**a**) Decentralized control allocation approach. (**b**) SVD-decoupling control allocation approach.

3.3. Quantitative Analysis

A quantitative summary of the experimental responses is presented in Table 5, which shows the *Mean Squared Error (MSE)* of the experimental responses of Figures 15 and 16.

Comparing both control allocation approaches when varying the pitch angle reference, there was a reduction of 20.4% in the MSE of the yaw angle cross-coupling when using the SVD-decoupling scheme. A reduction of 70.5% in the MSE of the pitch angle cross-coupling was also observed when the yaw angle reference was modified with the SVD-decoupling scheme.

On the other hand, Table 5 also reveals that, interestingly, the cross-coupling reductions of the SVD-based scheme came at the cost of a slight increase in MSE at the reference angle responses. In particular, increases of 2.15% and 1.59% were observed for the pitch and yaw angles, respectively.

This trade-off of cross-coupling reduction and main angle tracking performance was not observed in the simulated responses, which suggests again that actuator limitations and asymmetry may be a possible explanation.

Figures 17 and 18 present the error distribution of the experimental responses. A closer look at Figure 17, which corresponds to the experiment where the pitch angle reference is modified, shows that the error distribution for the yaw angle is considerably reduced with the SVD-decoupling approach, with much tighter quartiles overall, while the median is similarly close to zero in both cases. In contrast, for the pitch angle, the SVD-decoupling approach yields similarly large quartiles with a slightly larger range. In addition, the distribution of the inner quartiles with the SVD-decoupling approach is significantly skewed compared with the decentralized approach, while maintaining the same overall size, resulting in the median being farther from zero in the SVD-decoupling approach, which can explain the slight increase of MSE in this case.

In the case of the yaw varying experiment, Figure 18 shows that the error distribution of the pitch angle, which represents the cross-coupling in this experiment, is also considerably reduced with the SVD-decoupling approach, with all quartiles much closer to zero as well as the median. For the yaw angle, the SVD-decoupling approach yields a very similar error distribution to the decentralized approach, with the main differences being a median closer to zero and a slight skew to negative errors for the outer quartiles.

These figures confirm and extend the observations made with Table 5. That is, introducing the SVD-decoupling scheme reduces the cross coupling significantly while only slightly affecting the reference angle performance.

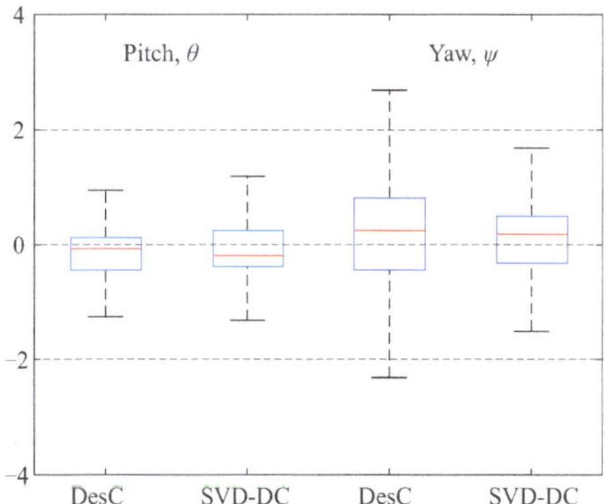

Figure 17. Error distribution of the experimental responses of the pitch θ varying experiment. DesC = decentralized control. SVD-DC = SVD decoupling control.

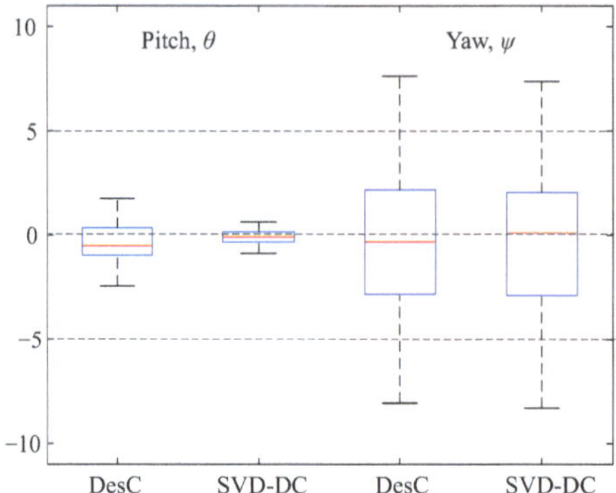

Figure 18. Error distribution of the experimental responses of the yaw ψ varying experiment. DesC = decentralized control. SVD-DC = SVD decoupling control.

Table 5. Mean Squared Error comparison between decentralized and decoupling controllers for pitch θ and yaw ψ reference variations.

Experiment	DoF	Decentralized	Decoupling	$\Delta\%$
Pitch θ	Pitch θ	37.4525	38.2600	2.1560
Variation	Yaw ψ	4.1196	3.2768	-20.4583
Yaw ψ	Pitch θ	2.3904	0.7042	-70.5405
Variation	Yaw ψ	118.6156	120.5088	1.5960

4. Discussion and Conclusions

In this article, a modified PVTOL configuration with an arbitrary number of co-linear tilting rotors was introduced as a baseline to study the control allocation problem found in many of the recently proposed VTOL aircraft designs. A general model for the resulting propulsion subsystem, as well as its linearization, was also developed. This linearization, which is necessary for many of the current control allocation strategies, is presented in a general and structured manner, to facilitate a better understanding of the effects of each of the physical parameters of the propulsion system.

Following the proposed modified PVTOL configuration, a two-tilting-rotor test bench was implemented experimentally. This test bench yields an over-actuated system which can be used to study the control allocation problem theoretically and experimentally. An SVD control allocation approach was developed for the resulting test bench. The proposed approach allows deriving an adequate input mixer algorithm for the over-actuated system, and it is shown that the remaining free degrees of freedom can be used to integrate physical insight and to solve general optimization goals (in this case, the minimization of error due to linearization).

The resulting SVD control allocation scheme is compared with a more traditional scheme derived by physical insight. It is shown that both approaches yield a similar use of the available control variables, with the SVD-based scheme introducing minor input modifications which allow obtaining a considerable decrease in cross-coupling in the test bench orientation angles.

Simulation and experimental results validate that the SVD control allocation scheme allows decreasing cross-coupling while maintaining the same overall performance as the traditional approach. The main deviations from the simulated and experimental results are due to the presence of friction damping in the experimental test bench, which was expected.

Interestingly, other smaller deviations suggest that actuator discrepancies can also have a significant effect on the resulting performance, that is, differences in actuator time responses, saturation levels, etc., due to the type of actuator, for instance, tilt-rotor servo dynamics vs. electric motor controller and propeller dynamics. Further research which incorporates better specific actuator properties is required to extend the benefits of decoupling control allocation schemes in general. This is an important finding, since many applications which require control allocation solutions use a mixture of dissimilar actuators.

Finally, the results support the continued use of the proposed modified PVTOL configuration, in either the presented two-rotor version or in a more complex configuration, for the study of the control allocation problem.

Author Contributions: Conceptualization, L.A.-B. and D.H.-A.; methodology, L.A.-B.; software, E.M.-M.; validation, L.A.-B., D.H.-A. and E.M.-M.; formal analysis, L.A.-B. and E.M.-M.; investigation, L.A.-B. and E.M.-M.; resources, L.A.-B. and D.H.-A.; data curation, E.M.-M.; writing—original draft preparation, L.A.-B. and E.M.-M.; writing—review and editing, D.H.-A.; visualization, E.M.-M.; supervision, L.A.-B. and D.H.-A.; project administration, L.A.-B. and D.H.-A.; funding acquisition, L.A.-B. and D.H.-A. All authors have read and agreed to the published version of the manuscript.

Funding: This research received no external funding.

Data Availability Statement: Data are contained within the article.

Conflicts of Interest: The authors declare no conflict of interest.

References

1. EASA. *MOC SC-VTOL, Proposed Means of Compliance with the Special Condition VTOL*; Technical Report; European Union Aviation Safety Agency: Cologne, Germany, 2022.
2. EASA. *RMT.0731, Terms of Reference for Rulemaking Task, New Air Mobility*; Technical Report; European Union Aviation Safety Agency: Cologne, Germany, 2021.
3. Wang, X.; Cai, L. Mathematical modeling and control of a tilt-rotor aircraft. *Aerosp. Sci. Technol.* **2015**, *47*, 473–492. [CrossRef]
4. Kong, Z.; Lu, Q. Mathematical Modeling and modal switching control of a novel tiltrotor UAV. *J. Robot.* **2018**, *2018*, 8641731. [CrossRef]
5. Lu, K.; Liu, C.; Li, C.; Chen, R. Flight Dynamics Modeling and Dynamic Stability Analysis of tilt-rotor aircraft. *Int. J. Aerosp. Eng.* **2019**, *2019*, 5737212. [CrossRef]
6. Sheng, H.; Zhang, C.; Xiang, Y. Mathematical Modeling and Stability Analysis of tiltrotor aircraft. *Drones* **2022**, *6*, 92. [CrossRef]
7. Çakıcı, F.; Leblebicioğlu, M.K. Modeling and simulation of a small-sized tiltrotor UAV. *J. Def. Model. Simul. Appl. Methodol. Technol.* **2011**, *9*, 335–345. [CrossRef]
8. Chen, C.; Shen, L.; Zhang, D.; Zhang, J. Mathematical Modeling and control of a tiltrotor UAV. In Proceedings of the 2016 IEEE International Conference on Information and Automation (ICIA), Ningbo, China, 1–3 August 2016. [CrossRef]
9. Liu, Z.; He, Y.; Yang, L.; Han, J. Control techniques of Tilt Rotor Unmanned Aerial Vehicle Systems: A Review. *Chin. J. Aeronaut.* **2017**, *30*, 135–148. [CrossRef]
10. Hegde, N.T.; George, V.; Nayak, C.G.; Kumar, K. Design, dynamic modelling and control of tilt-rotor uavs: A Review. *Int. J. Intell. Unmanned Syst.* **2019**, *8*, 143–161. [CrossRef]
11. Bacchini, A.; Cestino, E. Electric VTOL Configurations Comparison. *Aerospace* **2019**, *6*, 26. [CrossRef]
12. Amezquita-Brooks, L.; Liceaga-Castro, E.; Gonzalez-Sanchez, M.; Garcia-Salazar, O.; Martinez-Vazquez, D. Towards a standard design model for quad-rotors: A review of current models, their accuracy and a novel simplified model. *Prog. Aerosp. Sci.* **2017**, *95*, 1–23. [CrossRef]
13. Amezquita-Brooks, L.; Hernandez-Alcantara, D.; Santana-Delgado, C.; Covarrubias-Fabela, R.; Garcia-Salazar, O.; Ramirez-Mendoza, A.M.E. Improved Model for Micro-UAV Propulsion Systems: Characterization and Applications. *IEEE Trans. Aerosp. Electron. Syst.* **2020**, *56*, 2174–2197. [CrossRef]
14. Grau, S.; Kapitola, S.; Weiss, S.; Noack, D. Control of an over-actuated spacecraft using a combination of a fluid actuator and reaction wheels. *Acta Astronaut.* **2021**, *178*, 870–880. [CrossRef]
15. Qin, Z.; Liu, K.; Zhao, X. A smooth control allocation method for a distributed electric propulsion VTOL aircraft test platform. *IET Control Theory Appl.* **2023**, *17*, 925–942. [CrossRef]
16. Santos, M.; Honório, L.; Moreira, A.; Garcia, P.; Silva, M.; Vidal, V. Analysis of a Fast Control Allocation approach for nonlinear over-actuated systems. *ISA Trans.* **2022**, *126*, 545–561. [CrossRef] [PubMed]
17. Hamandi, M.; Usai, F.; Sablé, Q.; Staub, N.; Tognon, M.; Franchi, A. Design of multirotor aerial vehicles: A taxonomy based on input allocation. *Int. J. Robot. Res.* **2021**, *40*, 027836492110259. [CrossRef]
18. Zheng, X.; Yang, X. Improved adaptive NN backstepping control design for a perturbed PVTOL aircraft. *Neurocomputing* **2020**, *410*, 51–60. [CrossRef]
19. Bonilla, M.; Blas, L.; Azhmyakov, V.; Malabre, M.; Salazar, S. Robust structural feedback linearization based on the nonlinearities rejection. *J. Frankl. Inst.* **2020**, *357*, 2232–2262. [CrossRef]

20. Łakomy, K.; Madonski, R. Cascade extended state observer for active disturbance rejection control applications under measurement noise. *ISA Trans.* **2021**, *109*, 1–10. [CrossRef] [PubMed]
21. Offermann, A.; Castillo, P.; Miras, J.D. Control of a PVTOL with tilting rotors. In Proceedings of the 2019 International Conference on Unmanned Aircraft Systems (ICUAS), Atlanta, GA, USA, 11–14 June 2019. [CrossRef]
22. Harindranath, A.; Arora, M. A Systematic Review of User-Conducted Calibration Methods for MEMS-based IMUs. *Measurement* **2023**, 114001. [CrossRef]
23. Otegui, J.; Bahillo, A.; Lopetegi, I.; Díez, L.E. Performance Evaluation of Different Grade IMUs for Diagnosis Applications in Land Vehicular Multi-Sensor Architectures. *IEEE Sens. J.* **2021**, *21*, 2658–2668. [CrossRef]
24. de Alteriis, G.; Conte, C.; Lo Moriello, R.S.; Accardo, D. Use of Consumer-Grade MEMS Inertial Sensors for Accurate Attitude Determination of Drones. In Proceedings of the 2020 IEEE 7th International Workshop on Metrology for AeroSpace (MetroAeroSpace), Pisa, Italy, 22–24 June 2020; pp. 534–538. [CrossRef]
25. Villarreal Valderrama, J.F.; Takano, L.; Liceaga-Castro, E.; Hernandez-Alcantara, D.; Zambrano-Robledo, P.D.C.; Amezquita-Brooks, L. An integral approach for aircraft pitch control and instrumentation in a wind-tunnel. *Aircr. Eng. Aerosp. Technol.* **2020**, *92*, 1111–1123. [CrossRef]

Article

Intelligent PIV Fuzzy Navigation and Attitude Controller for an Octorotor Mini-UAV

Pablo A. Tellez-Belkotosky [1], Luis E. Cabriales-Ramirez [1], Manuel A. Gutierrez-Martinez [1] and Edmundo Javier Ollervides-Vazquez [1,2,*]

[1] Aerospace Engineering Research and Innovation Center, Faculty of Mechanical and Electrical Engineering, Autonomous University of Nuevo Leon, Apodaca 66616, Nuevo Leon, Mexico
[2] Technological Institute of La Laguna-TecNM, Torreon 27000, Coahuila, Mexico
* Correspondence: ejollervidesv@lalaguna.tecnm.mx

Abstract: In this research, a proportional plus integral plus velocity (PIV) fuzzy gain scheduling flight controller for an octorotor mini-unmanned aerial vehicle is developed. The designed flight controller scheme, with a PIV term, is combined with a fuzzy gain scheduling approach. The tracking controller PIV fuzzy gain scheduling is based on two controllers connected in cascade with a saturation approach. The Newton–Euler equations of motion are applied to obtain a mathematical model for the octorotor mini-unmanned aerial vehicle (mini-UAV). The flight controller approach is applied to obtain coupling moments and forces with interconnected attitude and navigation tracking trajectory. In the design of a flight navigation controller with two layers, the inner layer consists of a PIV fuzzy gain scheduling controller that is applied to the attitude dynamics, obtaining the references for the coupling outer layer PIV fuzzy gain scheduling controller, which manipulates the translational dynamics. The navigation PIV fuzzy gain scheduling controller is saturated for bounding in translational forces to avoid large deviations of commands to Euler angles pitch and roll, and another saturated controller is implemented for the bounded thrust rotor to avoid the excessive angular speed of these rotors. The octorotor mini-UAV flight navigation simulation is performed to validate the tracking control of a sequence of motions in each axis, which is presented as a validation for the proposed control scheme.

Keywords: fuzzy gain scheduling; octorotor; mini-UAV

Citation: Tellez-Belkotosky, P.A.; Cabriales-Ramirez, L.E.; Gutierrez-Martinez, M.A.; Ollervides-Vazquez, E.J. Intelligent PIV Fuzzy Navigation and Attitude Controller for an Octorotor Mini-UAV. *Machines* **2023**, *11*, 266. https://doi.org/10.3390/machines 11020266

Academic Editor: Antonios Gasteratos

Received: 30 December 2022
Revised: 26 January 2023
Accepted: 28 January 2023
Published: 10 February 2023

1. Introduction

In the last decade, multirotor UAVs have had important developed applications such as agriculture, construction, market, parcel, civil applications, surveillance, research, and rescue missions. However, as the needs of society grow, new challenges and demands for UAV applications also appear. Four-rotor configurations are the most widely used and have many uses; however, given the new demands, these configurations may be limited. Therefore, more robust vehicles with greater fault tolerance and stability are necessary. It is here where eight-rotor configurations called octorotors have been studied. On the other hand, the control techniques that allow vehicles to respond to faults quickly are just as important.

Multi-rotors are designed with different capacities and sizes that depend on the characteristics of the application. In [1], an octorotor UAV with a star-shaped way for radar applications was implemented for vertical take-off and landing, adding disturbance to be controlled with PID and LQR controllers. In other applications such as [2], an octorotor is used for heavy cargo transportation using different kinds of techniques such as magnetic sensor systems. In [3], heavy cargo transportation is applied in the medical field by transporting first-aid kits or external defibrillators. In agriculture [4–6], the authors used an octorotor for geolocation and thus detected diseased regions.

For example, in [7], they proposed an orthographic projection based on the Kalman filter to detect disease zones. The configuration of the octorotor that was used allowed a gimbal placed face-down and stability for the orthographic projection method. Furthermore, in [8], technologies for mapping and recollecting data, including thermographic cameras, were used to inspect and detect citrus trees and other types of fruit. Monitoring rivers [9] is another application which was implemented using a high-resolution camera in an octorotor UAV. In [10], an octorotor was developed for night-time visual infrastructure inspection. Here, the octorotor configuration was chosen because of its superiority in terms of payload, stability, and fault tolerance. In [11], a lightweight octocopter was developed for monitoring the indoor environment. The octorotor was designed with a weight of 486 (g), 308 (mm) in width, and 160 (mm) in height to get through narrow spaces in a complex indoor environment. Hover experiments which were conducted demonstrated the high-flying capabilities of the octorotor. Other papers [12,13] have described the advantages of using octorotors when a failure presents in one or more rotors because of unknown forces, since the redundancy of eight rotors possesses the skills to continue the flight without the failed rotor by detecting errors and using a recovery system to change the configuration.

According to the literature, octorotors are UAVs that are currently under research, meaning that there are some unknown dynamics that remain to be studied. For this reason, there exists a need for controllers that can respond to their uncertainty and disturbances. In [14], a controller for a coaxial octorotor with six degrees of freedom was proposed, and the proposed control was designed to respond in the presence of actuator faults. The technique of the control used an approach of a radial base function neural network (RBFNN), fuzzy logic control (FLC), and sliding mode control (SMC), named fault-tolerant control (FTC). The control technique avoids issues of modeling and attenuating the chattering due to the SMC. Moreover, the proposed controller allowed to reduce of the number of rulers of the fuzzy logic. The simulations showed that the FTC guarantees the stability and robustness of the system and reduces the chattering effect to have good tracking in the presence of faults of the octorotor. In [15], another controller with estimation and fault tolerance was designed. The controller presented three fault-tolerant control strategies for the octorotor. The first was a mixing strategy, the second was a robust adaptive sliding mode control, and the third was a new one based on the self-tuning sliding mode of the control system. The experiments for the octorotor were carried out indoors with self-imposed faults in the rotors. The comparison between the techniques showed efficient response fault tolerance results for the three proposed techniques. In [16], a controller based on backstepping with a fast terminal sliding mode was proposed for coaxial octorotor exposure to wind disturbance. The proposed controller works with a new learning technique and extended inverse multi-quadratic radial basis function network that estimates the unmodeled dynamics of the octorotor. The simulations showed improved performance when compared with PID and LQR controllers.

One can find the fuzzy controller among the most innovative techniques, which is used to improve the controllers, specifically the dynamic behavior of the octorotors. In [17], an adaptive fuzzy tracking control combining Lyapunov stability theory and backstepping was proposed. The fuzzy controller was used to estimate the unknown dynamics of the quadrotor; nonlinear disturbance observer was also applied to compensate for external disturbance and an approximation error for the fuzzy controller of the system. Lyapunov stability showed that the quadrotor was stable, and the experimental shows the effectiveness of the proposed controller. In [18], a new adaptative fuzzy terminal sliding mode controller and two proportional and derivative (PD) controllers were proposed. The PD controllers were proposed to determine the quadrotor desired attitude, and the sliding mode was used for the rotation of the motors. Furthermore, two Kalman filters estimated the quadrotor attitude and position. The simulations and implementation showed a good performance of the control for the quadrotor.

In this work, an octorotor mini-UAV is proposed to be used in pipe inspections, surveillance, and coordinated flight as a swarm in closed places. The proposed flight controller for the octorotor mini-UAV is a fuzzy gain scheduling that adapts the PIV controller gains. The dynamic model with the controller is performed using simulation experiments with the Newton–Euler model of the octorotor mini-UAV to validate the proposed system.

Main Contribution

In this work, the implementation of a PIV fuzzy gain scheduling control, integrated with the attitude, altitude, and navigation controllers, is presented, in which the attitude and altitude have saturation functions to avoid exceeding the established limits. In this sense, the rotors use the saturation functions that establish the speed limits for each rotor. The conceptual design of a small-sized UAV is carried out, which is developed based on the 3D print tools using the fused deposition modeling (FDM) technique. The contribution of this work is summarized as follows:

1. A PIV control with fuzzy gain scheduling is designed for the trajectory tracking of the attitude and navigation of the mini-UAV.
2. The design of the flight control system is validated by simulations using the Newton–Euler dynamic model of the mini-UAV.
3. The real parameters of the mini-UAV were obtained through CAD software and 3D-printed parts.

The work was structured as follows: The dynamic equations for the octorotor mini-UAV are presented in Section 2. The PIV controller is designed in Section 3. The gain scheduling controller design was developed in Section 4. The simulation results are shown in Section 5. Finally, the conclusions of this work are presented in Section 6.

2. Dynamic Equations for the Octorotor Mini-UAV

The dynamic model of the octorotor mini-UAV uses the configuration frame north–east–down (NED) for aerospace. The X–axis is coupled in the north direction, the Y–axis is located in the east direction, and the Z–axis is in the direction of the center of the Earth. An Earth-fixed inertial frame is defined as $\mathcal{I}=\{x_\mathcal{I}, y_\mathcal{I}, z_\mathcal{I}\}$, and a body frame attached to the center of gravity of the octorotor mini-UAV as $\mathcal{B}=\{x_\mathcal{B}, y_\mathcal{B}, z_\mathcal{B}\}$ [19,20], as can be seen in Figure 1. In order to formulate the dynamics equation, the Newton–Euler approach is used. The dynamic model is defined as

$$\dot{\zeta} = V \tag{1}$$

$$m\dot{V} = R(-T_T) + mge_3 \tag{2}$$

$$\dot{\eta} = E\Omega \tag{3}$$

$$J\dot{\Omega} = -\Omega \times J\Omega + \tau_a \tag{4}$$

where $\zeta = [x, y, z]^\top \in \mathbb{R}^3$ describes the position coordinates relative to the inertial frame and $\eta = [\phi, \theta, \psi]^\top \in \mathbb{R}^3$ describes the rotation coordinates for the octorotor mini-UAV. This is given by an orthogonal rotation matrix $R \in SO(3) : \mathcal{B} \to \mathcal{I}$ parameterized by the Euler angles, where ϕ is the roll angle, θ is the pitch angle, and ψ is the yaw angle, and then, the rotation matrix is defined as

$$R = \begin{bmatrix} c_\theta c_\psi & s_\phi s_\theta c_\psi - c_\phi s_\psi & c_\phi s_\theta c_\psi + s_\phi s_\psi \\ c_\theta s_\psi & s_\phi s_\theta s_\psi + c_\phi c_\psi & c_\phi s_\theta s_\psi - s_\phi c_\psi \\ -s_\theta & s_\phi c_\theta & c_\phi c_\theta \end{bmatrix} \tag{5}$$

$$E = \begin{bmatrix} 1 & \frac{s_\phi s_\theta}{c_\theta} & \frac{c_\phi s_\theta}{c_\theta} \\ 0 & c_\phi & -s_\phi \\ 0 & \frac{s_\phi}{c_\theta} & \frac{c_\phi}{c_\theta} \end{bmatrix} \tag{6}$$

where c_η, s_η are cosine and sine operators for each Euler angle, respectively. $\Omega = [p, q, r]^\top \in \mathbb{R}^3$ is the angular velocity vector in $\mathcal{B}$, $V = [v_x, v_y, v_z]^\top \subset \mathbb{R}^3$ is the translational velocity vector in $\mathcal{I}$, $T_T = [0, 0, f_T]^\top \in \mathbb{R}^3$ is the vertical thrust attached to the body frame considering the rotation speed of each rotor such that $f_T \geq 0$, and $\tau_a \subset \mathbb{R}^3$ is the moments acting on the center of gravity of the octorotor mini-UAV. In addition, the canonical basis vectors of $\mathbb{R}^3$ are considered; these are represented by $e_1 = [1, 0, 0]^\top$, $e_2 = [0, 1, 0]^\top$, and $e_3 = [0, 0, 1]^\top$. $E \in \mathbb{R}^{3 \times 3}$ is the Euler matrix. The term $m \in \mathbb{R}$ denotes the mass of the octorotor mini-UAV, while $J \in \mathbb{R}^{3 \times 3}$ contains moments of inertia such that:

$$J = \begin{bmatrix} J_{xx} & J_{xy} & J_{xz} \\ J_{xy} & J_{yy} & J_{yz} \\ J_{xz} & J_{yz} & J_{zz} \end{bmatrix} \tag{7}$$

To design the corresponding attitude, altitude and navigation controllers, models (1)–(4) are used.

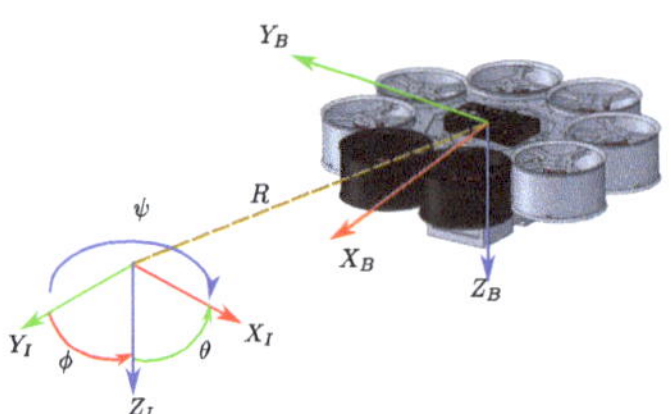

Figure 1. Reference frames and the direction of rotation of each rotor in the octorotor mini-UAV.

3. Flight Controller Design

3.1. Attitude and Altitude Controller

The thrust forces produced by the k-th individual rotor are given by $f_k = c_T \omega_k^2$, where ω_k is the angular velocity of the k-th rotor, and $c_T > 0$ is the thrust coefficient of the propeller, which can be determined by experimental tests from the measurement of thrust in the steady state.

Considering that the vector $\tau_a = [\tau_\phi, \tau_\theta, \tau_\psi]^\top$ represents the moments acting on the octorotor mini-UAV center of gravity and considering the allocation the pair of rotors. For pitch moment τ_θ actuator, rotors $m_1 - m_2$ and $m_5 - m_6$ are parallel to the X_B–axis with a separated angle of $\pi/4$ (rad) among each rotor. For roll moment τ_ϕ, the actuator rotors $m_3 - m_4$ and $m_7 - m_8$ are parallel to the Y_B–axis. The reactive moment produced by the propeller of each rotor of the octorotor mini-UAV is expressed as $Q_k = c_Q \omega_k^2$ with $k = 1, \ldots, 8$, where $c_Q > 0$ is the aerodynamic coefficient of the rotor propeller. The yaw moment τ_ψ must be assigned in the direction of rotation against each pair of rotors (see Figure 2), where the rotors that turn clockwise are positives and the rotors that turn counterclockwise are negative. Consequently, due to those mentioned above, the total thrust and moments applied to the octorotor mini-UAV center of gravity are grouped as follows:

$$\begin{bmatrix} f_T \\ \tau_\phi \\ \tau_\theta \\ \tau_\psi \end{bmatrix} = \begin{bmatrix} c_T & c_T & c_T & c_T & c_T & c_T & c_T & c_T \\ c_T l & -c_T l & -c_T L & -c_T L & -c_T l & c_T l & c_T L & c_T L \\ c_T L & c_T L & c_T l & -c_T l & -c_T L & -c_T L & -c_T l & c_T l \\ c_Q & c_Q & -c_Q & -c_Q & c_Q & c_Q & -c_Q & -c_Q \end{bmatrix} \begin{bmatrix} u_1 \\ u_2 \\ u_3 \\ u_4 \\ u_5 \\ u_6 \\ u_7 \\ u_8 \end{bmatrix} \tag{8}$$

where parameters $L = d\cos(\pi/8)$ and $l = d\sin(\pi/8)$ are the distance d of the octorotor mini-UAV center of gravity with respect to the axis of rotation of the propellers. The compact expression of the Equation (8) is shown as:

$$\Gamma = C_A U \tag{9}$$

where $C_A \in \mathbb{R}^{4\times 8}$ represents the matrix of the aerodynamic coefficients of the octorotor mini-UAV. Vector $U = [\omega_1^2, \cdots, \omega_8^2] = [u_1, \cdots, u_8]$ is the vector of quadratic rotor speed, and $\Gamma = [f_T, \tau_\phi, \tau_\theta, \tau_\psi]^\top$ represents the vector that connects the total scalar thrust f_T with the octorotor mini-UAV moments vector. Since the C_A matrix is not square, the inverse matrix is obtained using a pseudo-inverse equation that is expressed as follows

$$C_A^{-1} = C_A^\top (C_A C_A^\top)^{-1} \tag{10}$$

Then, the inverse expression of the Equation (9) is shown below.

$$U = C_A^{-1}\Gamma \tag{11}$$

Thus, the vector of the desired controller variables is obtained, which corresponds to each of the rotors. Below, such a developed matrix expression is shown:

$$
\begin{bmatrix} u_1 \\ u_2 \\ u_3 \\ u_4 \\ u_5 \\ u_6 \\ u_7 \\ u_8 \end{bmatrix}
=
\begin{bmatrix}
\frac{1}{8c_T} & \frac{1}{4c_T l} & \frac{1}{4c_T L} & \frac{1}{8c_Q} \\[4pt]
\frac{1}{8c_T} & \frac{-1}{4c_T l} & \frac{1}{4c_T L} & \frac{1}{8c_Q} \\[4pt]
\frac{1}{8c_T} & \frac{-1}{4c_T L} & \frac{1}{4c_T l} & \frac{-1}{8c_Q} \\[4pt]
\frac{1}{8c_T} & \frac{-1}{4c_T L} & \frac{-1}{4c_T l} & \frac{-1}{8c_Q} \\[4pt]
\frac{1}{8c_T} & \frac{-1}{4c_T l} & \frac{-1}{4c_T L} & \frac{1}{8c_Q} \\[4pt]
\frac{1}{8c_T} & \frac{1}{4c_T l} & \frac{-1}{4c_T L} & \frac{1}{8c_Q} \\[4pt]
\frac{1}{8c_T} & \frac{1}{4c_T L} & \frac{-1}{4c_T l} & \frac{-1}{8c_Q} \\[4pt]
\frac{1}{8c_T} & \frac{1}{4c_T L} & \frac{1}{4c_T l} & \frac{-1}{8c_Q}
\end{bmatrix}
\begin{bmatrix} f_T \\ \tau_\phi \\ \tau_\theta \\ \tau_\psi \end{bmatrix}
\tag{12}
$$

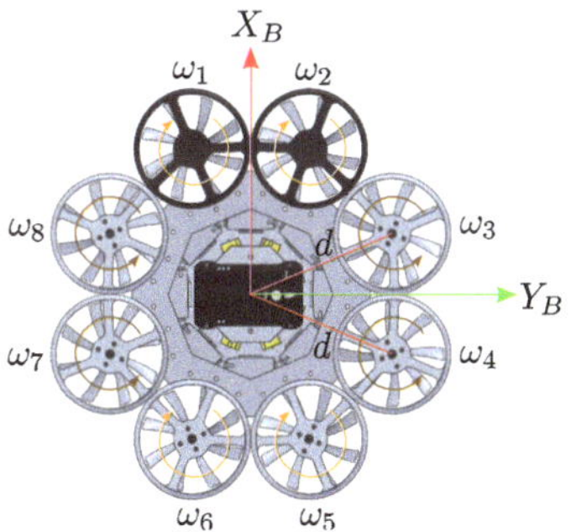

Figure 2. Reference frames and direction of rotation of each rotor in the octorotor mini-UAV.

3.1.1. Altitude Controller

The thrust T_T, which is aligned with Z_B, is not always in the direction of the inertial frame e_3, and the component of the desired inertial force vector F_{z_d} associated with the altitude controller can be calculated in terms of the roll angle ϕ and the pitch angle θ such that

$$F_{z_d} = m\ddot{z}_d = -f_{T_d}\cos(\theta)\cos(\phi) + mg \tag{13}$$

Then, the dynamics of the desired thrust force is proposed f_{T_d} for the actuators, with the purpose of complying with the stationary or ascending flight regime, obtaining the following expression

$$f_{T_d} = \frac{1}{\cos(\theta)\cos(\phi)}(mg - F_{z_d}) \tag{14}$$

where $\phi < \pi/2$ (rad) and $\theta < \pi/2$ (rad) are constrained to avoid a singularity of altitude compensation, and the desired thrust force F_{z_d} is reached by an altitude controller *PIV*, such that this controller is defined as

$$F_{z_d} = k_{pz}\tilde{z} + k_{vz}\tilde{v}_z + k_{iz}\int \tilde{z}\,dt \tag{15}$$

where $\tilde{z} = z_d - z$ represents the translational position error, z_d represents the desired position, and z is the actual value of the translational position, with both variables in e_3. $\tilde{v}_z = v_{zd} - v_z$ represents the translational velocity error, where v_{zd} is the desired velocity, and v_z is the actual value of the translational velocity. A saturation is added to improve trajectory tracking with the desired thrust force. The expression of the thrust-saturated controller is shown below.

$$F_{z_d}^{sat} = \text{Sat}(F_{z_d}) = \begin{cases} F_{z_{max}} & \text{if} \quad F_{z_d} \geq F_{z_{max}} \\ F_{z_d} & \text{if} \quad F_{z_{min}} < F_{z_d} < F_{z_{max}} \\ F_{z_{min}} & \text{if} \quad z \leq F_{z_{min}} \end{cases} \tag{16}$$

The *PIV* controller with thrust compensation is described as follows

$$F_{z_d}^{sat} = \text{Sat}\left(k_{pz}\tilde{z} + k_{vz}\tilde{v}_z + k_{iz}\int \tilde{z}\,dt\right) \tag{17}$$

Combining the Equations (14) and (17), the thrust controller is described below

$$f_{T_d}^{sat} = \frac{1}{\cos(\theta)\cos(\phi)}(mg - F_{z_d}^{sat}) \tag{18}$$

3.1.2. Attitude Controller

For the formulation of the attitude control problem, the desired attitude vector $\eta_d = [\phi_d, \theta_d, \psi_d]^\top$ is considered, such that $\tilde{\eta} = \eta_d - \eta$ giving $\tilde{\eta} = [\tilde{\phi}, \tilde{\theta}, \tilde{\psi}]^\top$ representing the attitude errors. η_d is the vector containing the desired angles, and η are the actual angles. $\tilde{\Omega} = \Omega_d - \Omega$ represents the angular velocity error, where vector Ω_d is the desired angular velocity, and vector Ω is the actual value of the angular velocity. The control problem is strictly the proposal of a controller vector $\tau_{a_d} = [\tau_{\phi_d}, \tau_{\theta_d}, \tau_{\psi_d}]^\top$. For attitude control, a *PIV* controller is proposed. Taking into account the definition established in vector (8), the external moments are rewritten to be expressed in terms of the proposed controller law, as shown below as

$$\begin{bmatrix} \tau_{\phi_d} \\ \tau_{\theta_d} \\ \tau_{\psi_d} \end{bmatrix} = \begin{bmatrix} k_{p\phi}\tilde{\phi} + k_{v\phi}\tilde{p} + k_{i\phi}\int \tilde{\phi}\,dt \\ k_{p\theta}\tilde{\theta} + k_{v\theta}\tilde{q} + k_{i\theta}\int \tilde{\theta}\,dt \\ k_{p\psi}\tilde{\psi} + k_{v\psi}\tilde{r} + k_{i\psi}\int \tilde{\psi}\,dt \end{bmatrix} \tag{19}$$

A saturation is added to improve the attitude in the torques, and it is expressed below

$$\tau_{\eta_d}^{sat} = \text{Sat}(\tau_{\eta_d}) = \begin{cases} \tau_{\eta_{max}} & \text{if} \quad \tau_{\eta_d} \geq \tau_{\eta_{max}} \\ \tau_{\eta_d} & \text{if} \quad \tau_{\eta_{min}} < \tau_{\eta_d} < \tau_{\eta_{max}} \\ \tau_{\eta_{min}} & \text{if} \quad \tau_{\eta_d} \leq \tau_{\eta_{min}} \end{cases} \tag{20}$$

Then, the developed attitude controller equations are expressed as follows

$$\tau_{\phi_d}^{sat} = \mathrm{Sat}\left(k_{p\phi}\tilde{\phi} + k_{v\phi}\tilde{p} + k_{i\phi}\int \tilde{\phi}\,dt\right) \tag{21}$$

$$\tau_{\theta_d}^{sat} = \mathrm{Sat}\left(k_{p\theta}\tilde{\theta} + k_{v\theta}\tilde{q} + k_{i\theta}\int \tilde{\theta}\,dt\right) \tag{22}$$

$$\tau_{\psi_d}^{sat} = \mathrm{Sat}\left(k_{p\psi}\tilde{\psi} + k_{v\psi}\tilde{r} + k_{i\psi}\int \tilde{\psi}\,dt\right) \tag{23}$$

3.2. Navigation Controller

The attitude controller is integrated into the flight control of the octorotor mini-UAV in such a way that the desired setpoints of the attitude vector η_d are calculated from the translational position controller in the $X - Y$ plane and using the reference rotation matrix $R(\psi) \in SO(2)$. To obtain the desired forces of the translational position controller in the $X - Y$ plane, a controller of type *PIV* is proposed as follows

$$\begin{bmatrix} F_{x_d} \\ F_{y_d} \end{bmatrix} = \begin{bmatrix} k_{px}\tilde{x} + k_{dx}\tilde{v}_x + k_{ix}\int \tilde{x}\,dt \\ k_{py}\tilde{y} + k_{dy}\tilde{v}_y + k_{iy}\int \tilde{y}\,dt \end{bmatrix} \tag{24}$$

where $\tilde{x} = x_d - x$ and $\tilde{y} = y_d - y$ represent the translational position error, x_d and y_d represent the desired position, and x, and y are the actual value of the translational position, with both variables in e_1 and e_2, respectively. $\tilde{v}_x = v_{xd} - v_x$ and $\tilde{v}_y = v_{yd} - v_y$ represent the translational velocity errors, where v_{xd} and v_{yd} are the desired velocities, and v_x and v_y are the actual value of the translational velocities. The equations developed by the navigation controller in the navigation plane $X - Y$ are expressed as follows.

$$\begin{aligned} F_{x_d} &= k_{px}\tilde{x} + k_{vx}\tilde{v}_x + k_{ix}\int \tilde{x}\,dt \\ F_{y_d} &= k_{py}\tilde{y} + k_{vy}\tilde{v}_y + k_{iy}\int \tilde{y}\,dt \end{aligned} \tag{25}$$

The proposed *PIV*-type navigation controller is linked with the external controller layer by the rotation matrix $R(\psi) \in SO(2)$ to generate the desired reference angles θ_d and ϕ_d of the attitude controller, in such a way that the following connection between both attitude and translational controllers are obtained.

$$\begin{aligned} \phi_d &= F_{x_d}\sin(\psi) - F_{y_d}\cos(\psi) \\ \theta_d &= F_{x_d}\cos(\psi) + F_{y_d}\sin(\psi) \end{aligned} \tag{26}$$

Equation (26) is proposed for the navigation controller of the octorotor mini-UAV to track the trajectories in the plane $X - Y$.

Mixer Rotor Controllers

As a final step, the control laws corresponding to the motors of the octorotor mini-UAV, considering the mixing matrix (12), are presented as follows

$$
\begin{aligned}
u_{c_1} = \text{Sat}(u_1) &= \text{Sat}\left(\frac{f_{T_d}^{sat}}{8c_T} + \frac{\tau_{\phi_d}^{sat}}{4c_T l} + \frac{\tau_{\theta_d}^{sat}}{4c_T L} + \frac{\tau_{\psi_d}^{sat}}{8c_Q} \right) \\
u_{c_2} = \text{Sat}(u_2) &= \text{Sat}\left(\frac{f_{T_d}^{sat}}{8c_T} - \frac{\tau_{\phi_d}^{sat}}{4c_T l} + \frac{\tau_{\theta_d}^{sat}}{4c_T L} + \frac{\tau_{\psi_d}^{sat}}{8c_Q} \right) \\
u_{c_3} = \text{Sat}(u_3) &= \text{Sat}\left(\frac{f_{T_d}^{sat}}{8c_T} - \frac{\tau_{\phi_d}^{sat}}{4c_T L} + \frac{\tau_{\theta_d}^{sat}}{4c_T l} - \frac{\tau_{\psi_d}^{sat}}{8c_Q} \right) \\
u_{c_4} = \text{Sat}(u_4) &= \text{Sat}\left(\frac{f_{T_d}^{sat}}{8c_T} - \frac{\tau_{\phi_d}^{sat}}{4c_T L} - \frac{\tau_{\theta_d}^{sat}}{4c_T l} - \frac{\tau_{\psi_d}^{sat}}{8c_Q} \right) \\
u_{c_5} = \text{Sat}(u_5) &= \text{Sat}\left(\frac{f_{T_d}^{sat}}{8c_T} - \frac{\tau_{\phi_d}^{sat}}{4c_T l} - \frac{\tau_{\theta_d}^{sat}}{4c_T L} + \frac{\tau_{\psi_d}^{sat}}{8c_Q} \right) \\
u_{c_6} = \text{Sat}(u_6) &= \text{Sat}\left(\frac{f_{T_d}^{sat}}{8c_T} + \frac{\tau_{\phi_d}^{sat}}{4c_T l} - \frac{\tau_{\theta_d}^{sat}}{4c_T L} + \frac{\tau_{\psi_d}^{sat}}{8c_Q} \right) \\
u_{c_7} = \text{Sat}(u_7) &= \text{Sat}\left(\frac{f_{T_d}^{sat}}{8c_T} + \frac{\tau_{\phi_d}^{sat}}{4c_T L} - \frac{\tau_{\theta_d}^{sat}}{4c_T l} - \frac{\tau_{\psi_d}^{sat}}{8c_Q} \right) \\
u_{c_8} = \text{Sat}(u_8) &= \text{Sat}\left(\frac{f_{T_d}^{sat}}{8c_T} + \frac{\tau_{\phi_d}^{sat}}{4c_T L} + \frac{\tau_{\theta_d}^{sat}}{4c_T l} - \frac{\tau_{\psi_d}^{sat}}{8c_Q} \right)
\end{aligned}
\tag{27}
$$

where $u_{c_k} = \text{Sat}(u_k)$ for $k = 1, \ldots, 8$. It is necessary to remark that the saturation functions consider the actual limits of the angular velocity of the motors. This saturation function is defined as

$$
u_{c_k} = \text{Sat}(u_k) = \begin{cases} u_{max} & \text{if} \quad u_k \geq u_{max} \\ u_k & \text{if} \quad u_{min} < u_k < u_{max} \\ u_{min} & \text{if} \quad u_k \leq u_{min} \end{cases}
\tag{28}
$$

where u_{max} represents the maximum quadratic rotor speed and u_{min} is the minimum quadratic rotor speed. The complete control scheme and dynamics are represented in a diagram in Figure 3.

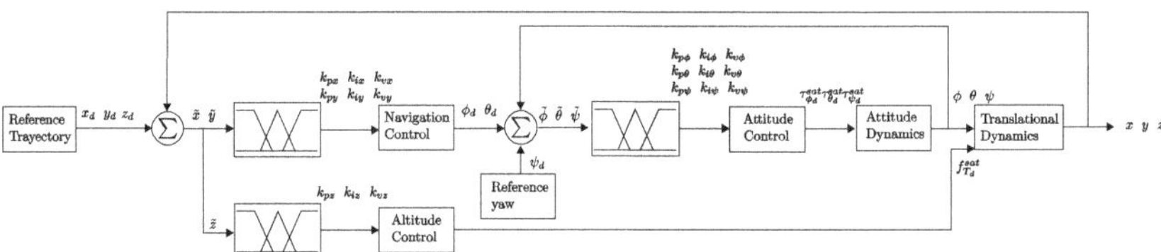

Figure 3. Diagram of attitude and navigation controllers applied to a dynamic model of an octorotor mini-UAV.

4. Gain Scheduling Controller Design

The PIV controller is widely used due to its fast solution, and some researchers use tuning applications to find the optimal constant gains for the PIV controller. Meanwhile, the gain scheduling combines fuzzy and PIV controllers to autotune the gains in real time. A control strategy using gain scheduling with a PIV controller is proposed for the octorotor mini-UAV. As explained above, the objective of PIV control is to avoid deviations during the dynamic motion. The controller can adjust the gains depending on the error size.

The fuzzy logic controller is connected to the PIV gains of the attitude, altitude, and navigation controller with a total of six fuzzy logic controllers; each fuzzy has a membership function, as is shown in Figure 4.

$$
\begin{aligned}
Z &= \begin{bmatrix} a1 & a1 & a2 & a3 \end{bmatrix} \\
SP &= \begin{bmatrix} b1 & b2 & b3 \end{bmatrix} \\
BP &= \begin{bmatrix} c1 & c2 & c3 & c3 \end{bmatrix}
\end{aligned}
\tag{29}
$$

The membership functions have the error and velocity error as inputs while proportional, velocity, and integral gains as outputs. The fuzzy always considers the inputs as positive to determine whether the error is zero (Z), small positive (SP), or big positive (BP) value.

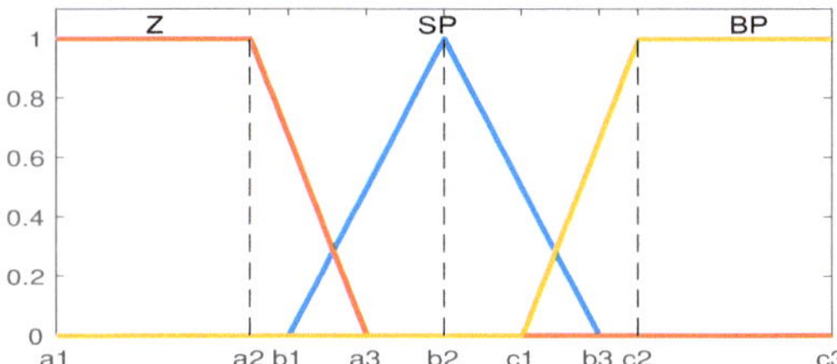

Figure 4. Membership function for input and output of each controller.

Nine rules are defined in Table 1 by combining the inputs, where $\mu(A)$ and $\mu(B)$ are the input position error and input velocity error, respectively. These rules are set to the six fuzzies in each controller. The Tables 2–5 are the values for the fuzzy determined by simulating an octorotor mini-UAV using only PIV control to estimate the range of parameters.

Table 1. Fuzzy rules.

$\mu(A)$ \\ $\mu(B)$	Z	SP	BP
Z	Z	Z	SP
SP	Z	SP	BP
BP	SP	BP	BP

Table 2. Parameters for roll and pitch in membership function.

	Z	SP	BP
$\tilde{\phi}, \tilde{\theta}$	[0 0 0.094 0.14]	[0.11 0.1571 0.2042]	[0.17 0.22 0.3142 0.3142]
$\tilde{p}, \tilde{q}$	[0 0 0.061 0.13]	[0.095 0.1571 0.22]	[0.19 0.25 0.3142 0.3142]
kp	[0 0 64.8 129.6]	[97.2 162 226.8]	[194.4 259.2 324 324]
kv	[0 0 6.48 12.96]	[9.72 16.2 22.68]	[19.44 25.92 32.4 32.4]
ki	[0 0 0.45 1.3]	[0.81 1.62 2.43]	[1.91 2.75 3.24 3.24]

Table 3. Parameters for yaw in membership function.

	Z	SP	BP
$\tilde{\psi}$	[0 0 0.78 1.25]	[0.94 1.57 2.2]	[1.99 2.35 3.142 3.142]
$\tilde{r}$	[0 0 0.074 0.12]	[0.083 0.157 0.23]	[0.19 0.24 0.314 0.314]
kp	[0 0 90 180]	[135 225 315]	[270 360 450 450]
kv	[0 0 31.5 49.5]	[31.5 67.5 103.5]	[85.5 103.5 135 135]
ki	[0 0 0.86 1.67]	[0.99 2.25 3.51]	[2.84 3.69 4.5 4.5]

Table 4. Parameters for X and Y navigation in membership function.

	Z	**SP**	**BP**
$\bar{x},\ \bar{y}$	[0 0 0.75 1.2]	[0.9 1.5 2.1]	[1.8 2.25 3 3]
$\bar{v}_x,\ \bar{v}_y$	[0 0 0.6 1.18]	[0.73 1.5 2.24]	[1.8 2.4 3 3]
kp	[0 0 5.76 11.52]	[8.64 14.4 20.16]	[17.28 23.04 28.8 28.8]
kv	[0 0 2.09 3.31]	[1.74 3.6 5.34]	[5.18 6.48 8.64 8.64]
ki	[0 0 0.28 0.5]	[2.09 4.32 6.41]	[0.94 1.2 1.44 1.44]

Table 5. Parameters for Z altitude in membership function.

	Z	**SP**	**BP**
$\bar{z}$	[0 0 1.25 2]	[1.5 2.5 3.5]	[3 3.75 5 5]
$\bar{v}_z$	[0 0 2.5 4]	[2.5 5 7.5]	[6 7.5 10 10]
kp	[0 0 240 480]	[360 600 840]	[720 960 1200 2400]
kv	[0 0 54 86]	[54 120 180]	[150 180 240 480]
ki	[0 0 0.12 0.21]	[0.14 0.3 0.45]	[0.38 0.49 0.6 0.6]

5. Flight Controller Simulation

5.1. Parameters of the Octorotor Mini-UAV Controller

The parameters used for the numerical simulations are obtained through the octorotor mini-UAV model in Solidworks. In Table 6, the numerical values of the model parameters are shown.

Table 6. Parameters of the octorotor mini-UAV.

Parameter	**Value**	**Parameter**	**Value**
c_T	4.8853×10^{-7} (N-s^2)	J_{xx}	0.0048 (kg-m^2)
c_Q	1.4181×10^{-8} (N-m-s^2)	J_{yy}	0.0051 (kg-m^2)
m	1.772 (kg)	J_{zz}	0.0079 (kg-m^2)
g	9.81 (m/s^2)	J_{xy}	2.3420×10^{-8} (kg-m^2)
d	0.115 (m)	J_{yz}	1.9100×10^{-8} (kg-m^2)
		J_{xz}	8.8301×10^{-7} (kg-m^2)

The parameters used for the saturation are set as follows in Table 7.

Table 7. Saturation parameters.

Parameter	**Value**	**Parameter**	**Value**
$F_{z_{min}}$	-11.76 (N)	$F_{z_{max}}$	11.76 (N)
$\tau_{\eta_{min}}$	-13 (N-m)	$\tau_{\eta_{max}}$	13 (N-m)
u_{min}	1000 (RPM)	u_{max}	34,500 (RPM)

5.2. Control Surface

The control surface is produced by the inference logic fuzzy mechanism for a fuzzy PIV attitude controller with programming gains. It is generated with the tuned parameters in Tables 2 and 3 to compute the variation range of the gain sets ($k_{p\phi}$, $k_{v\phi}$ and $k_{i\phi}$), ($k_{p\theta}$, $k_{v\theta}$ and $k_{i\theta}$) and ($k_{p\psi}$, $k_{v\psi}$ and $k_{i\psi}$) of the attitude controller for roll, pitch, and yaw, (as can be seen in Figure 5). While the control surface is produced by the inference logic fuzzy mechanism for a fuzzy PIV navigation controller with programming gains. It is generated with the tuned parameters in Tables 4 and 5 to compute the variation range of the gain sets (k_{px}, k_{vx} and k_{ix}), (k_{py}, k_{vy} and k_{iy}) and (k_{pz}, k_{vz} and k_{iz}) of navigation and altitude controllers in the X, Y, and Z axes, (see Figure 6). The geometric shape of the control surface (Figures 5 and 6) is the result of the properties that achieve the tuning of fuzzy rules implemented with the Mamdani inference method. It is used to set up the fuzzy PIV controllers with programming gains in the attitude, altitude, and navigation dynamics of the octorotor mini-UAV.

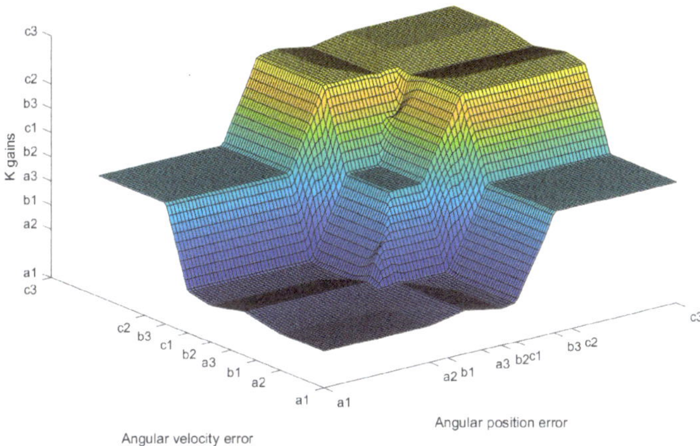

Figure 5. Attitude control surface.

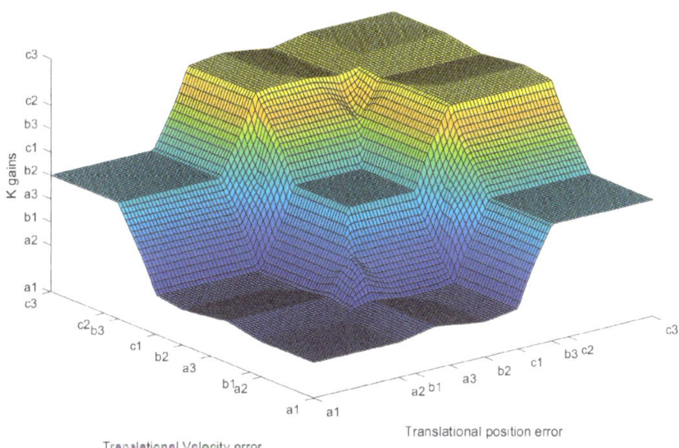

Figure 6. Translational control surface.

5.3. Trajectory Tracking Navigation

The results of the simulation were obtained using Matlab/Simulink. In previous works, similar algorithms were implemented. In [21], a quadrotor with a two-layer saturated *PIV* controller was shown. The quadrotor follows an ascending helical trajectory which is decreasing. In [22], an octorotor is proposed using a nonlinear PV plus a linear integral; in this work, the octorotor has a mass of 8 (kg), and its inertial matrix is simplified. The octorotor follows a helical trajectory, which remains constant during time simulation. The octorotor mini-UAV performs a sequence of movements through the axes traveling 1 m in each axis. The sequence is $-Z, X, Y, -X, Z$, and the negative part of Z-axis is the take-off. Meanwhile, the positive part is the landing, in the X-axis, and the positive part is the forward movement, and the negative part is the backward movement. Finally, in the Y-axis, the positive part indicates the movement from left to right.

A disturbance is added in Equation (4), such that $J\dot{\Omega} = -\Omega \times J\Omega + \tau_a - D$, where D is the disturbance vector that contains disturbances in each torque. The vector $D = [D_{\tau_\phi}, 0, 0]^\top$ with $D_{\tau_\phi} = ae^{(-(t-b)^2/2c^2)}$ is a Gauss peak equation acting as a disturbance, where $a = 4$ is the height of the peak curve, $b = 9$ is the time where the peak is in the center, $c = 1$ is the standard deviation, and t is the time, see Figure 7.

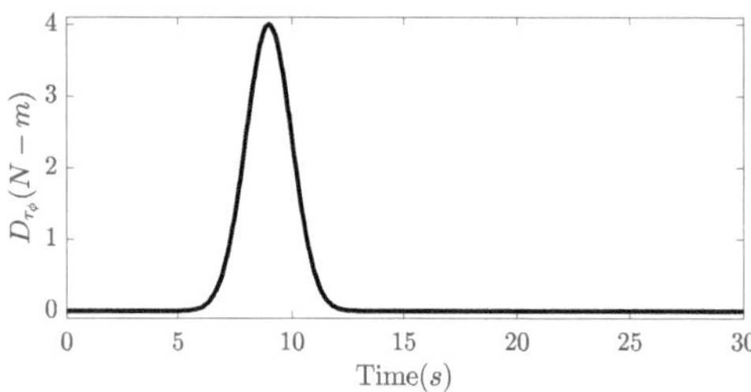

Figure 7. Disturbance D_{τ_ϕ} for the desired torque of roll.

In the simulation, Figure 8 shows the tracking of the desired (red line) and actual (black line) trajectories of the octorotor mini-UAV. A disturbance is added to analyze the behavior of the gain scheduling controller that autotunes the gains to recover stability in the octorotor mini-UAV. The following Figures 9 and 10 show the octorotor mini-UAV in attitude, altitude, and navigation behavior. Then, the rotor speed and gains parameters are shown. Finally, the moments and total thrust of the octorotor mini-UAV are presented.

Figure 9 shows the response of Euler angles; in the roll angle (Figure 9a), the disturbance recalculated the desired value (red line) of the roll angle for a stabilization. While the actual value (black line) continues using less of an angle to recover the trajectory. For the pitch angle (Figure 9b), there is an absence of disturbance due to this simulation considering the disturbance only in τ_ϕ which indirectly affects the Y-axis due to the roll angle. There is a small disturbance in the yaw angle (Figure 9c) when the disturbance in the torque is presented at a time of 9 (s), the following two disturbances are related to the change in direction between the X and Y axes at the times of 12 (s) and 18 (s).

Figure 10 shows the sequence of movements in each axis. In the X-axis (Figure 10a), as can be seen, the controller is performs stabilization each time the direction changes due to the attitude controller in τ_θ. In the Y-axis (Figure 10b), there is a disturbance with a peak at 9 (s) due to the disturbance in τ_ϕ. Finally, the Z-axis (Figure 10c) presents rapid stabilization during ascending and descending.

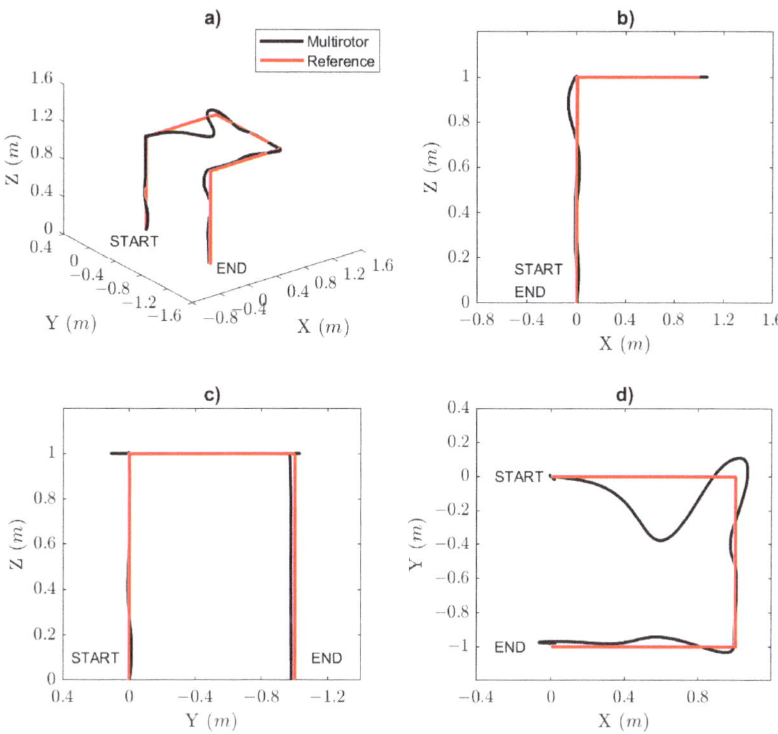

Figure 8. Trajectory tracking of the octorotor mini-UAV, (**a**) isometric view, (**b**) side plane, (**c**) front plane, and (**d**) top view.

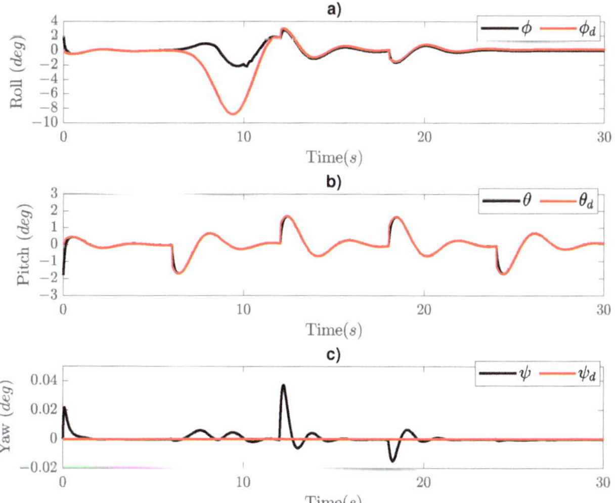

Figure 9. Euler angle responses in: (**a**) ϕ, (**b**) θ, and (**c**) ψ.

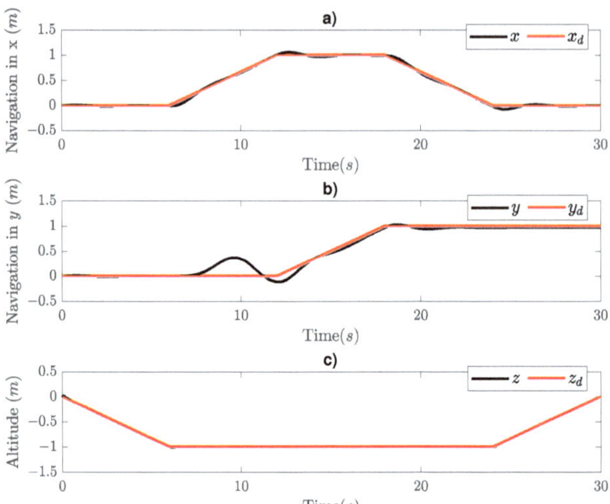

Figure 10. Translational motion responses in: (**a**) X–axis, (**b**) Y–axis, and (**c**) Z–axis as the altitude.

Figure 11 shows the angular velocity in RPM of all rotors, which have a maximum speed limit set to 34,500 (RPM), as obtained in the datasheet of the motor, and a minimum speed limit set to 1000 (RPM), to avoid turning off the rotors. The regulated speed after stabilization is 16,570 (RPM) in each rotor. All rotors are paired in two in the sense that $\omega_1 \simeq \omega_2$, $\omega_3 \simeq \omega_4$, $\omega_5 \simeq \omega_6$, and $\omega_7 \simeq \omega_8$ present a similar behavior with a small difference when the octorotor mini-UAV is changing direction. When the octorotor mini-UAV ascends, the whole rotors are stabilized at 16,570 (RPM), and then, in the X–axis forward direction, rotors ω_1 and ω_2 (Figure 11a) increase their speed and ω_5 and ω_6 (Figure 11c) decrease their speed. When the octorotor mini-UAV changes direction towards the east, rotors ω_7 and ω_8 (Figure 11d) increase their speed and ω_3 and ω_4 (Figure 11b) decrease their speed.

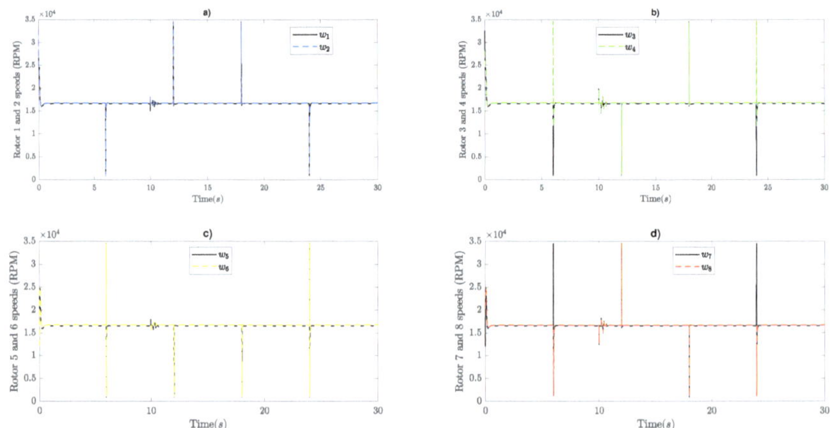

Figure 11. Rotor speed responses: (**a**) ω_1 and ω_2, (**b**) ω_3 and ω_4, (**c**) ω_5 and ω_6, and (**d**) ω_7 and ω_8.

Figure 12 shows the gain behavior in the attitude controller. Figure 12a illustrates the gains of τ_ϕ controller, where the proportional gain changes by the disturbance set. For each change in direction, the proportional and velocity gains change. The gains of the τ_θ controller (Figure 12b) present the same behavior when the octorotor mini-UAV changes direction, except that there is no disturbance in τ_θ. In the τ_ψ controller, the gains (Figure 12c) remain constant due to the desired yaw set to zero.

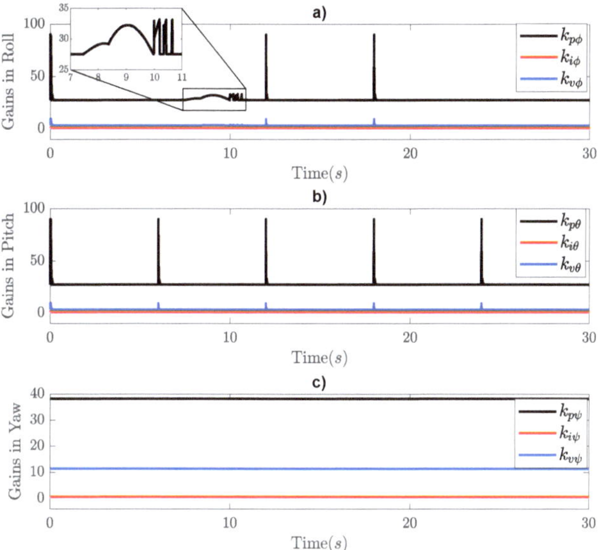

Figure 12. PIV gain responses in attitude controller: (**a**) $k_{h\phi}$, (**b**) $k_{h\theta}$, and (**c**) $k_{h\psi}$ with $h = p, i, v$.

Figure 13 shows the gain behavior in the translation controller. Due to the gain in attitude adjusting, the gains in the translation controller remain constants without changes.

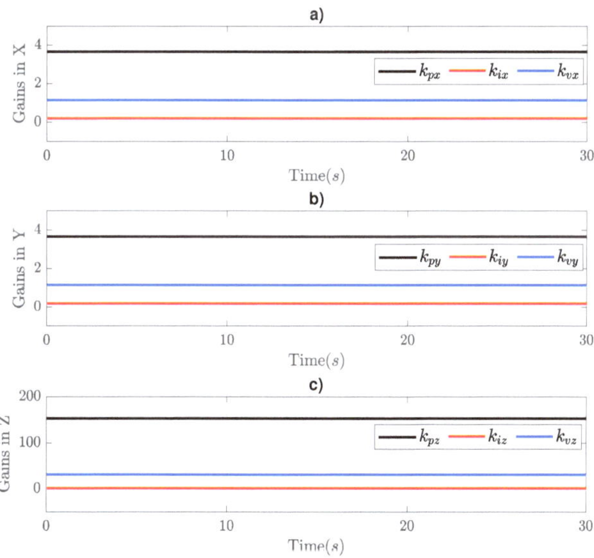

Figure 13. PIV Gain responses in the translation controller: (**a**) k_{hx}, (**b**) k_{hy}, and (**c**) k_{hz}, with $h = p, i, v$.

Figure 14 shows the total thrust of all rotors in the octorotor mini-UAV. The maximum thrust shown is 22.55 (N) at the beginning, and then the stable thrust is 11.77 (N), which is the minimum to maintain the octorotor mini-UAV hovering during flight. The minimum total thrust shown is 5.45 (N) due to the finalization of ascending and when it starts descending.

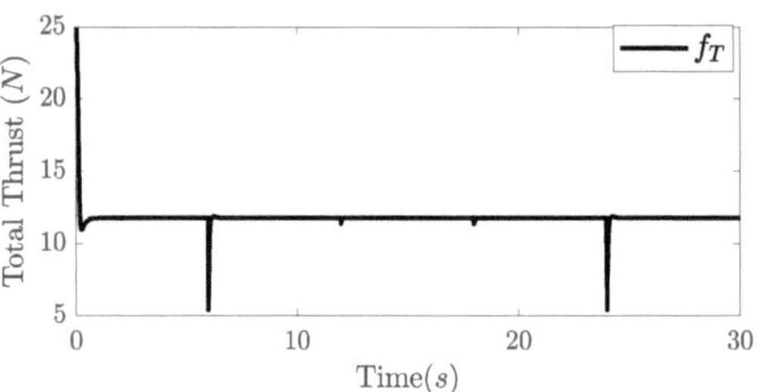

Figure 14. Response of total thrust of the octorotor mini-UAV.

Figure 15 shows moment actuators in the octorotor mini-UAV, where τ_ϕ has the disturbance added at 9 (s) to a disturbated Y–axis as it can be seen in Figure 15a. Then, Figure 15b shows the moment τ_θ when the octorotor mini-UAV performs the X–axis movements. Then, Figure 15c illustrates the τ_ψ signal with a minimum actuation when the octorotor mini-UAV stops ascending when the octorotor mini-UAV performs the Y–axis movement. Finally, when the octorotor mini-UAV starts descending, these actions affect the yaw angle between -0.02 (deg) and 0.03 (deg).

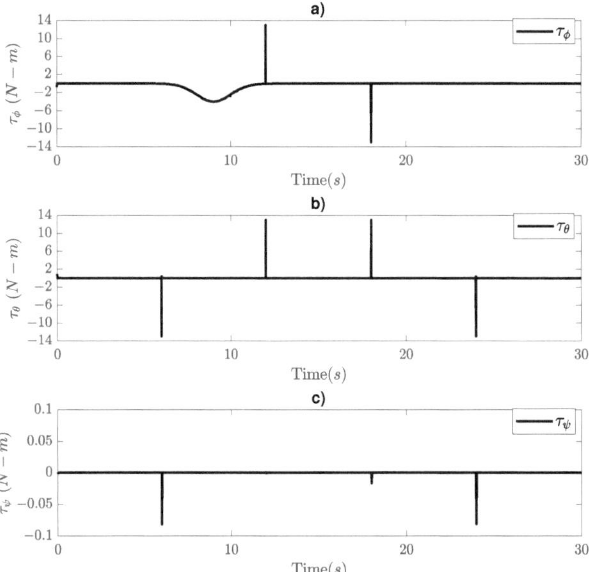

Figure 15. Response of moments: (**a**) τ_ϕ, (**b**) τ_θ, (**c**) τ_ψ.

5.4. Octorotor Mini-UAV Manufacturing

The octorotor mini-UAV frame was manufactured using the FDM (fused deposition modeling) technique with PLA (polylactide) as the primary material. The 3D printer used was an Ultimaker. Figure 16 shows eight separately printed ducts with two ducts printed using the black PLA to identify the front of the aircraft. The base and top of the fuselage were printed separately due to the volume capacity of Ender 6. Finally, Figure 17 shows the octorotor mini-UAV battery support, which is printed in two parts to avoid future installation problems with the battery.

Figure 16. Top view of the octorotor mini-UAV manufactured fuselage.

Figure 17. Bottom view of the octorotor mini-UAV manufactured fuselage.

The electronic power propulsion and avionic system are proposed using eight 2205 Brushless Direct Current (BLDC) motors with 2300 (KV or RPM/V), eight Electronic Speed Controller (ESC) with 30A-S and BLHeli_ S firmware, as well as an eight-blade propeller mounted on each motor. A power distribution board capable of supporting 200 (A) is selected. A PixHawk PX4 ver. 2.4.8 with an STM32F103 processor that works with 160 (MHz), and 256k of RAM is selected. A Turnigy lipo battery with 4000 (mAh), 4S configuration, and a weight of 396 (g) was selected to power up all electronic components.

6. Conclusions

The flight navigation controller PIV fuzzy gain scheduling proposed for the trajectory tracking of the octorotor mini-UAV with attitude inner PIV fuzzy gain scheduling controller was designed. In addition, the PIV fuzzy gain scheduling navigation controller of the octorotor mini-UAV fulfillment a satisfactory performance for tracking trajectories. In effect, it is mentioned that the capacity of the PIV fuzzy gain scheduling flight controller can be tested against heavy cargo transportation and enduring gust wind disturbances. The validation of the attitude and navigation controllers in a closed-loop system with the dynamic model of an octorotor mini-UAV is obtained via numerical simulations. In the simulations, a saturated PIV attitude controller with fuzzy gain scheduling tuning for a fast response was performed. The membership functions of the fuzzy flight controller were tuned on a heuristic procedure to obtain an underdamped response for translational and angular positions with a low overshoot and a fast settling time. For the flight controller tuning procedure, the output membership function of the proportional gain is tuned for fast output response of attitude and navigation positions, then the output membership function of the velocity gain is tuned to reduce the system oscillations; finally, the output membership function of the integral gain is tuned to reduce the steady-state error to a value close to zero. Furthermore, this PIV gain scheduling controller was designed to reject wind gust disturbances fulfilling satisfactory performance of the trajectory tracking responses of the octorotor mini-UAV. Finally, the octorotor mini-UAV fuselage was manufactured using polylactic acid (PLA) material by the fused deposition modeling (FDM) technique, which provides a robust and lightweight platform.

Author Contributions: P.A.T.-B. worked on the mathematical model, controllers, and wrote the first draft. L.E.C.-R. worked on the conceptual design, controllers, and simulation design. M.A.G.-M. worked on the tuning methodology, controller analysis, and the simulation results. E.J.O.-V. proposed the octorotor UAV design, improved the manuscript and supervised the overall project. All authors have read and agreed to the published version of the manuscript.

Funding: This research work is supported by The Technological Institute of La Laguna-TecNM through the award number 14492.22-P. and 10929.21-P.

Institutional Review Board Statement: Not applicable.

Informed Consent Statement: Not applicable.

Data Availability Statement: All research data supporting this study are included in the manuscript.

Acknowledgments: The authors would like to thank The Technological Institute of La Laguna-TecNM and Autonomous University of Nuevo Leon (CIIIA-FIME-UANL).

Conflicts of Interest: The authors declare no conflict of interest.

References

1. Makarov, M.; Maniu, C.S.; Tebbani, S.; Hinostroza, I.; Beltrami, M.M.; Kienitz, J.R.; Menegazzi, R.; Moreno, C.S.; Rocheron, T.; Lombarte, J.R. Octorotor UAVs for radar applications: Modeling and analysis for control design. In Proceedings of the 2015 Workshop on Research, Education and Development of Unmanned Aerial Systems (RED-UAS), Cancun, Mexico, 23–25 November 2015; pp. 288–297.
2. Phelps, G.; Bracken, R.; Spritzer, J.; White, D. Achieving sub-nanoTesla precision in multirotor UAV aeromagnetic surveys. *J. Appl. Geophys.* **2022**, *206*, 104779. [CrossRef]
3. Baumgarten, M.C.; Röper, J.; Hahnenkamp, K.; Thies, K.C. Drones delivering automated external defibrillators—Integrating unmanned aerial systems into the chain of survival: A simulation study in rural Germany. *Resuscitation* **2022**, *172*, 139–145. [CrossRef] [PubMed]
4. Sankey, T.; Donager, J.; McVay, J.; Sankey, J.B. UAV lidar and hyperspectral fusion for forest monitoring in the southwestern USA. *Remote Sens. Environ.* **2017**, *195*, 30–43. [CrossRef]
5. Panagiotidis, D.; Abdollahnejad, A.; Slavik, M. 3D point cloud fusion from UAV and TLS to assess temperate managed forest structures. *Int. J. Appl. Earth Obs. Geoinf.* **2022**, *112*, 102917. [CrossRef]
6. Garcia, C.A.; Xu, Y. Multitarget geolocation via an agricultural octorotor based on orthographic projection and data association. *Proc. Inst. Mech. Eng. Part J. Aerosp. Eng.* **2018**, *232*, 2076–2090. [CrossRef]
7. Garcia, C.A.; Xu, Y. Target Geolocation for agricultural applications via an Octorotor. *IFAC-PapersOnLine* **2016**, *49*, 27–32. [CrossRef]
8. Donmez, C.; Villi, O.; Berberoglu, S.; Cilek, A. Computer vision-based citrus tree detection in a cultivated environment using UAV imagery. *Comput. Electron. Agric.* **2021**, *187*, 106273. [CrossRef]
9. Rivas, Casado, M.; Ballesteros, Gonzalez, R.; Kriechbaumer, T.; Veal, A. Automated identification of river hydromorphological features using UAV high resolution aerial imagery. *Sensors* **2015**, *15*, 27969–27989. [CrossRef] [PubMed]
10. Madokoro, H.; Woo, H.; Sato, K.; Shimoi, N. Development of Octo-Rotor UAV Prototype with Night-vision Stereo Camera System Used for Nighttime Visual Inspection. In Proceedings of the 2019 19th International Conference on Control, Automation and Systems (ICCAS), Jeju, Republic of Korea, 15–18 October 2019; pp. 998–1003.
11. Zhang, B.; Tsuchiya, S.; Lim, H. Development of a Lightweight Octocopter Drone for Monitoring Complex Indoor Environment. In Proceedings of the 2021 6th Asia-Pacific Conference on Intelligent Robot Systems (ACIRS), Tokyo, Japan, 7–9 July 2021; pp. 1–5.
12. Walter, A.; McKay, M.; Niemiec, R.; Gandhi, F. Trim Analysis of a Classical Octocopter After Single-Rotor Failure. In Proceedings of the 2018 AIAA/IEEE Electric Aircraft Technologies Symposium (EATS), Cincinnati, OH, USA, 9–11 July 2018; pp. 1–20.
13. Marks, A.; Whidborne, J.F.; Yamamoto, I. Control allocation for fault tolerant control of a VTOL octorotor. In Proceedings of the 2012 UKACC International Conference on Control, Cardiff, UK, 3–5 September 2012; pp. 357–362.
14. Zeghlache, S.; Mekki, H.; Bouguerra, A.; Djerioui, A. Actuator fault tolerant control using adaptive RBFNN fuzzy sliding mode controller for coaxial octorotor UAV. *Isa Trans.* **2018**, *80*, 267–278. [CrossRef] [PubMed]
15. Hamadi, H.; Lussier, B.; Fantoni, I.; Francis, C.; Shraim, H. Comparative study of self tuning, adaptive and multiplexing FTC strategies for successive failures in an Octorotor UAV. *Robot. Auton. Syst.* **2020**, *133*, 103602. [CrossRef]
16. ul Amin, R.; Inayat, I.; Jun, L.A. Finite time position and heading tracking control of coaxial octorotor based on extended inverse multi-quadratic radial basis function network and external disturbance observer. *J. Frankl. Inst.* **2019**, *356*, 4240–4269. [CrossRef]
17. Li, C.; Wang, Y.; Yang, X. Adaptive fuzzy control of a quadrotor using disturbance observer. *Aerosp. Sci. Technol.* **2022**, *128*, 107784. [CrossRef]
18. Nekoukar, V.; Dehkordi, N.M. Robust path tracking of a quadrotor using adaptive fuzzy terminal sliding mode control. *Control. Eng. Pract.* **2021**, *110*, 104763. [CrossRef]

19. Lozano, R. *Unmanned Aerial Vehicles Embedded Control*; John Wiley-ISTE Ltd.: Hoboken, NJ, USA, 2010.
20. Stengel, R.F. *Flight Dynamics*; Princeton University Press: Princeton, NJ, USA, 2004.
21. Ollervides-Vazquez, E.J.; Rojo-Rodriguez, E.G.; Rojo-Rodriguez, E.U.; Cabriales-Ramirez, L.E.; Garcia-Salazar, O. Two-layer saturated PID controller for the trajectory tracking of a quadrotor UAV. In Proceedings of the 2020 International Conference on Mechatronics, Electronics and Automotive Engineering (ICMEAE), Cuernavaca, Mexico, 16–21 November 2020; pp. 85–91.
22. Tellez-Belkotosky, P.A.; Ollervides-Vazquez, E.J.; Rojo-Rodriguez, E.G.; Santillan-Avila, J.L.; Cabriales-Ramirez, L.E.; Gutierrez-Martinez, M.A.; Garcia-Salazar, O. Nonlinear flight navigation controller for an octorotor unmanned aerial vehicle. In Proceedings of the 2021 International Conference on Mechatronics, Electronics and Automotive Engineering (ICMEAE), Cuernavaca, Mexico, 22–26 November 2021; pp. 52–59.

Article

The Coupled Wing Morphing of Ornithopters Improves Attitude Control and Agile Flight

Yu Cai, Guangfa Su , Jiannan Zhao * and Shuang Feng *

Guangxi Key Laboratory of Intelligent Control and Maintenance of Power Equipment, School of Electrical Engineering, Guangxi University, Nanning 530004, China; caiy@gxu.edu.cn (Y.C.); guangfasu@163.com (G.S.)
* Correspondence: jzhao@gxu.edu.cn (J.Z.); fshuang@gxu.edu.cn (F.S.)

Abstract: Bird wings are exquisite mechanisms integrated with multiple morphological deformation joints. The larger avian species are particularly adept at utilizing their wings' flapping, folding, and twisting motions to control the wing angle and area. These motions mainly involve different types of spanwise folding and chordwise twisting. It is wondered whether the agile maneuverability of birds is based on the complex coupling of these wing morphing changes. To investigate this issue, we designed a two-section wing structure ornithopter capable of simultaneously controlling both spanwise folding and chordwise twisting and applied it to research on heading control. The experimental data collected from outdoor flights describe the differing flight capabilities between the conventional and two-section active twist wing states, indicating that incorporating an active twist structure enhances the agility and maneuverability of this novel flapping aircraft. In the experiments on yaw control, we observed some peculiar phenomena: although the twisting motion of the active twist ornithopter wings resembles that of a fixed-wing aileron control, due to the intricate coupling of the wing flapping and folding, the ornithopter, under the control of active twist structures, exhibited a yaw direction opposite to the expected direction (directly applying the logic assumed by the fixed-wing aileron control). Addressing this specific phenomenon, we provide a plausible model explanation. In summary, our study with active twist mechanisms on ornithopters corroborates the positive impact of active deformation on their attitude agility, which is beneficial for the design of similar bio-inspired aircraft in the future.

Keywords: ornithopter; wing morphing coupling; UAV attitude control; two-section wing

1. Introduction

Birds can adjust the geometric morphology of their wings to alter their posture [1]. Compared to insect wings, avian wings possess multiple joints and exhibit greater maneuverability [2]. Therefore, investigating avian wing deformation offers inspiration for aircraft attitude control [3–5]. Birds can alter their wing morphology through various muscular and joint movements, including spanwise folding [6], spanwise extension–retraction [7], and chordwise twist [8], as depicted in Figure 1a–d. These complex muscular joint configurations and wing deformation capabilities ensure the agility of avian flight [9]. Figure 1 depicts the four representative geometric morphologies of avian wings during flight. Among these, the flapping motion typified by hummingbirds does not apparently alter the static geometric morphing of the wings [10], generating lift solely through up-and-down flapping (Figure 1a,e). As shown in Figure 1b,f, the spanwise extension–retraction exhibited by high-speed avian species (such as falcons) manifests as a slight forward and aft sweeping motion of the wings [11]. It is reported that varying this in-plane morphing parameter enhances flight maneuverability [12]. In Figure 1c,g, represented by albatrosses and other large-wingspan birds, the spanwise folding deformation involves an "M"-shaped up-and-down flapping motion of the wings; it is reported that it can enhance the flight stability during deceleration [13,14]. Figure 1d,h illustrate the chordwise twist movement,

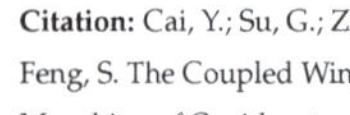

Citation: Cai, Y.; Su, G.; Zhao, J.; Feng, S. The Coupled Wing Morphing of Ornithopters Improves Attitude Control and Agile Flight. *Machines* **2024**, *12*, 486.
https://doi.org/10.3390/machines12070486

Academic Editors: Octavio Garcia-Salazar, Anand Sanchez-Orta and Aldo Jonathan Muñoz-Vazquez

Received: 12 June 2024
Revised: 11 July 2024
Accepted: 12 July 2024
Published: 19 July 2024

which is typically studied in pigeons; the research has focused on the effect of chordwise twist on agile attitude control [8].

The study of morphological parameters and their impact on flapping flight is a research hotspot. Song pointed out that research on the static morphological characteristics of avian wings has been extensive in non-flapping states. However, there has been relatively little theoretical research on the morphological deformation coupled with flapping states [15].

The morphological research in non-flapping states can be summarized such that (1) altering the wing shape parameters at different angles of attack can affect the lift coefficient [16]; (2) actions similar to the spanwise folding movements observed in large-winged birds (Figure 1a) can enhance flight stability [13]; and (3) pigeons can improve flight agility through chordwise twist [8]. However, the morphological variables of these studies are not coupled with flapping, thus underestimating the complex aerodynamic situations induced by the coupling of flapping motion and wing deformation.

Does the torsional ability of wing muscles along the chord differ during flapping motions (such as extension–folding or extension–retraction) compared to non-flapping states?

The research on the morphological variables of wings during the flapping states remains scarce [4]. Only a few studies have focused on the influence of the aspect ratio on lift and thrust performance [15]. For instance, seagull wings exhibit excellent lift characteristics during flapping flight due to spanwise folding [17]. Some studies have effectively explained the advantages of large-winged birds in generating lift and saving power during flapping by describing the effects of wing folding on leading-edge vortices and Strouhal numbers [18]. Thielicke, W. [19] and the team led by Song [20] have conducted relevant studies on active chordwise twist, demonstrating that chordwise twist can reduce the constraint of the wing on vortex streets, decrease the magnitude of the lift, and alter the distribution of the effective angles of attack. However, this literature solely comprises aerodynamic simulations without investigating the coupling effects of chordwise twists and the flapping motion on the attitude of flapping-wing aircraft. To the best of the authors' knowledge, this paper represents the first study regarding the coupling effect of chordwise twist (Figure 1c) and spanwise folding (Figure 1d) on attitude control during flapping.

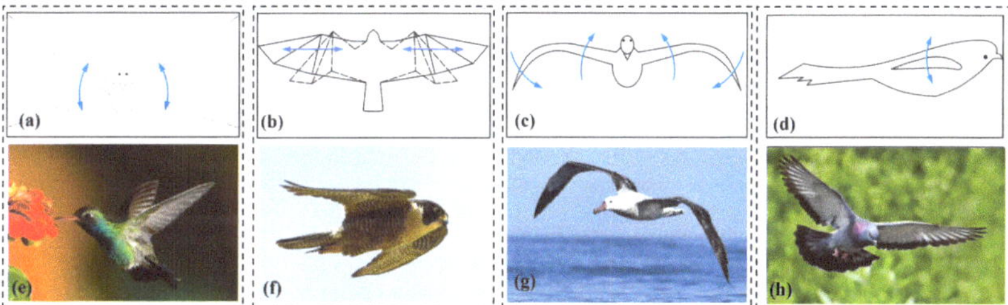

Figure 1. Geometric morphology of bird wing motion, where the first row depicts simplified schematic diagrams of motion, and the second row shows corresponding bird flight forms. (**a**) The up-and-down flapping of the wings without static geometric deformation. (**b**) Spanwise extension–retraction of the wings. (**c**) Spanwise folding of the wings, folding upwards and downwards. (**d**) Chordwise twisting of the wings. (**e**) Flapping flight mode, exemplified by hummingbirds, where the wings do not undergo static deformation. (**f**) Spanwise extension–retraction flight mode, exemplified by peregrine falcons. (**g**) Spanwise folding flight mode, represented by albatrosses with large wing spans. (**h**) Flight mode of pigeons, enhancing agility through chordwise twisting during flight [8].

In the literature on biomimetic mechanical design, there are numerous studies focusing on the mechanical structures mimicking each of the three wing deformation modes depicted in Figure 1b–d individually. For instance, in studies on wing morphology resembling spanwise extension–retraction (achieving the motion depicted in Figure 1b), Chan from Stanford University improved upon a fixed-wing prototype by creating a drone with foldable wings. The research indicates that altering the wing area and angle of fixed-wing aircraft through joint unfolding movements can significantly enhance the operability, agility, and stability of the aircraft [12]. Their study utilized a fixed-wing aircraft platform (with the lift generated by propellers rather than flapping) to investigate the effect of asymmetric spanwise retraction on aircraft attitude control. However, constrained by the nature of the fixed-wing platform, their wing deformations were not coupled with a flapping motion. In contrast, researchers at Northwestern Polytechnical University designed a mechanical structure that couples wing deformation with flapping motion. Their developed RoboFalcon flapping-wing aircraft replicated the effect of the asymmetric spanwise folding of wings on attitude control during the flapping states (an asymmetric form of the extension–folding depicted in Figure 1b). RoboFalcon achieves flight and attitude control through asymmetric forward and aft sweeping motions [21]. Their study indicates that the morphing-coupled flapping mode exhibits higher lift effects compared to the single-wing flapping mode with unchanged static characteristics when increasing the downstroke duration [11].

The studies on wing morphology resembling spanwise folding (emulating the motion depicted in Figure 1c) began in 2011, and the two-section wing ornithopter mimicking the avian wing deformation mechanisms was first developed by the Festo Corporation. The flagship model, known as Smartbird, has gathered significant attention due to its remarkably realistic flight posture resembling that of actual birds [22]. Based on the structure of Smartbird, Zhang et al. developed an avian ornithopter with feather covers and achieved reliable flight [23].

In the studies on wing morphology resembling chordwise twist (emulating the motion depicted in Figure 1d), both theoretical and experimental research have been limited to single-section wing ornithopters. For example, the team led by Xiao et al. designed a differential control mechanism for active chordwise twist, verifying that lateral control using wing differential torsion yields superior performance compared to rudder control [24].

The aforementioned studies have independently investigated the influence of altering certain morphological parameters of wings on their aerodynamic performance or attitude control. However, during avian flapping flight, various types of geometric parameters of wings should be coupled [15]. Therefore, when studying flapping-wing aircraft, considering multiple geometric parameters of wings and studying the coupling effects of the morphology and flapping (morphing-coupled flapping) are essential. Due to the complexity of the flapping motion, it is currently impractical to fully replicate the combined actions of all the muscles in birds [9]. However, it is feasible to gradually observe the combined effects of several morphological parameters on flapping-wing aircraft. We note the lack of studies investigating the coupling effects of spanwise folding and chordwise twisting, which becomes the focus of our study. Given the complexity of the flapping motion, the addition of each new mechanism may potentially render the existing motion model ineffective. Therefore, we have adopted a cautious approach and designed a two-section ornithopter that retains a vertical tail mechanism, active chordwise twist mechanisms, and spanwise folding mechanisms. Through extensive experimentation (initially encountering numerous uncontrollable phenomena due to insufficient analysis of the aerodynamic model), we ultimately propose a model and control principles that enhance the agility of attitude control.

To summarize, inspired by the control strategies of the complex morphing strategy of avians, this paper investigates the impact of chordwise active twist coupled with spanwise folding-induced flapping through a two-section wing ornithopter. The main contributions are as follows:

- We designed a two-section wing ornithopter with a spanwise folding and chordwise active twist mechanism. This flapping vehicle achieved agile flight and attitude control in the open air.
- We experimentally analyzed the impact of the chordwise active twist structure on the attitude control during flapping. The results demonstrated that the spanwise active

twist structure can provide independent direction control for headings. Therefore, it can enhance the maneuverability and agility of the ornithopter.

- The experimental findings revealed that, due to the complex coupling of chordwise folding and spanwise active twist, the attitude control phenomenon could not be predicted by the traditional fixed-wing aileron control mechanism. Therefore, a novel model analysis and explanation is proposed for this coupled wing-flapping pattern. The proposed "M"-shaped model describes the particularly special phenomenon that the attitude control effect of a single spanwise active twist is contrary to that when it is coupled with chordwise folding.

2. Design of Active Twist Two-Section Wing Ornithopter

To investigate the influence of wing deformation on avian posture control for potential application in unmanned aerial vehicle (UAV) design, we developed a two-section wing with an active twist structure inspired by the morphology of seagulls, as illustrated in Figure 2a and parameter specifications are listed in Table 1. The propulsion architecture of this flapping wing mechanism resembles Festo's Smartbird design. This model's outer and inner wing segments are composed of two four-bar linkage structures; the outer segment and the inner segment are driven by a single motor, resulting in a coupled motion between the outer segment and the inner segment. Therefore, it is an actuated system with one degree of freedom [22]; the motor only provides lift and thrust without posture control capability.

Table 1. Main properties of ornithopter.

Property	Value
Wing span/m	1.5
Aspect ratio	4.26
Weight/g	400
Wing length (outer)/m	0.60
Wing length (inner)/m	0.45

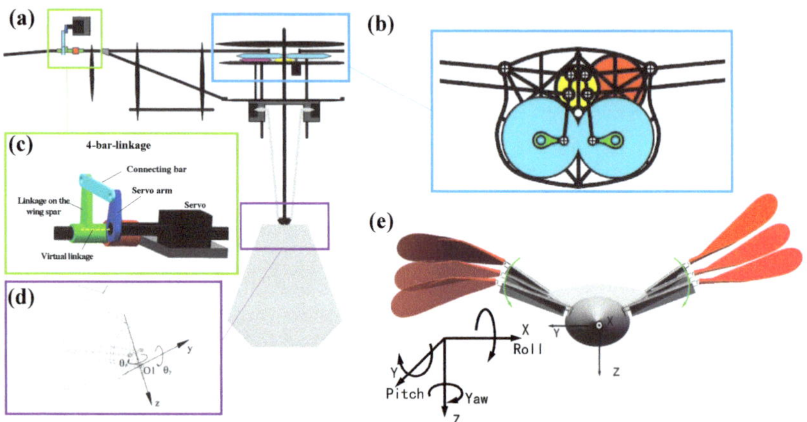

Figure 2. (**a**) Skeletal structure of an ornithopter. The power system motor and variable speed gears form a set of variable speed systems. The left wing has a three-stage transmission, and the right wing has a four-stage transmission. (**b**) Diagram of the effect of the twisted wing structure in action (with fuselage coordinates, originating at the center of mass of the ornithopter). (**c**) Twisted wing structure, a four-bar linkage servo arm (blue), linkage on the wing spar (green), connecting bar (light blue), and virtual linkage between wing spar and servo arm root. (**d**) Diagram of the change in the center position of the flapping wing craft with the flapping of the wings. The blue color is for a two-section wing flap ornithopter, and the red color is for a single-ended wing flap of the same size. (**e**) Definition of the coordinates of the tail–fuselage junction.

During flight, birds can generate rolling moments by altering the posture and shape of their two wings differently. To achieve a similar effect, we propose a mechanism wherein servo motors are positioned at the joints of the outer and inner wing segments (as depicted in Figure 2b). The rotation of the outer wing's twist servo motor can induce the rotation of the wing spar. Under remote control, the servo motors of the left and right wings can execute opposite directional twists, thereby inducing different deformation effects in the outer wing segments on either side (as illustrated in Figure 2d). This generates an asymmetric moment on the fuselage, thereby achieving posture control. To obtain controlled experimental results without active twist, our aircraft retained the commonly observed vertical tail structure for posture control, as depicted in Figure 2c. Through simulation analysis conducted using SolidWorks 2024 and MATLAB R2024a software, it has been observed that, compared to single-wing flapping-wing aircraft, the two-section wing structure exhibits minimal variation in the center of gravity position, as illustrated in Figure 2e, which is advantageous for the design of flapping-wing aircraft controllers. This observation is consistent with the conclusions drawn by Lee JS and colleagues regarding the research on folding-wing and flapping-wing aircraft [25]. A dynamic demonstration of the special wing-morphing (flapping–twisting) mechanism, can be found in the video of the suplymentary material.

3. Aerodynamic Analysis

The factors influencing the lift and thrust of an aircraft are numerous, including the air density coefficient ρ, the horizontal projected wing area S, the incoming flow velocity v_L, the cruising velocity v_T, the lift coefficient C_L, and the thrust coefficient C_T. The mathematical expressions are represented as shown in Equations (1) and (2):

$$F_L = \frac{1}{2}C_L\rho Sv_L^2 \tag{1}$$

$$F_T = \frac{1}{2} C_T \rho S v_T^2 \tag{2}$$

Thus, the lift and thrust coefficients can be represented by Equations (3) and (4):

$$C_L = \frac{2F_L}{\rho S v_L^2} \tag{3}$$

$$C_T = \frac{2F_T}{\rho S v_T^2} \tag{4}$$

The horizontal projected area of the flapping wing dominates the trend of lift variation, where the local tangential velocity (v) and horizontal projected area determine the instantaneous lift of the flapping wing mechanism, which is closely related to the development of leading-edge vortices [6].

Experimental testing of the aerodynamic performance of the bi-plane flapping-wing aircraft was conducted without yaw control. The experiments involved measuring the lift and thrust of the aircraft using an ANIPRO RL4 turntable system [26], as illustrated in Figure 3a. From Figure 3b, it can be observed that, at a certain relative airspeed, the thrust coefficient of the flapping-wing aircraft is directly proportional to the wing flapping frequency [27]. Furthermore, as shown in Figure 3c, the thrust coefficient of the two-section flapping-wing aircraft is maximally affected by the relative airspeed at low frequencies. These results and analysis guarantee the basic flight safety of our outdoor real-flight experiments.

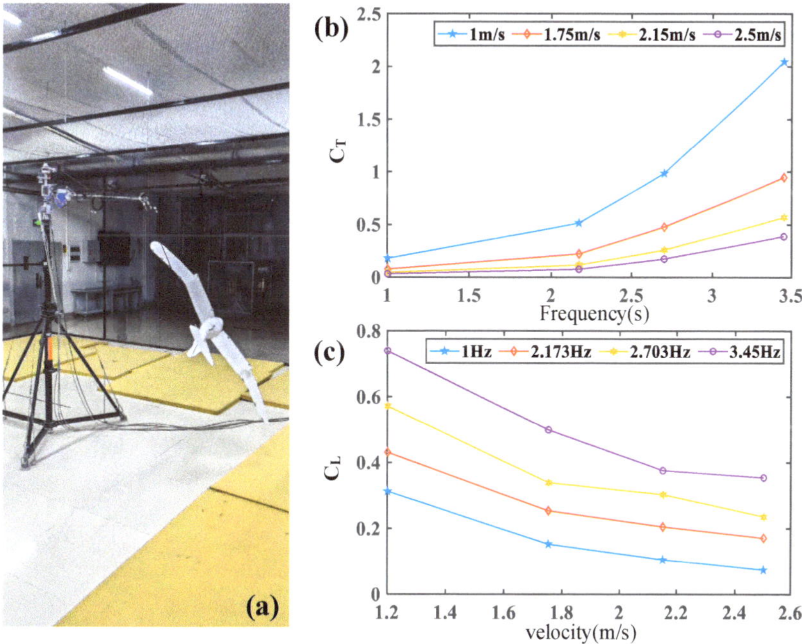

Figure 3. (a) Experimental force analysis of ornithopters under ANIPRO RL4 turntable system allows for the measurement of the lift and thrust forces experienced by the ornithopters under specific conditions, accompanied by a dynamic capture system to measure the flapping frequency of the ornithopters. (b) The relationship between flapping frequency and thrust coefficient (C_T) of ornithopters at different airspeeds. (c) The relationship between relative airspeed and lift coefficient (C_L) of the ornithopter at various flapping frequencies.

4. Outdoor Real Flight Experiments on Horizontal Maneuverability

As depicted in Figure 4, multiple stable flight tests were conducted outdoors, during which the attitude data and control signals were recorded using onboard sensors (more information of the outdoor flight can be found in the suplymentary material). These experiments encompassed both pre-modification flight trials and flights after installing the active twist mechanism. Due to the lack of historical literature and prior knowledge on the aeroelastic testing of wingtip twist, to ensure flight safety, we incorporate an active wingtip twist structure onto a relatively mature scheme with a vertical tail. During the experimental exploration of the influence of the active twist structure on the flight attitudes, the control of the vertical tail is disabled to ensure the uniqueness of the variable.

Figure 4. (**a**,**b**) The outdoor flight experiments of two-section wing ornithopter.

The ability of an aircraft to change its flight direction is referred to as horizontal maneuverability [28,29]. During the actual flight, it was observed that discrepancies between the left and right wings were inevitable due to manufacturing variations, resulting in an imbalance of the forces on both sides of the aircraft. In this case, relying solely on the tail fin cannot effectively control the aircraft's roll to a neutral position, leading to the torpid climbing performance of the aircraft. Human pilots have repeatedly observed this occurrence during outdoor experiments. It indicates that relying solely on tail control for the heading direction control of a flapping-wing aircraft can lead to a deficiency in the aircraft's horizontal maneuverability.

To evaluate the control performance, the correlation between the roll angle and the control signals (from the remote controller) is calculated using the root mean squared error (RMSE). This evaluation is feasible because the control signals provided by the remote controller are proportional to the mechanism's rotational angles, and its effects on attitude control can be measured by the Inertial Measurement Unit directly. Data points are recorded by the onboard processor at 25 Hz and represented in a two-dimensional coordinate system as $P(controlinput, attitudeangle)$.

The calculation of RMSE is provided as follows:

$$RMSE_1 = \sqrt{\frac{\sum_{i=1}^{n}(Roll_i - \widehat{Roll_i})^2}{n}}, \tag{5}$$

$$RMSE_2 = \sqrt{\frac{\sum_{i=1}^{n}(Pitch_i - \widehat{Pitch_i})^2}{n}}. \tag{6}$$

where n is the total number of recorded data points, $Roll$ represents the roll angle of the aircraft, $\widehat{Roll}$ is the mean obtained from the regression equation, $Pitch$ represents the pitch angle of the aircraft, and $\widehat{Pitch}$ is the mean obtained from the regression equation.

4.1. Horizontal Maneuverability without Active Twist

Before installing the active twist structure, the aircraft's attitude control relies solely on the T-tail. The tail is capable of rotation about the Z-axis and the Y-axis (as shown in Figure 2a,c). The control principle is as follows: rotation about the Z-axis increases the force on the horizontal surface of the tail, thereby increasing the aircraft's angle of attack; rotation about the Y-axis alters the force on the vertical surface of the tail, thereby modifying the aircraft's yaw and roll angles. During the actual flight experiment, the relationship between the T-tail control input and the aircraft's attitude is illustrated in Figure 5. It can be observed that there is a strong correlation between the aircraft's roll angle and the control signal along the Z-axis of the tail surface (Figure 5a), as well as the correlation between the pitch angle and the control signal along the Y-axis (Figure 5c). In comparison to the linear regression equations between the attitude angles and the control signals, the root mean square errors (RMSEs) are 1.1530 for the roll angle and 3.1161 for the pitch angle, as depicted in Figure 5b,d.

In this configuration, the aircraft cannot fully control all three degrees of freedom due to only two servos acting as active control mechanisms on two degrees of freedom. However, the control signal along the Z-axis influences both the yaw and roll angles. According to [30], this coupling phenomenon in the control effectiveness of roll and yaw angles is attributed to the T-tail configuration, which causes the center of pressure of the horizontal control surface to deviate from the xoz plane where the aircraft's center of mass is located, resulting in a rolling moment. Consequently, the tilting lift generates a lateral force in the horizontal direction, providing the centripetal force required for the aircraft's yaw rotation [28]. Comparing the data for roll and yaw (Figure 6), it is evident that there is a significant coupling effect between them in terms of control.

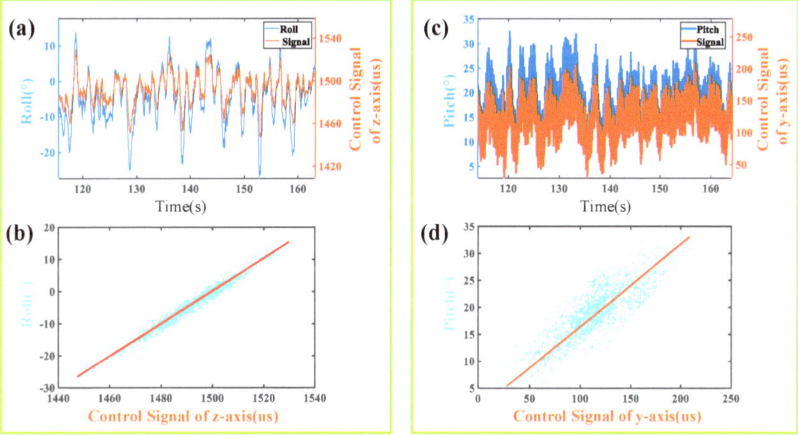

Figure 5. Graph of the relationship between the attitude of an ornithopter and control signals. (**a**) Roll angle and tail wing Z-axis control signal for the ornithopter. (**b**) The linear regression plot of roll angle against Z-axis control signal. (**c**) Pitch angle and tail wing Y-axis control signal for the ornithopter. (**d**) The linear regression plot of pitch angle against the Y-axis control signal.

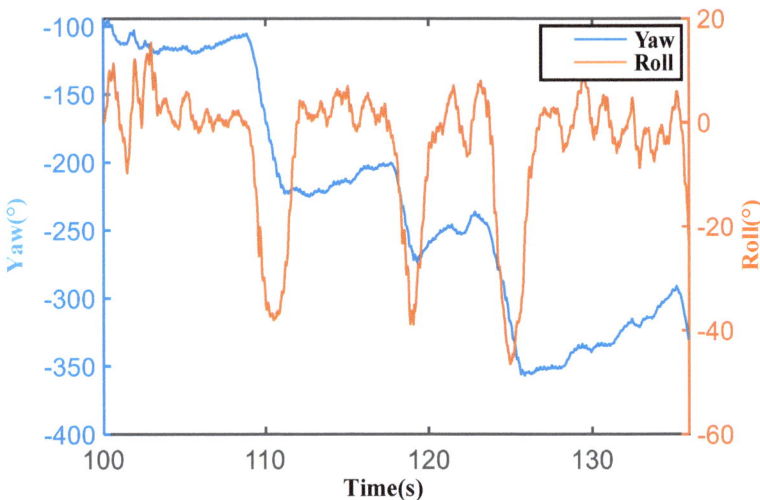

Figure 6. The coupling relationship diagram between pitch angle and roll angle under tail control.

4.2. Horizontal Maneuverability with Active Twist Mechanism

Birds adjust their flight posture by flexibly changing the shape of their wings during flight. Wing deformation causes the displacement of the vortices on the wing surface, so the deformation of the outer wing of a two-section ornithopter can result in changes in lift and thrust [31]. We aim to install an active twist wing structure to enable the ornithopter to also have the capability to adjust its aircraft attitude through imbalanced wing deformation.

In actual flight experiments, the control effect of the active twist mechanism is illustrated in Figure 7. The magnitude of the control signal (*Signal*) is directly proportional to the active twisting angle of the ornithopter's outer wing sections. This mechanical structure is somewhat similar to the ailerons on fixed-wing aircraft. On a fixed-wing aircraft, when the ailerons undergo differential deflection as indicated by the signal, different lift distributions are generated on the two wing surfaces, resulting in a rolling moment, as depicted in Figure 7a. According to the theory of aileron control on fixed-wing aircraft [32], the roll angle change of the aircraft should be as shown in Figure 7a (a fake curve generated by inverting the roll data in Figure 7b), demonstrating that the higher side of the aileron on the fixed-wing aircraft rolls downward. However, as shown in Figure 7b, the experimental data contradict this expectation. In comparison to the linear regression equation between attitude angles and control signals, the root mean square error (RMSE) is $RMSE_3 = 9.0975$, as depicted in Figure 7e. The active twist wing mechanism causes the roll situation of the ornithopter to be the opposite, with the side of the outer wing twisting upward, resulting in the aircraft rolling upward. During actual flight, it was observed that, when the ornithopter's wings are higher than the fuselage, the reversal effect of the roll angle becomes significant. When the wings are level with or lower than the fuselage, the roll angle control effect tends to resemble that of a fixed-wing aircraft.

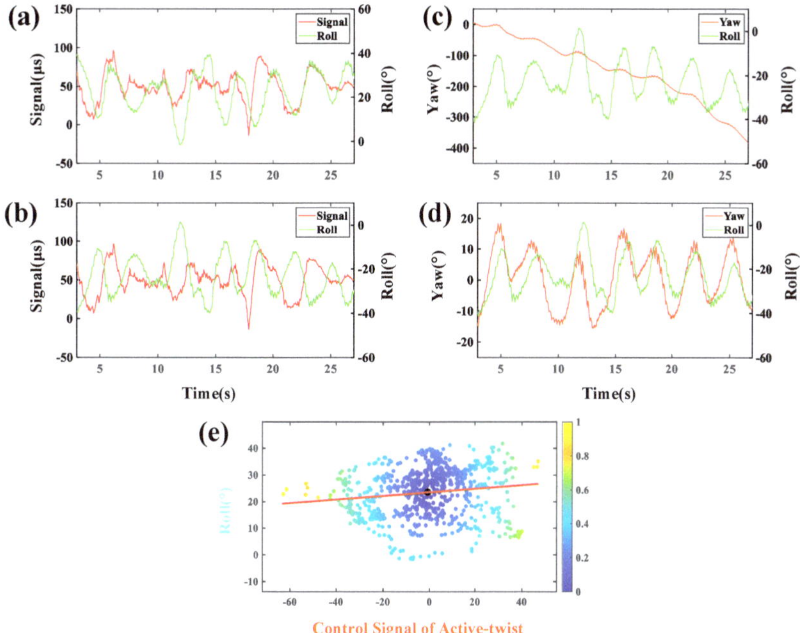

Figure 7. The control relationship between the wing twist control signal and the aircraft's roll and yaw angles. (**a**) The twist wing signal's anticipated control effect on the ornithopter's roll angle (similar to the aileron's control effect on fixed-wing aircraft). (**b**) The twist wing signal's actual control effect on the ornithopter's roll angle is completely opposite to the anticipated effect. (**c**) The coupling relationship between the yaw and roll angles during actual flight. (**d**) The coupling relationship between the yaw angle signal filtered by low frequency and the roll angle. (**e**) The linear regression curve between the wing twist signal and the roll angle.

In the experiments with the installed active twist wing mechanism, the rolling induced by wing deformation has alleviated the previously encountered climbing difficulties and the manufacturing asymmetry issue. The ornithopter can now achieve left or right turns or adjust the bias caused by left–right asymmetry through the imbalance active twisting of the outer wing. Comparative experiments indicate that, by solely controlling the twisting of the ailerons, it is possible to change the aircraft's yaw angle, as shown in Figure 7c. In Figure 7c, the roll angle deviates from the direction away from $0°$, causing the aircraft to continuously circle in one direction, while the value of the yaw angle increases in the negative direction in an integrated form. Under these circumstances, does the coupling phenomenon still exist between the roll angle and yaw angle control? By extracting the high-frequency components of the yaw angle, filtering out the low-frequency fundamental signal of yaw (0.0185 Hz), and comparing it with the roll angle control signal, as shown in Figure 7d, it can be observed that the high-frequency part of the yaw angle changes synchronously with the roll angle, indicating that the coupling effect still exists in the control. This result suggests that the active wing-warping mechanism can provide the required torque for attitude changes, achieving control over both the roll and yaw angles. The experiential feedback from pilots is that, during horizontal directional control, the phenomenon of aircraft center of mass drift is significantly reduced, making turning more flexible and easier. Therefore, the significance of the active twist wing in ornithopters lies in adjusting the imbalance wing deformation, enhancing the control strategies, compensating for the shortcomings of tail control, and making ornithopter flight more agile.

5. Kinetic Model of the Active Twist Wing Ornithopter

As described above, during flight experiments, it was observed that the principle of fixed-wing aileron control could not explain the performance of the yaw control of the ornithopter with active twist. The actual yaw deviation of the ornithopter is opposite to the expected deviation direction based on the principle of fixed-wing aileron control. Moreover, this reverse control phenomenon becomes more pronounced as the outer wing's folding degree increases. Additionally, it was found during experiments that, the higher the flapping position of the ornithopter, the more significant its inverted turning control effect becomes.

The inverted control phenomenon induced by outer segment flapping, which was unexpected at the outset of the experiments, was not explained in the literature citations. Additionally, the existing studies have not provided a kinetic model for two-section wing twisting, necessitating a new model to explain the current control phenomenon and its underlying principles. This paper proposes an M-shaped model to analyze and elucidate this control phenomenon, guiding the controller in understanding this new control characteristic. As illustrated in Figure 8a, considering the limit position when the ornithopter is folded by 90°, the inverted control effect can be explained intuitively at this limit position.

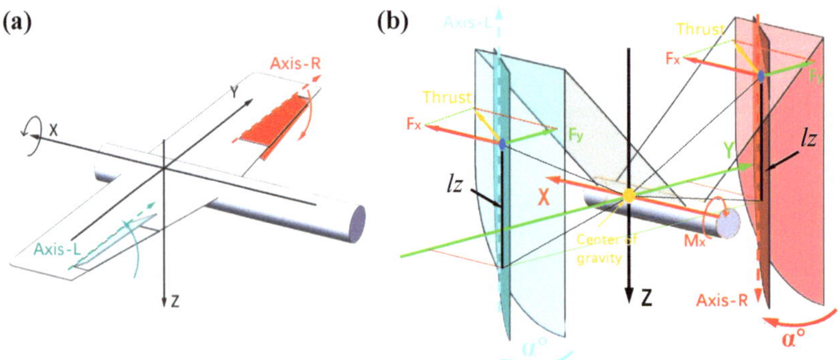

Figure 8. Comparison diagram between fixed-wing aircraft and two-section wing ornithopter. (**a**) Schematic illustration of aileron twisting in fixed-wing aircraft. (**b**) Schematic illustration of outer wing twisting in two-section wing ornithopter, along with its corresponding force analysis.

The thrust acting on the outer wing is denoted as $Thrust(F_T)$, expressed by Equation (2), which can be decomposed into forces in the X and Y directions:

$$\begin{bmatrix} F_X \\ F_Y \end{bmatrix} = F_T \begin{bmatrix} sin\alpha \\ cos\alpha \end{bmatrix}, \tag{7}$$

Comparing Figure 8b with Figure 8a, it is noted that the rotation axis of the ailerons on a fixed-wing aircraft remains aligned with the Y-axis throughout the flight process, and, when the wings of the ornithopter do not fold, the direction of the wing warping rotation axis aligns with that of the ailerons on a fixed-wing aircraft. Additionally, the rotation axes on both sides of the wings are aligned ($Axis - L = Axis - R$). However, when the wings of the ornithopter fold by 90° during flapping, the direction of the wing warping rotation axis changes, becoming oriented in opposite directions along the Z-axis ($Axis - L = -Axis - R$). Under the influence of differential control signals, the initially opposite rotations of the wing warping on both sides around the Y-axis gradually transform into rotations in the same direction around the Z-axis.

As the center of pressure of the outer wing is consistently positioned above and aft of the aircraft's center of mass, the resulting torque exerted on the fuselage can be expressed as follows:

$$\begin{bmatrix} M_X \\ M_Y \\ M_Z \end{bmatrix} = \begin{bmatrix} 2F_Y \times l_Z \\ 2F_X \times l_Z \\ 0 \end{bmatrix}, \tag{8}$$

In the above expressions, M_X represents the rolling moment caused by wing twisting, M_Y denotes the pitching moment, and M_Z signifies the yawing moment. The agility of an aircraft is employed to quantify its capability for attitude rotation. The control of the rotational angular acceleration of a flapping-wing aircraft by wing twisting is as follows:

$$\dot{\Omega} = \begin{bmatrix} \dot{p} \\ \dot{q} \\ \dot{r} \end{bmatrix} = \begin{bmatrix} \frac{M_X}{I_{xx}} \\ \frac{M_Y}{I_{yy}} \\ \frac{M_Z}{I_{zz}} \end{bmatrix}, \tag{9}$$

where $\dot{\Omega}$ represents the angular acceleration vector of the airframe influenced by the wing twisting, $\dot{p}$ denotes the roll acceleration, $\dot{q}$ denotes the pitch acceleration, and $\dot{r}$ denotes the yaw acceleration. I_{xx} is the moment of inertia of the fuselage about the X-axis, I_{yy} is the moment of inertia of the fuselage about the Y-axis, and I_{zz} is the moment of inertia of the fuselage about the Z-axis.

The yawing moment is not directly influenced by wing warping. However, the rolling motion induced by wing twisting causes a change in the lift direction of the ornithopter, resulting in a lateral force in the horizontal direction, leading to a change in the heading of the ornithopter. Additionally, due to the transition of the active twist wing rotation axis from parallel to opposite directions during the flapping process of the ornithopter, the aircraft exhibits a directional control characteristic that is completely opposite to that of fixed-wing aircraft, as analyzed in the model.

6. Conclusions

(1) The active twist wing ornithopter, redesigned based on the prototype of a vertical-tail ornithopter, has been systematically analyzed and reliably validated through actual flight experiments.

(2) The field flight experiment results demonstrate that the active twist wing can provide independent horizontal control capability and enhance the maneuverability and agility of the ornithopter.

(3) The observed phenomenon in the experiment, where the roll angle and yaw control effect of the ornithopter after using the active twist wing function are opposite to the common control phenomenon of the fixed-wing aileron control, indicates that the active twist wing structure introduces complex aeroelastic changes that the existing models cannot describe to explain its control principles. Therefore, this paper presents an initial model to analyze and describe this new yaw control principle.

Overall, measuring the attitude control effect of the active twist on the two-section wing ornithopter during actual flight is of great significance for studying the agility and maneuverability of folding-wing birds, providing some insights for future experiments investigating this behavior. To further explore the flight attitude control of birds, our laboratory will establish wind tunnel experiments to investigate the magnitude of the rolling moment under the active deformation of the two-section wing ornithopter. In the future, we aim to develop a two-section wing ornithopter that is completely independent of vertical tail control, aiming to more closely replicate the flight mode of birds.

Supplementary Materials: The following supporting information including original experimental data and video of manual controled out-door flight can be downloaded at: https://www.mdpi.com/article/10.3390/machines12070486/s1, Video S1: Outdoor Flight.

Author Contributions: Conceptualization, Y.C. and J.Z.; methodology, Y.C. and J.Z.; software, G.S.; validation, G.S., J.Z. and Y.C.; formal analysis, G.S. and Y.C.; investigation, Y.C., G.S. and J.Z.; resources, J.Z. and S.F.; data curation, G.S.; writing—original draft preparation, Y.C. and G.S.; writing—review and editing, J.Z.; visualization, G.S.; supervision, J.Z.; project administration, J.Z.; funding acquisition, J.Z. All authors have read and agreed to the published version of the manuscript.

Funding: This research was funded by the National Natural Science Foundation of China, grant number 62206065, and the Bagui Scholar Program of Guangxi.

Data Availability Statement: The data that support the findings of this study are available in the Supplementary Materials of this article.

Acknowledgments: The authors would like to thank Chen Taiqing, Xie Quansheng, Pei Gao, Jiong Zheng, Linkun Song, and Delin Pang for helping to build the dataset.

Conflicts of Interest: The authors declare no conflicts of interest.

References

1. Abas, M.F.B.; Rafie, A.S.B.M.; Yusoff, H.B.; Ahmad, K.A.B. Flapping wing micro-aerial-vehicle: Kinematics, membranes, and flapping mechanisms of ornithopter and insect flight. *Chin. J. Aeronaut.* **2016**, *29*, 1159–1177. [CrossRef]
2. Alexander, D.E.; Wang, Z.J. Nature's Flyers: Birds, Insects, and the Biomechanics of Flight. *Phys. Today* **2003**, *56*, 358.
3. Ajanic, E.; Paolini, A.; Coster, C.; Floreano, D.; Johansson, C. Robotic avian wing explains aerodynamic advantages of wing folding and stroke tilting in flapping flight. *Adv. Intell. Syst.* **2023**, *5*, 2200148. [CrossRef]
4. Marques, P.; Spiridon, E. Adaptive wing technology, aeroelasticity and flight stability: The lessons from natural flight. In Proceedings of the 2013 Maui International Engineering Education Conference, Kuala Lumpur, Malaysia, 4–5 December 2013; pp. 25–34.
5. Harvey, C.; Gamble, L.L.; Bolander, C.R.; Hunsaker, D.F.; Joo, J.J.; Inman, D.J. A review of avian-inspired morphing for UAV flight control. *Prog. Aerosp. Sci.* **2022**, *132*, 100825. [CrossRef]
6. Lang, X.; Song, B.; Yang, W.; Yang, X. Effect of spanwise folding on the aerodynamic performance of three dimensional flapping flat wing. *Phys. Fluids* **2022**, *34*, 021906. [CrossRef]
7. Stowers, A.K.; Matloff, L.Y.; Lentink, D. How pigeons couple three-dimensional elbow and wrist motion to morph their wings. *J. R. Soc. Interface* **2017**, *14*, 20170224. [CrossRef] [PubMed]
8. Warrick, D.R.; Dial, K.P. Kinematic, aerodynamic and anatomical mechanisms in the slow, maneuvering flight of pigeons. *J. Exp. Biol.* **1998**, *201*, 655–672. [CrossRef]
9. Chin, D.D.; Matloff, L.Y.; Stowers, A.K.; Tucci, E.R.; Lentink, D. Inspiration for wing design: How forelimb specialization enables active flight in modern vertebrates. *J. R. Soc. Interface* **2017**, *14*, 20170240. [CrossRef] [PubMed]
10. Keennon, M.; Klingebiel, K.; Won, H. Development of the nano hummingbird: A tailless flapping wing micro air vehicle. In Proceedings of the 50th AIAA Aerospace Sciences Meeting including the New Horizons Forum and Aerospace Exposition, Nashville, Tennessee, 9–12 January 2012; p. 588.
11. Chen, A.; Song, B.; Wang, Z.; Liu, K.; Xue, D.; Yang, X. Experimental Study on the Effect of Increased Downstroke Duration for an FWAV with Morphing-coupled Wing Flapping Configuration. *J. Bionic Eng.* **2024**, *21*, 192–208. [CrossRef]
12. Ajanic, E.; Feroskhan, M.; Mintchev, S.; Noca, F.; Floreano, D. Bioinspired wing and tail morphing extends drone flight capabilities. *Sci. Robot.* **2020**, *5*, eabc2897. [CrossRef]
13. Sachs, G. Speed stability in birds. *Math. Biosci.* **2009**, *219*, 1–6. [CrossRef] [PubMed]
14. Grant, D.T.; Abdulrahim, M.; Lind, R. Design and analysis of biomimetic joints for morphing of micro air vehicles. *Bioinspir. Biomimetics* **2010**, *5*, 045007. [CrossRef] [PubMed]
15. Song, B.; Lang, X.; Xue, D.; Yang, W.; Bao, H.; Liu, D.; Wu, T.; Liu, K.; Song, W.; Wang, Y. A review of the research status and progress on the aerodynamic mechanism of bird wings. *Sci. Sin. Technol.* **2022**, *52*, 893–910. [CrossRef]
16. Lees, J.J.; Dimitriadis, G.; Nudds, R.L. The influence of flight style on the aerodynamic properties of avian wings as fixed lifting surfaces. *PeerJ* **2016**, *4*, e2495. [CrossRef] [PubMed]
17. Han, C. Investigation of unsteady aerodynamic characteristics of a seagull wing in level flight. *J. Bionic Eng.* **2009**, *6*, 408–414. [CrossRef]
18. Taylor, G.K.; Nudds, R.L.; Thomas, A.L. Flying and swimming animals cruise at a Strouhal number tuned for high power efficiency. *Nature* **2003**, *425*, 707–711. [CrossRef] [PubMed]
19. Thielicke, W.; Stamhuis, E.J. The effects of wing twist in slow-speed flapping flight of birds: trading brute force against efficiency. *Bioinspir. Biomimetics* **2018**, *13*, 056015. [CrossRef]
20. Dong, Y.; Song, B.; Yang, W.; Xue, D. A numerical study on the aerodynamic effects of dynamic twisting on forward flight flapping wings. *Bioinspir. Biomimetics* **2024**, *19*, 026013. [CrossRef] [PubMed]
21. CCTV. Northwestern Polytechnical University Achieves New Breakthrough in Biomimetic Flapping Wing Aircraft. *Shaanxi Educ.* **2024**, *9*.

22. Send, W.; Fischer, M.; Jebens, K.; Mugrauer, R.; Nagarathinam, A.; Scharstein, F. Artificial hinged-wing bird with active torsion and partially linear kinematics. In Proceedings of the 28th Congress of the International Council of the Aeronautical Sciences, Brisbane, Australia, 23–28 September 2012; Volume 10.

23. Zhang, Z.; Wei, A.; Cai, Y. A flapping feathered wing-powered aerial vehicle. *arXiv* **2021**, arXiv:2102.12687.

24. Duan, W.B.; Ang, H.S.; Xiao, T.H. Design and wind tunnel test of an ornithopter with differential twist wings. *Shiyan Liuti Lixue/J. Exp. Fluid Mech.* **2013**, *27*, 35–40.

25. Lee, J.S.; Han, J.H. Indoor flight testing and controller design of bioinspired ornithopter. In Proceedings of the Intelligent Autonomous Systems 12: Volume 1 Proceedings of the 12th International Conference IAS-12, Jeju Island, Republic of Korea, 26–29 June 2012; Springer: Berlin/Heidelberg, Germany, 2012; pp. 825–834.

26. Send, W.; Scharstein, F.; GbR, A. Thrust measurement for flapping-flight components. In Proceedings of the 27th ICAS Congress, Nice, France, 19–24 September 2010; pp. 19–24.

27. Xin, Z.; Feng, S.; Yu, C.; Taiqing, C. Multi-stage Ornithopter Aerodynamic Characteristics. *Chin. Hydraul. Pneum.* **2023**, *47*, 1–7.

28. Thomas, A.L. The flight of birds that have wings and a tail: Variable geometry expands the envelope of flight performance. *J. Theor. Biol.* **1996**, *183*, 237–245. [CrossRef]

29. Liefer, R.K. *Fighter Agility Metrics*; Technical Report; University of Kansas, Aerospace Engineering: Lawrence, KS, USA, 1990.

30. Hu, M.L.; Wei, R.X.; Cui, X.F.; Kong, T. Attitude Stabilization of Bird-like Flapping Wing Air Vehicle. *Robot* **2008**, 481 –485.

31. Rbiger, H.; Holst, E. *Lift during Wing Upstroke, Horst Räbiger, Nuremberg, 2015*, Version 10.1; 2018. Available online: http://www.ornithopter.de/english/data/wing_upstroke.pdf (accessed on 11 July 2024).

32. Stanewsky, E. Aerodynamic benefits of adaptive wing technology. *Aerosp. Sci. Technol.* **2000**, *4*, 439–452. [CrossRef]

Article

Q-Learning with the Variable Box Method: A Case Study to Land a Solid Rocket

Alejandro Tevera-Ruiz, Rodolfo Garcia-Rodriguez, Vicente Parra-Vega and Luis Enrique Ramos-Velasco

[1] Robotics and Advanced Manufacturing Department, Research Center for Advanced Studies (CINVESTAV), Ramos Arizpe 25900, Mexico

[2] Aeronautical Engineering Program and Postgraduate Program in Aerospace Engineering, Univ. Politécnica Metropolitana de Hidalgo, Tolcayuca 43860, Mexico

* Correspondence: rogarcia@upmh.edu.mx

Abstract: Some critical tasks require refined actions near the target, for instance, steering a car in a crowded parking lot or landing a rocket. These tasks are critical because failure to comply with the constraints near the target may lead to a fatal (unrecoverable) condition. Thus, a higher resolution action is required near the target to increase maneuvering precision. Moreover, completing the task becomes more challenging if the environment changes or is uncertain. Therefore, novel approaches have been proposed for these problems. In particular, reinforcement learning schemes such as Q-learning have been suggested to learn from scratch, subject to exploring action–state causal relationships aimed at action decisions that lead to an increase in the reward. Q-learning refines iterative action inputs by exploring state spaces that maximize the reward. However, reducing the (constant) resolution box needed for critical tasks increases the computational load, which may lead to the tantamount curse of the dimensionality problem. This paper proposes a variable box method to maintain a low number of boxes but reduce its resolution only near the target to increase action resolution as needed. The proposal is applied to a critical task such as landing a solid rocket, whose dynamics are highly nonlinear, underactuated, non-affine, and subject to environmental disturbances. Simulations show successful landing without leading to a curse of dimensionality, typical of the classical (constant box) Q-learning scheme.

Keywords: variable box method; Q-learning; rocket landing; thrust vector control; underactuated systems

Citation: Alejandro Tevera-Ruiz, Rodolfo Garcia-Rodriguez, Vicente Parra-Vega and Luis Enrique Ramos-Velasco Q-Learning with the Variable Box Method: A Case Study to Land a Solid Rocket. *Machines* **2023**, *11*, 214. https://doi.org/10.3390/machines11020214

Academic Editor: Tao Li

Received: 2 December 2022
Revised: 22 January 2023
Accepted: 30 January 2023
Published: 2 February 2023

1. Introduction

Space launches have improved autonomous landing technology for reusable spacecraft in the last few years. These aim to recover the rocket's main booster for analysis and to retrofit, saving about 60% of total mission costs [1]. Why does it take 60 years of rocket technology development to achieve landing technology? We need to revise this briefly. First, the physics involved in a rocket landing is difficult to command; this stems from the fact that descending a vertical longitudinal rocket's dynamics are highly nonlinear, underactuated, non-affine, and subject to environmental disturbances. This critical task requires extreme maneuverability when using thrust vector control (TVC). Furthermore, unlike the fixed nozzle configuration that commands two controllers (variable thrust and surface control), non-affine underactuation arises when using TVC because it implements one control (constant thrust with a variable nozzle placed at the rocket's bottom end, controlling it within a small angle range). Second, TVC is used because the low approaching velocities make the surface control input irrelevant, exacerbating maneuvering toward the landing spot [2,3]. Over the years, model-based optimization schemes subject to constraints have been addressed. However, due to changing environmental conditions, the lack of exact knowledge jeopardizes the implementation of conservative model-based approaches. In these circumstances, Machine Learning (ML) tools, such as Reinforcement Learning

(RL), have materialized as an alternative to learning the actions required to produce the desired trajectories.

RL algorithms exploit the reward evaluation to optimize a value function through the Bellman Equation that governs the learning process under an optimal policy. RL has solved many problems, such as automatic translation, image recognition, medical diagnosis, and text and speech recognition, to name a few. However, further algorithmic improvement is needed for physical systems due to RL iterates. Thus, critical tasks such as landing a rocket may lead to fatal failure (where the system cannot recover operation) for certain iterations. For example, RL has been used for image classification to identify landing spots for aerospace missions [4–6], including deep RL [7]; however, there are no studies for landing dynamical rockets.

Q-learning is a salient scheme of RL algorithms that has proven successful in uncertain dynamical systems (see Appendix A). It associates a discrete state with a discrete box corresponding to a discrete action that eventually leads to the optimal outcome after exploring the whole state space; such discretizations handle advantageous uncertainties. Classical Q-learning relies on the box method, which assigns boxes of a constant resolution (CR) corresponding to a discretized state. When a better input resolution is required, the conventional solution is to increase the number of boxes, which may lead to the curse of dimensionality [8]. Moreover, for goal-oriented tasks, the Q-algorithm's learning architecture allows for the modification of the resolution of the boxes as the system state approaches the goal state. Thus, we aim to use a low number of boxes to maintain the low state-space dimension and reduce only the box size near the goal, avoiding the curse of dimensionality.

This paper proposes a box method with a variable resolution (VR) to increase the learning resolution where needed without increasing the computational costs or the resolution for each state variable. The VR method is applied to the Q-learning algorithm to maneuver the landing of a rocket. Representative dynamic simulations using the complex solid rocket dynamics using the real parameters of a NASA rocket are presented. It is shown that the learned policy (controls) produces admissible trajectories that comply with this critical task, even when the rocket is subject to disturbances.

This paper is organized as follows. The problem statement is presented in Section 2, followed by the proposed variable box method. Then, a brief revision of RL algorithms is given in Section 3. Section 4 introduces the TVC rocket dynamics and the simulation results are shown. Finally, the conclusions are given in Section 5.

2. Problem Statement

We consider the following nonlinear, non-affine, underactuated, disturbed, and state-constrained system in the continuous state-space form given by

$$\dot{\mathbf{x}} = \mathbf{f}(\mathbf{x}, t) + \mathbf{g}(\mathbf{x}, u, t) + \boldsymbol{\eta}(t) \tag{1}$$

$$\mathbf{y} = \mathbf{h}(\mathbf{x}) = \mathbf{x} \tag{2}$$

where $\mathbf{x}, \mathbf{y} \in \Re^n$ are the vector state and system output, respectively, $\mathbf{f}(\mathbf{x}, t)$ is the flow of the nonlinear ODE (1), $\mathbf{g}(\mathbf{x}, u, t)$ is the input matrix, and $\boldsymbol{\eta}(t)$ is a Liptchitz disturbance. Let the Euler method be used to obtain the following difference equation of (1) and (2),

$$\mathbf{x}_{t+1} = \mathbf{x}_t + h\left[\mathbf{f}(\mathbf{x}_t) + \mathbf{g}(\mathbf{x}_t, u_t) + \boldsymbol{\eta}_t\right] \tag{3}$$

$$\mathbf{y}_t = \mathbf{x}_t \quad \Rightarrow \mathbf{y}_{t+1} = \mathbf{x}_{t+1} \tag{4}$$

where $h > 0$ is the step size and $\mathbf{y}_t$ corresponds to the output at the t time step. Note that Systems (1) and (2), or (3) and (4), are:

1. nonlinear (superposition theorem does not apply);
2. underactuated (there exist more degrees of freedom than control inputs);

3. non-affine in the control input u (since $\mathbf{g}$ cannot be written as $\mathbf{g} = \bar{\mathbf{g}}u$ for a given $\bar{\mathbf{g}}$ input matrix).

Since 1–3 account for a complex system and assuming that the task is complex for traditional control techniques, a Q-algorithm is a feasible solution using a variable resolution box method to refine the state space that allows learning from scratch, avoiding increasing the computational load. Then, assuming full access to the output vector $\mathbf{y}_t$ of system (3) and (4), the problem statement is how to apply the Q-Algorithm using a variable box method to produce a set of admissible trajectories with a feasible (without incurring the curse of dimensionality) iterative learning policy u_t that eventually converges to its optimal $u_t^* = \pi_*$ that maximizes a reward policy.

3. Box Methods in RL algorithms

3.1. Brief Background of Reinforcement Learning

Unlike supervised or unsupervised learning, reinforcement learning is characterized as the learning process carried out from scratch, taking into account environmental information and reward assignment, typically using a Markov Decision Process (MDP) [9]. As a result, undesirable actions and states are allowed throughout the learning process. Moreover, RL avoids them in subsequent trials to improve learning. The RL idea is that the agent applies action u_t to the environment, consequently producing a measurable state s_t at time t, which is evaluated by a reward policy, see Figure 1. The agent's goal is to learn the set of actions that optimizes a value cost function $V(s_t)$ throughout a long-term reward process r_{t+1}, which at $t + 1$, leads to complying with the reward policy until s_{t+1} reaches the (desired) goal state s_{t+1}^d. The optimization process minimizes $V(s_t)$ by assigning a *value* to each state, followed by a *policy* $\pi(s_t)$ that defines the agent's behavior. Thus, the agent learns the behavior that leads to an optimal policy $\pi_*(s_t)$, and, therefore, to optimal states of an optimal value function $V_*(s_t)$. The optimization problem is solved using Dynamic Programming (DP) to satisfy the fundamental Bellman Equation.

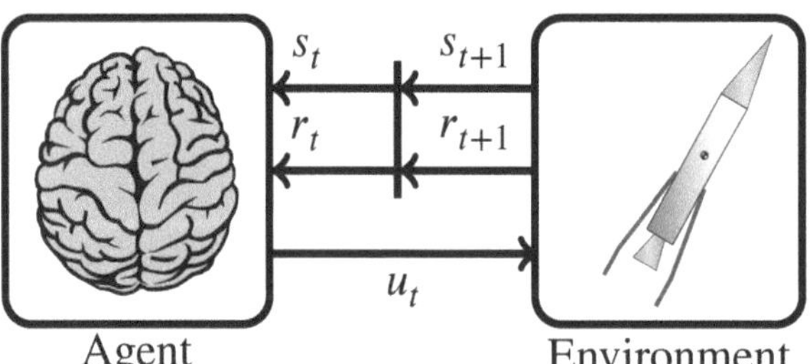

Figure 1. Conceptual reinforcement learning scheme.

3.2. Classical Box Method

Q-learning encloses continuous state and action values into finite sets by indexing intervals into so-called boxes [10]. The classical boxing method, called the box method, uses two user parameters, the set of finite fence limits and the constant resolution $\epsilon = b_{up} - b_{lw}$, which determine the size of each box based on its upper b_{up} and lower b_{lw} limits. Then, the i-th element of the output vector $\mathbf{y}_t$ is divided into n_i boxes of a constant resolution ϵ, where t is the time step and $\mathbf{i}_t \in \mathbb{N}^N$ is the index vector. There arise N-elements of $\mathbf{y}_t$ as components of a new set $\mathcal{B}(\cdot)$, where $\mathcal{B}(\cdot)$ maps the state space to index the vector space $\mathbf{y}_t \mapsto \mathbf{i}_t$. In this way, one has the following two spaces:

-   Finite State-Number Space. Let $\mathbf{i}_t$ be the index vector of the finite state-number space $\mathcal{S}$ for $\dim(\mathcal{S}) = \prod_{i=1}^{N}(n_i + 1)$, whose elements are the state numbers s_t codified by $\mathbf{i}_t$'s elements. Then, $\mathcal{S}$ models a vector of the integer numbers' range $[0, \dim(\mathcal{S}) - 1]$ as elements of a tensor $\mathfrak{T}$ of N-axes, coding the state number s_t into $\mathfrak{T}$ to finally relate the indexed output vector $\mathbf{y}_t$ to $\mathbf{i}_t$ and then to $s_t = \mathfrak{T}(\mathbf{i}_t) : \mathbf{i}_t = \mathcal{B}(\mathbf{y}_t)$.

- Finite Action Space. Aiming at obtaining a finite action space $\mathcal{A} = \{u_1, u_2, \ldots, u_k, \ldots, u_m\}$ for the bounded action (control) variable $u \in [u_{lw}, u_{up}]$ represented by m-boxes of $\epsilon > 0$ width each, then $dim(\mathcal{A}) = m$. Each element u_k is indexed by the m-th action agent from an arbitrary policy. Note that a high number of u values is suggested to enable more state-number explorations. However, this may lead to an unfeasible explosion of dimensions, or the curse of dimensionality.

In recent years, research has focused on improving learning algorithms such as deep Q-learning or actor–critic, rather than the box method, which is fundamental for many of these algorithms.

Limitations of the Classical Box Method. Methods have been proposed to deal with one of the main drawbacks of the classical box method: tuning ϵ without leading to an unfeasible explosion of dimensions. Several variable boxing methods have been proposed. Ref. [11] proposes an interpolation approach using a coarse grid for the learning process and then increases the accuracy with Kuhn Triangulation around the interested regions. In [12], feudal reinforcement learning is proposed by dividing the state space into regions, where each is iterative-divided to reduce state-space searching using the manager concept, similar to the demons in neuron-like adaptive elements [13]. Ref. [14] introduced an adaptive resolution boxes approach by approximating the value function using interpolation and tree structures to refine the grid, similar to [15], using kd-trees to identify attractive state-space regions. Recently, a fuzzy action assignment method has been proposed to generate continuous control based on an optimal trained Q-table [16]. In these approaches, the resolution is adaptive online because the goal box is unknown, as in a chess game. However, there are applications where the goal is known a priori, allowing an offline variable box resolution.

3.3. The Proposed Variable Box Method

Let the i_b box be variable with a resolution ϵ_i given by

$$\epsilon_i = \begin{cases} \epsilon_{i_b-1} + \frac{1}{2}\epsilon_{gb} M(|\Delta i_b|) & \text{if } \Delta i_b > 0 \\ \epsilon_{i_b+1} + \frac{1}{2}\epsilon_{gb} M(|\Delta i_b|) & \text{if } \Delta i_b < 0 \\ \epsilon_{gb} & \text{otherwise} \end{cases} \tag{5}$$

where $|\Delta i_b| = |i_b - i_{gb}|$, with i_b the current box index and i_{gb} the goal-box index. The scalar function $M(|\Delta i_b|) \in \mathcal{C}^2$ is monotonous away from the origin $M(0) = 0$ such that it increases (decreases) the accuracy in the vicinity of larger (smaller) distances to the goal-box resolution ϵ_{gb}. As an example of function $M(|\Delta i_b|)$, the discrete *Fibonacci Serie* $\{1, 1, 2, 3, 5, \ldots\}$ can be considered. The proposed algorithm for the variable box method is shown in Algorithm 1. The box resolution ϵ_i yields:

a. a faster reaction for larger errors;
b. a constant box dimension; the remaining boxes are enlarged (shrank) when ϵ_{i_b-1} is reduced (expanded), like an accordion of a fixed length;
c. continuous action-taken history and influence [14], with smooth transitions between state numbers.

Note that for each i-th element of the output vector, $y_i \in \mathbf{y}_t$, there exists a goal box with an index box i_{gb} and a resolution ϵ_{gb} defined by prior knowledge of the desired state-variable value $y_i^d \in [y_{i,lw}, y_{i,up}]$; each box resolution can be increased (or decreased) according to the proximity of the goal box. Additionally, ϵ_{i_b} should be truncated to ϵ_s for safety resolution to ensure sufficient accuracy in edged boxes; otherwise, many state vectors would be assigned the same state number. Finally, note that the classical box method is a particular case of our proposed variable box method since it is included in the third condition in (5).

Algorithm 1 The proposed variable box method enables higher resolution near the goal but lower resolution away from it to avoid exploiting dimensionality.

Input: $y_i^d \in \Re$; $\epsilon_{gb} > 0$; $(y_{i,lw}, y_{i,up})$ and $\epsilon_s >> 0$;
Result: A set $\mathcal{B}$ with n_i-boxes.
Initialize;
$\mathcal{B} \leftarrow$ empty list;
Calculate initial lower limit as $b_i \leftarrow y_i^d - \frac{1}{2}\epsilon_{gb}$;
while $b_i > y_{i,lw}$ **do**
$\quad$ Add b_i value to $\mathcal{B}$ set;
$\quad b_i \leftarrow b_i - \epsilon_{i_b}$;
$\quad$ **if** $\epsilon_i < \epsilon_s$ **then** update ϵ_{i_b} by (5) **else** $\epsilon_{i_b} = \epsilon_s$;
end
Calculate initial upper limit as $b_i \leftarrow y_i^d + \frac{1}{2}\epsilon_{gb}$;
while $b_i < y_{i,up}$ **do**
$\quad$ Add b_i value to $\mathcal{B}$ set;
$\quad b_i \leftarrow b_i + \epsilon_{i_b}$;
$\quad$ **if** $\epsilon_i < \epsilon_s$ **then** update ϵ_{i_b} by (5) **else** $\epsilon_{i_b} = \epsilon_s$;
end
Add $y_{i,lw}$ and $y_{i,up}$ values to $\mathcal{B}$ set;
return *Shorted* $\mathcal{B}$.

3.4. How to Deal with the Curse of Dimensionality

A comparative grid example of the classical box and variable box methods is shown in Figure 2, considering two state variables with $y_{i,lw} = -1$ and $y_{i,up} = 1$. Figure 2a uses a constant resolution $\epsilon_i = 1 \times 10^{-2}$, whereas the variable box method considers $\epsilon_{gb} = 1 \times 10^{-2}$ and $\epsilon_s = 0.1$. Note that both methods have the same box resolution and accuracy around the goal box. This way, finer action is taken with a greater reward when it is near the goal.

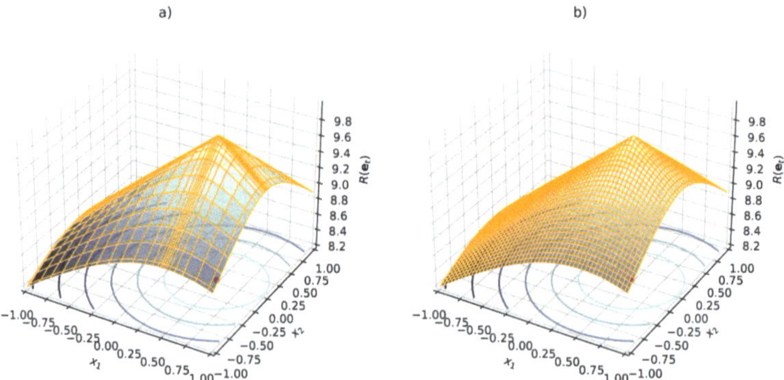

Figure 2. Level sets for vertical rocket landing were obtained with (**a**) the variable box method using 576 boxes, and (**b**) the classical box method using 40,000 boxes, both tuned to obtain similar final results. The dramatic reduction in the number of boxes achieved using our proposed variable method results from the symmetrical higher resolution around the goal box, which yields a higher reward in areas with zero errors.

Otherwise, the variable box method reduces the curse of dimensionality in a Q-learning implementation using only 14.4% of boxes to render similar results to the classical box method without changing their attributes around the interested region.

3.5. Reward Assignment

A critical element of RL algorithms is the design of the reward policy because it codifies the policy that leads to the learning goal [17]. In this work, the reward is defined as an evaluation function $R_t(\mathbf{e}_t)$ that evaluates the error of the goal of the current agent. Thus, instead of increasing the reward, which may lead to an aggressive action, we propose to increase the box resolution when needed by using the variable box method for the same reward policy to guarantee the learning goal.

The following section describes a case study showing how the proposed variable box method with Q-learning can successfully land a complex plant, as depicted in Figure 3.

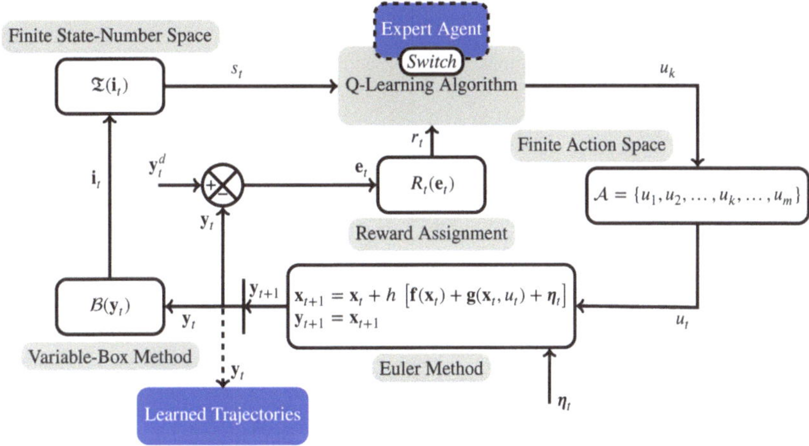

Figure 3. Schematic diagram of the proposed variable box method with Q-learning. Two feedback loops reinforce the intertwined closed-loop dynamic interplay of the expert agent to guarantee $\mathbf{y}_t \to \mathbf{y}_t^d$, whereas the cumulative reward r_t is maximized

4. A Case Study: Rocket Landing

The proposed variable box method is synthesized to learn the descent trajectories of a solid rocket driven by TVC using the Q-learning algorithm. First, a brief description of rocket dynamics is given, followed by the concise Lagrangian model. Then, the comparative simulation results of the classical box and variable box methods are presented.

4.1. Dynamical Model

4.1.1. Translational Dynamics

Applying the second law of Newton, the forces acting on the rocket are defined as

$$\mathbf{a} = \frac{1}{m_r}\mathbf{F} = \frac{1}{m_r}[T_{ECI}^B(\mathbf{F}_{aer} + \mathbf{F}_{thrust}) + \mathbf{F}_g] \tag{6}$$

where m_r is the rocket mass, $\mathbf{F}_{aer}$ are the aerodynamic forces, $\mathbf{F}_{thrust}$ is the thrust force, and $\mathbf{F}_g$ is the gravity force, with $T_R^B = R(x,\phi)R(y,\theta)R(z,\psi)$ the rotation matrix between the reference frames B and R, see Figure 4a. Note that the thrust and aerodynamic forces are defined with respect to the body reference frame B. In contrast, the gravity force is defined with respect to the inertial reference frame R.

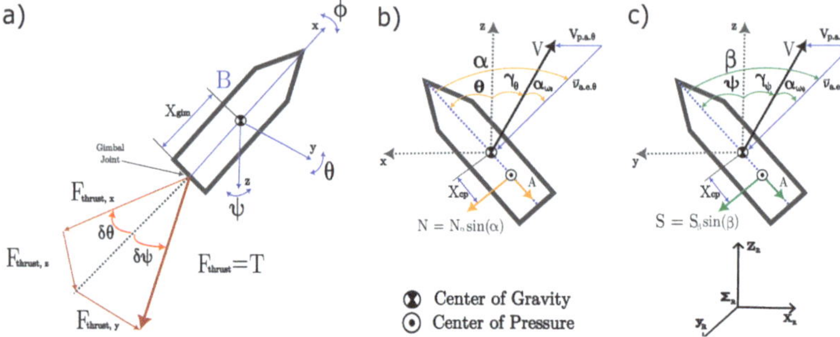

Figure 4. (**a**) Rocket with an integrated TVC, with a bounded angle nozzle, and the aerodynamics forces shown in (**b**) the xz-plane, and (**c**) the yz-plane. Notice that CoG and CoP are at different locations, which yields a rather highly nonlinear coupled non-affine underactuated system.

Aerodynamic Forces. Two components of aerodynamic forces arise, the lift force acting in the normal direction and the dissipative drag forces, both acting throughout the center of pressure. The normal force is longitudinal to the principal axis of the rocket, $N = N_\alpha sin\alpha$, whereas the axial force resulting from drag force is given by $A = \frac{1}{2}C_a\rho V_{air}^2\varrho$. The lateral slip force is defined as $S = S_\beta sin\beta$, where N_α is the normal force depending on the angle of attack $\alpha = \theta + \gamma_\theta + \alpha_{w\theta}$, where C_a is the aerodynamic axial force coefficient, ρ is the density of the air, V_{air} is the airspeed, ϱ is the coincidence surface of air, and S_β is the lateral slip force dependent on the angle of slip $\beta = \psi + \gamma_\psi + \alpha_{w\psi}$, with θ and ψ as the rocket angles, γ the drift angle, and α_w the wind disturbance, see Figure 4b,c. Finally, the normal and lateral slip forces are affected by the angle of attack, thus defined as $N_\alpha = \frac{1}{2}C_N\rho V_{air}^2\varrho$ and $S_\beta = \frac{1}{2}C_S\rho V_{air}^2\varrho$, where C_N and Cs represent the aerodynamic coefficients from the normal force (N) and lateral slip force (S), respectively. Because the center of pressure and the center of gravity are not generally located at the same point on the rocket, the aerodynamic forces can cause the rocket to rotate in flight. In this way, the aerodynamic forces $\mathbf{F}_{aer}$ are defined as

$$\mathbf{F}_{aer} = \begin{bmatrix} F_{aer,x} \\ F_{aer,y} \\ F_{aer,z} \end{bmatrix} = \begin{bmatrix} -A \\ S \\ -N \end{bmatrix} = \begin{bmatrix} -A \\ S_\beta s_\beta \\ -N_\alpha s_\alpha \end{bmatrix} \tag{7}$$

where $s_a = \sin(a)$, $c_a = \cos(a)$.

Thrust Forces. Thrust is produced by the engine acting in the opposite direction to the exhaust combustion gases and is given by $\mathbf{F}_{thrust} = T = \dot{m}V_e + (P_e - P_0)A_e$, where $\dot{m}V_e$ is the combustion of the propellant that burns and escapes at a constant rate and $(P_e - P_0)$ is the difference in pressure between the inside and outside of the nozzle at an exhaust surface A_e. Assuming that the thrust is constant (T), the components of the TVC, as shown in Figure 4a, are given as

$$\mathbf{F}_{thrust} = \begin{bmatrix} F_{thrust,x} \\ F_{thrust,y} \\ F_{thrust,z} \end{bmatrix} = \begin{bmatrix} Tc_{\delta_\psi}c_{\delta_\theta} \\ Ts_{\delta_\psi} \\ Ts_{\delta_\theta} \end{bmatrix} \tag{8}$$

Gravity Force. Consider the Earth's geometry as a WGS84 ellipsoid called Geoid and the gravity force vector is defined as [8,18]

$$\mathbf{F}_g = m_r \begin{bmatrix} g_x \\ g_y \\ g_z \end{bmatrix} = m_r \begin{bmatrix} -\frac{\mu}{r_e^2}\left(1 + \frac{3}{2}J_2(\frac{R_0}{r_e})^2(1 - 5(\frac{z}{r_e})^2)\right)\frac{x}{r_e} \\ -\frac{\mu}{r_e^2}\left(1 + \frac{3}{2}J_2(\frac{R_0}{r_e})^2(1 - 5(\frac{z}{r_e})^2)\right)\frac{y}{r_e} \\ -\frac{\mu}{r_e^2}\left(1 + \frac{3}{2}J_2(\frac{R_0}{r_e})^2(3 - 5(\frac{z}{r_e})^2)\right)\frac{z}{r_e} \end{bmatrix} \tag{9}$$

where μ represents the universal gravitational constant, R_0 is the distance between the surface location from the Earth to its center, J_2 is the oblateness term, and r_e is the distance between the rocket and the Earth's center.

4.1.2. Rotational Dynamics

Let the angular acceleration $\dot{\omega}$ be given by:

$$\dot{\omega} = [\dot{p}, \dot{q}, \dot{r}]^T = \hat{J}^{-1}[M - \omega \times (\hat{J} \cdot \omega)] \tag{10}$$

where M are the moments acting on the rocket that are produced by the thrust force and the aerodynamic force and $M = M_{aer} + M_{thrust} = r_{c.p.} \times F_{aer} + r_{gim} \times F_{thrust}$. Note that M_{aer} represents the moment of the aerodynamic force acting on the center of pressure (*c.p.*), whereas M_{thrust} is the moment from the center of gravity (*c.g.*) to the gimbal of the engine. Given that the gravity force acts uniformly on the rocket, it does not create momentum. Now, assuming that the axis of reference frame B is aligned to reference frame R, the inertia tensor is given as $\hat{J} = diag[J_{xx}, J_{yy}, J_{zz}]$. Substituting the rocket moments M and $\hat{J}$ in (10), there arises the rotational dynamics, which are defined as

$$\begin{bmatrix} \dot{p} \\ \dot{q} \\ \dot{r} \end{bmatrix} = \begin{bmatrix} \frac{-qr(J_{zz} - J_{yy})}{J_{xx}} \\ \frac{[-X_{cp} \cdot N_\alpha s_\alpha + X_{gim} T s_{\delta\theta} - pr(J_{xx} - J_{zx})]}{J_{yy}} \\ \frac{[-X_{cp} \cdot S_\beta s_\beta - X_{gim} T s_{\delta\psi} - pq(J_{yy} - J_{zx})]}{J_{zz}} \end{bmatrix} \tag{11}$$

4.1.3. Rotational Kinematics

The rotational kinematics of the rocket with respect to the R frame are defined as a function of the angular velocity as follows:

$$\begin{bmatrix} \dot{\phi} \\ \dot{\theta} \\ \dot{\psi} \end{bmatrix} = \frac{1}{c_\theta} \begin{bmatrix} c_\theta & s_\phi s_\theta & c_\phi s_\theta \\ 0 & c_\phi c_\theta & -s_\phi c_\theta \\ 0 & s_\phi & c_\phi \end{bmatrix} \begin{bmatrix} p \\ q \\ r \end{bmatrix} \tag{12}$$

where ϕ, θ, and ψ are the Euler angles, the roll, pitch, and yaw, respectively. Thus, the TVC rocket model is given in (6), (11) and (12).

4.2. Two-Dimensional Aerodynamic Rocket Model

Consider that the rocket landing is performed on the vertical plane xz; thus, the angles ϕ and ψ and their derivatives are omitted. Additionally, the slip S in this direction is zero so let the nozzle control δ_ψ be zero. In this way, the rocket model becomes

$$m_r \ddot{z} - A s_\theta + N_\alpha s_\alpha c_\theta + m_r g_z = -T(c_u + c_\theta s_u) + \eta_1(t) \tag{13}$$

$$J_{yy} \ddot{\theta} + X_{cp} N_\alpha s_\alpha = -T X_{gim} s_u + \eta_2(t) \tag{14}$$

where z and θ stand for the rocket altitude and the pitch angle, respectively, with $u = \delta\theta$ the nozzle angle (control input), X_{gim} the distance between the nozzle gimbal joint and the center of gravity (CoG) in the rocket body frame, J_{yy} the principal moment of inertia on the y-axis, X_{cp} the distance between the center of pressure and the gravity center, and α the angle of attack. In addition, A and N_α stand for the axial and normal forces, respectively, with $\alpha = \theta + \gamma_\theta + \alpha_{w_\theta}$, where γ_θ is the drift angle and α_{w_ψ} corresponds to an unknown wind disturbance. Finally, $\eta_1(t)$ and $\eta_2(t)$ represent the exogenous forces as Lipchitz disturbances applied to each dynamic component, respectively.

Equations (13) and (14) can be written in a state-space form as in (1) and (2) or (3) and (4), where $\mathbf{x} = [x_1, x_2, x_3, x_4]^T = [z, \dot{z}, \theta, \dot{\theta}]^T \in \Re^4$ is the vector state, $\mathbf{y} = \mathbf{x}$ is the system output, and $\mathbf{f}(\mathbf{x}, t) = [f_1, f_2]^T$ is the flow of the nonlinear ODE in (13) and (14), where

$f_1 = \frac{1}{m_r}(-m_r g x_1 - N_\alpha s_\alpha c_{x_3} + A s_{x_3})$ and $f_2 = -J_{yy}^{-1} X_{cp} N_\alpha s_\alpha$; $\mathbf{g} = \mathbf{g}(\mathbf{x}, u, t) = [g_1, g_2]^T$ is the input matrix, where $g_1 = -\frac{T}{m_r}(c_u + c_{x_3} s_u)$ and $g_2 = J_{yy}^{-1} X_{gim} T s_u$; and $\boldsymbol{\eta}(t) = [\eta_1, \eta_2]^T$.

Remark 1. *Note that the angle* $|x_3| < x_{3_M}$ *must remain bounded to avoid a stall. In addition, the height* $x_1 > 0$ *to avoid crashing against the landing pad. Thus, it is no surprise that it is so difficult to maneuver a landing rocket, even when accounting for a plethora of resources in large enterprises such as SpaceX, Blue Origin, or NASA.*

4.3. Simulation Study

Three conditions are simulated for comparative purposes to show the performance of the Q-learning algorithm in learning the rocket landing. The simulation:

1. shows the comparative performance of the classical and variable box methods for the same output limits, action space, and number states but with different resolutions.
2. shows the comparative performance considering the same goal-box resolution and state-space dimension for the variable box method.
3. shows the rocket landing trajectories when the rocket is subject to a Gaussian disturbance.

The simulator was written in Python, with a step size of $h = 0.01$ [s]. The principal modules were numpy to compute the math operations, matplotlib to depict the results, and our Q-learning algorithm, which was embedded in the "temporal difference" module.

The task was to learn the vertical rocket landing from an arbitrary initial condition $\mathbf{y}_0$ to $\mathbf{y}_t^d = [0.1, 0, 0, 0]$, see Figure 5. Note that offset was set due to the landing train and the command was implemented using only the nozzle angle.

The reward assignment was designed as follows:

$$R_t(\mathbf{e}_t) = \begin{cases} 10 - ||\mathbf{e}_t|| & \text{if } \mathbf{y}_t \text{ is within limits} & \text{(it awards)} \\ -1000 & \text{otherwise} & \text{(it penalizes)} \end{cases} \tag{15}$$

where $\mathbf{e}_t = \mathbf{y}_t^d - \mathbf{y}_t$ is the error vector and $|| \cdot ||$ is the Euclidean norm. Function $R_t(\mathbf{e}_t)$ increases as $||\mathbf{e}_t|| \to 0$, maximizing the reward, i.e., $\mathbf{y}_t \to \mathbf{y}_t^d$.

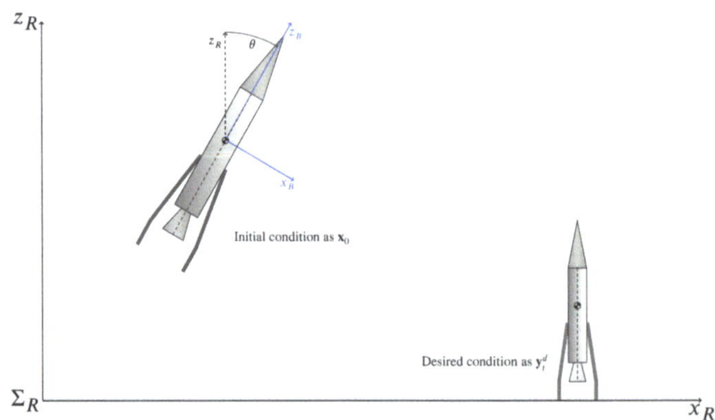

Figure 5. Schematic showing the task goal: learn the optimal admissible rocket trajectories for landing from an arbitrary initial condition $\mathbf{x}_0$ to a (constant) desired landing spot $\mathbf{y}_d^t$.

4.4. Rocket and Learning Parameters

The rocket parameters used were those of the Saturn V rocket from NASA [19], where $N_\alpha = 4.46477$ [N], $A = 6.09525$ [N], $m_r = 570 \times 10^3$ [kg], $T = 7.605 \times 10^6$ [N], $X_{gim} = 21$ [m], $X_{cp} = 10$ [m], $J_{yy} = 3.2 \times 10^7$ [kg m^2], $J_2 = 1.0826 \times 10^{-3}$, $\mu = 3.986 \times 10^{14} \left[\frac{m^3}{s^2}\right]$, $R_0 = 6.371 \times 10^6$ [m], $r_e = 6.372 \times 10^6$ [m], $\alpha_{\omega_\theta} = 0$ [rad], and $\gamma_\theta = \dot{z}/V$, with $V = 400 \left[\frac{m}{s}\right]$. The learning parameters were $\gamma = 0.7$ and $\alpha_Q = 0.01$ for 5×10^6 episodes. The first and second simulations were disturbed with $\boldsymbol{\eta}_t = [0,0]^T$.

4.5. Simulation A: Comparative Performance

Consider the output limits $[y_1, y_2] \in [-1,1]$ [m, m/s] and $[y_3, y_4] \in [-0.5, 0.5]$ [rad, rad/s], the action space $u_t \in [-0.2, 0.2]$ [rad], and 9 number states for both the classical and variable box methods. For the classical box method *CBM1*, the constant resolutions were $\epsilon_{y_1} = \epsilon_{y_2} = 3 \times 10^{-2}$ and $\epsilon_{y_3} = \epsilon_{y_4} = 8 \times 10^{-2}$. The first and last box indexes were considered open range ($\pm\infty$) as in membership functions in classical fuzzy logic. Then, there arose 28,224 number states. On the other hand, two sets of variable box parameters were tuned for the variable box method: (a) *VBM1*: The goal-box resolution was $\epsilon_{y_1} = \epsilon_{y_2} = \epsilon_{y_3} = \epsilon_{y_4} = 1 \times 10^{-2}$, with 28,224 number states as in CBM1. The action space used 21 action boxes. (b) *VBM2*: The goal-box resolution was the same as VBM1 but the output limits were set at $[y_1, y_2] \in [-10, 10]$ [m, m/s] and $[y_3, y_4] \in [-0.5, 0.5]$ [rad, rad/s], producing 51,984 number states. The action space also used 21 action boxes.

Figure 6 shows the results using CBM1, VBM1, and VBM2. The initial conditions for all cases were $\mathbf{x}_0 = [0.99 \text{ m}, 0 \text{ m/s}, 0.02 \text{ rad}, 0 \text{ rad/s}]^T$, for $\epsilon_{gb} = 1 \times 10^{-2}$. Simulations CBM1 and VBM1 needed $\dim(\mathcal{S}) = 28{,}224$ and VBM2 requires $\dim(\mathcal{S}) = 51{,}988$ to converge faster ($t_{land} = 3.7$ [s]); however, the highest reward ($r_{max} = 9.8618$) with the lower error norm ($\|\mathbf{e}_t\|_{min} = 0.1381$) is obtained with VBM1, showing that in any case, the proposed variable method performs better than the constant method, see Table 1.

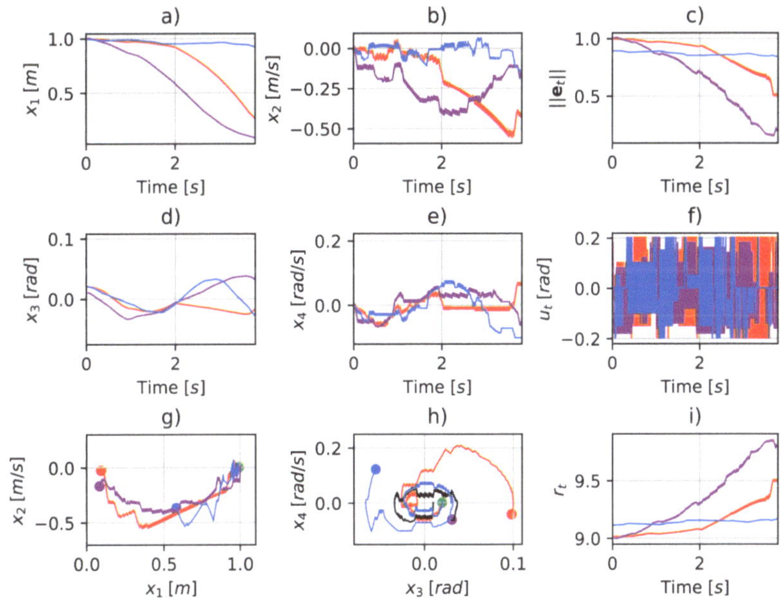

Figure 6. Comparative results for landing a rocket: classical box method CBM1 (*blue*), variable box method VBM1 (*red*), and VBM2 (*purple*), where the purple and black dots represent the initial and desired conditions, respectively. Subfigures depict the following: (**a,b,d,e**) show the rocket descending phase, for height velocity x_2, angle x_3, and angle velocity x_4 remain bounded while tracking errors remain near zero; (**a**) shows the better performance of our proposal, (**c**) shows how the error norm decreases while the reward increases (**i**), the control signal performance u_t is show in (**f**), and phase portraits are show in (**g,h**).

In Figure 6a,b,d,e, it can be seen that while the rocket is descending, the height velocity x_2, angle x_3, and angle velocity x_4 remain bounded. In addition, it can be seen that the state-variable errors are near zero, which is typical of reinforcement learning algorithms. Additionally, Figure 6a shows that the set of simulations using the variable box method had better landing performance than the classical box method. Furthermore, Figure 6f shows that the control signal u_t of the classical box method had a limited correlation between actions, states, and rewards, which limited the adequate exploration of the state space. Overall, the second set of simulation parameters with VBM2 performed better (as indicated in purple in Figure 6) at the expense of a significant increase in the number states. Finally, Figure 6g,h presents the portrait phase for each set of simulations, showing the convergence for the three cases; however, in the variable box method, the rocket approached smoothly due to the mesh's density.

4.6. Simulation B: Comparative Performance

We aimed to compare the landing performance using the classical and variable box methods based on three parameters: the minimum norm error $||\mathbf{e}_t||_{\min}$; maximum reward obtained $r_{\max}$; and time taken to land t_{land}. The parameters were the goal-box resolution ϵ_{gb} and state-space dimension $\dim(\mathcal{S})$ tuned by safety resolution ϵ_s, see Table 1. The initial conditions were $\mathbf{x}_0 = [0.99 \text{ m}, 0 \text{ m/s}, 0.02 \text{ rad}, 0 \text{ rad/s}]^T$.

Table 1. Comparative study results between CBM and VBM under different conditions of the goal-box resolution and state-space dimension. The best metrics ($\epsilon_{gb}, ||\mathbf{e}_t||_{\min}, r_{max}, t_{land}$ [s]) were obtained with VBM4 at the expense of a higher computational load ($\dim(\mathcal{S}) = 51{,}984$).

Box Method	Code	ϵ_{gb}	$\dim(\mathcal{S})$	$||\mathbf{e}_t||_{\min}$	r_{max}	t_{land} [s]
Classical	CBM1	1×10^{-2}	28,224	0.6304	9.3787	4.88
Variable	VBM1	1×10^{-2}	28,224	0.1381	9.8618	4.82
	VBM3	5×10^{-2}	**18,225**	0.1519	9.8480	4.93
Variable	VBM2	1×10^{-2}	51,984	0.1458	9.8541	3.7
	VBM4	1×10^{-3}	51,984	**0.0867**	**9.9132**	**3.22**

The first two rows in Table 1 correspond to the results presented in Figure 6, where *VBM1* achieved better performance than *CBM1* using the same state-space dimension (28,224) and box resolution. *VBM3* showed acceptable performance with only 18,225 (in bold) number states, which caused increasing errors and a poor reward; moreover, the landing time was the worst. Additionally, in *VBM2*, the state-space dimension increased to 51,984 due to the loose-fitting grid around the learning goal, the errors decreased slightly, and the landing took place in 3.7 [s].

If the goal-box resolution was defined as 1×10^{-3}, it yielded 51,984 number states and the best performance was achieved (in bold) in terms of the error, reward, and landing time, see *VBM4* in Table 1. Note that when applying the variable box method, it is suggested to establish a compromise among the learning goal, goal-box resolution, and state-space dimension on a trial-and-error basis.

4.7. Rocket Subject to Disturbances

Consider the rocket subject to a Gaussian disturbance $N_G(\cdot)$ such that $\eta_{t,1} = \eta_{t,2} = \kappa N_G(\mu_G, \sigma^2)$ for $\mu_G = 0$, $\sigma = 1$, and $\kappa = 8 \times 10^{-2}$ [N]. Figure 7 shows the performance of the rocket landing with and without perturbations using the policy of the VBM1 case defined in Section 4.5. The initial conditions were $\mathbf{x}_0 = [0.99 \text{ m}, 0 \text{ m/s}, 0.02 \text{ rad}, 0 \text{ rad/s}]^T$. Figure 7a,b show the rocket's portrait phase reaching the goal. Figure 7c shows the convergence under the previous learning process conditions, which were sufficient to compensate for disturbances that were unknown a priori.

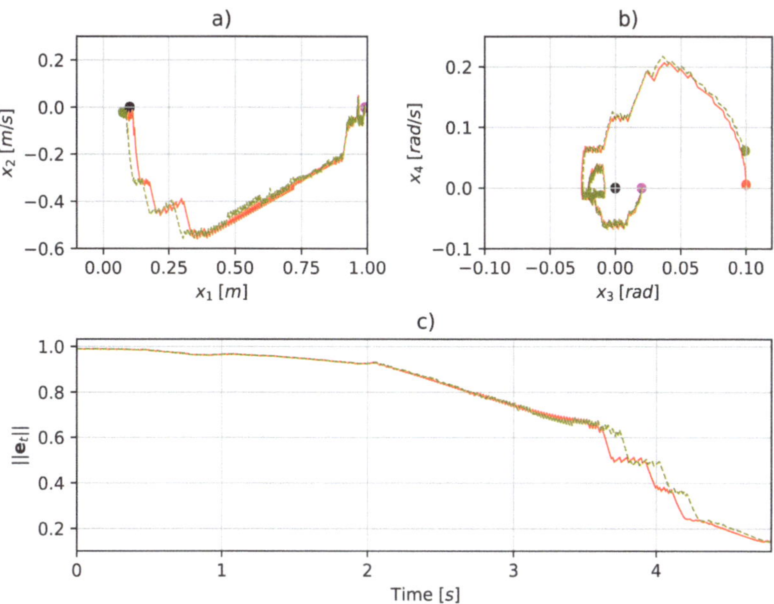

Figure 7. Rocket landing trajectories shown in phase portraits ((**a**) height position versus its velocity, and (**b**) angular position versus its velocity) using the proposed variable box method without disturbances (*red*) and subject to a disturbance η_t (*olive*). The purple and black dots represent the initial and desired conditions, respectively. Remarkably, L_2 norm of errors (**c**) yields similar plots, showing trajectories are learned for both cases, despite the strong disturbance (*olive*) at the expense of a slight (admissible) final deviation.

5. Conclusions

The classical Q-learning scheme guarantees iteratively reaching the goal states with a quasi-optimal policy. This powerful scheme seems promising for complex dynamical tasks such as landing a massive rocket but requires improved maneuverability. The variable box method is proposed for the Q-learning algorithm to yield higher resolution near the target region, where higher accuracy is required, such as the landing pad. In addition, this method suggests that asymmetric refinement may be introduced at different state regions in terms of the error threshold depending on the available computational power. This way, the learning process emerges with the refinement of boxes to actions without significantly increasing the computational costs. Simulations show how convenient it is to explore such refinements of box-sizing instead of increasing the number of boxes. In addition, the proposed variable box method can be introduced into other RL algorithms since the Q-learning architecture is maintained.

Due to the Q-learning algorithm being an offline learning scheme, a sample of the whole system behavior throughout the finite state-number space, which an expert user should tune on a trial-and-error basis, is needed to asymptotically compute the quasi-optimal policy rewarded by r_t. Once that quasi-optimal policy is learned offline, it is necessary to switch to an expert agent who already knows a local quasi-optimal policy $\pi_*(s_t)$. Moreover, caution is advised since, in practice, slight changes in states and actions lead to a slightly different policy, i.e., $\pi(s_t) \neq \pi_*(s_t)$. Thus, it is customary to run many simulation tests beforehand to "average" the policy $\hat{\pi}(s_t)$ tested in real systems, where $\hat{\pi}(s_t) \approx \pi_*(s_t)$. In our case, the proposed variable box method may assist in practice since it produces smoother but finer actions for larger errors.

Author Contributions: Methodology, A.T.-R.; Investigation, R.G.-R., V.P.-V. and L.E.R.-V.; Writing—original draft, A.T.-R.; Writing—review and editing, R.G.-R., V.P.-V. and L.E.R.-V. All authors have read and agreed to the published version of the manuscript.

Funding: This research received no external funding.

Institutional Review Board Statement: This study did not require ethical approval.

Informed Consent Statement: Not applicable.

Data Availability Statement: No new data were created.

Acknowledgments: The authors acknowledge the scholarship support granted to the first author from the National Council of Science and Technology (CONACyT) of Mexico.

Conflicts of Interest: The authors declare no conflict of interest.

Appendix A. The Standard Q-Learning Algorithm

The algorithm presented below merges the value iteration of the classical Dynamic Programming approach with a random sampling of the Monte Carlo method to approximate the value function asymptotically. In the algorithm, the temporal difference error $\delta_t = r_t + \gamma \max_{u_k} Q(s_{t+1}, \cdot) - Q(s_t, u_k)$, where $\gamma \in [0, 1)$ is the discount factor, which is fundamental. Subsequently, the current state value is updated as follows: $Q(s_t, u_k) \leftarrow Q(s_t, u_k) + \alpha_Q \delta_t$, where the learning rates $\alpha_Q \in (0, 1]$ and $Q(s_t, u_k)$ stand for the value function that is updated iteratively throughout δ_t. The optimal policy is given by $\pi_*(s_t) = \arg\max_{u_k} Q(s_t, u_k)$, which converges to the desired state vector $\mathbf{y}_t^d = \mathbf{x}_t^d$ by maximizing the reward r_t as the number of episodes $K_e \rightarrow \infty$. Note that $Q(s_t, u_k) \rightarrow Q_*(s_t, u_k)$ as the exploration improves. Particularly, all state numbers $s_t = \Im(\mathcal{B}(\mathbf{y}_t^d))$ are elements of the terminal-state-number subspace $\mathcal{S}^+ \subset \mathcal{S}$. When Q-learning is applied to dynamic systems, their state spaces and actions are mapped to a finite state-number space $\mathcal{S}$ and a finite action space $\mathcal{A}$, respectively; then, the box method that is applied to the instantaneous reward evaluates the new action, which yields a new state.

Algorithm A1 The classical Q-learning

Input: $\alpha_Q \in (0, 1]$; $\gamma \in [0, 1)$; and K_e.
Result: $\pi_*(s_t) \approx \pi(s_t)$.
Initialize;
$Q(s_t, u_k) \leftarrow 0 \; \forall s_t \in \mathcal{S}, \; u_k \in \mathcal{A}$;
Episode counter, $k_e \leftarrow 1$;
repeat for each episode
    Initialize randomly s_t;
    Select any action u_k according to state s_t;
    repeat for each step in current episode
        Apply action u_k and observe system response r_t, s_{t+1};
        if $s_t \notin \mathcal{S}^+$ **then**
            Compute *TD error*, $\delta_t \leftarrow r_t + \gamma \max_{u_k} Q(s_{t+1}, \cdot) - Q(s_t, u_k)$;
            Update value function $Q(s_t, u_k) \leftarrow Q(s_t, u_k) + \alpha_Q \delta_t$
        end
        $s_t \leftarrow s_{t+1}$;
    until $s_t \in \mathcal{S}^+$;
    $k_e \leftarrow k_e + 1$
until $k_e < K_e$;
$\pi(s_t) \leftarrow \arg\max_{u_K} Q(s_t, u_k)$;
return $\pi_*(s_t) \approx \pi(s_t)$

References

1. Nebylov, A.; Nebylov, V. Reusable Space Planes Challenges and Control Problems. *IFAC-PapersOnLine* **2016**, *49*, 480–485. [CrossRef]
2. Ünal, A.; Yaman, K.; Okur, E.; Adli, M.A. Design and Implementation of a Thrust Vector Control (TVC) Test System. *J. Polytech. Politek. Derg.* **2018**, *21*, 497–505. [CrossRef]
3. Oates, G.C. *Aerothermodynamics of Gas Turbine and Rocket Propulsion*, 3rd ed.; American Institute of Aeronautics and Astronautics: Washington, DC, USA, 1997; ISBN 978-1-56347-241-1.
4. Chen, Y.; Ma, L. Rocket Powered Landing Guidance Using Proximal Policy Optimization. In Proceedings of the 4th International Conference on Automation, Control and Robotics Engineering, Shenzhen, China, 19–21 July 2019; pp. 1–6. [CrossRef]
5. Yuan, H.; Zhao, Y.; Mou, Y.; Wang, X. Leveraging Curriculum Reinforcement Learning for Rocket Powered Landing Guidance and Control. In Proceedings of the China Automation Congress, Beijing, China, 22–24 October 2021; pp. 5765–5770. [CrossRef]
6. Sánchez-Sánchez, C.; Izzo, D. Real-Time Optimal Control via Deep Neural Networks: Study on Landing Problems. *J. Guid. Control Dyn.* **2018**, *41*, 1122–1135. [CrossRef]
7. Zhang, L.; Chen, Z.; Wang, J.; Huang, Z. Rocket Image Classification Based on Deep Convolutional Neural Network. In Proceedings of the 10th International Conference on Communications, Circuits and Systems, Chengdu, China, 22–24 December 2018; pp. 383–386. [CrossRef]
8. Stengel, R.F. *Flight Dynamics*; Princeton University Press: Princeton, NJ, USA, 2004; ISBN 978-1-40086-681-6.
9. Sutton, R.S.; Barto, A.G. *Reinforcement Learning: An Introduction*, 2nd ed.; A Bradford Book: Cambridge, MA, USA, 2018; ISBN 978-0-262-03924-6.
10. Michie, D.; Chambers, R.A. Boxes: An Experiment in Adaptive Control. *Mach. Intell.* **1968**, *2*, 137–152.
11. Davies, S. Multidimensional Triangulation and Interpolation for Reinforcement Learning. In Proceedings of the 9th International Conference on Neural Information Processing Systems, Denver, CO, USA, 3–5 December 1996; pp. 1005–1011. [CrossRef]
12. Dayan, P.; Hinton, G.E. Feudal Reinforcement Learning. In Proceedings of the Advances in Neural Information Processing Systems 5, Denver, CO, USA, 30 November–3 December 1992; pp. 271–278. [CrossRef]
13. Barto, A.G.; Sutton, R.S.; Anderson, C.W. Neuronlike Adaptive Elements that Can Solve Difficult Learning Control Problems. *IEEE Trans. Syst. Man Cybern.* **1983**, *SMC-13*, 834–846. [CrossRef]
14. Munos, R.; Moore, A. Variable Resolution Discretization in Optimal Control. *Mach. Learn.* **2002**, *49*, 291–323. [CrossRef]
15. Moore, A.W. Variable Resolution Dynamic Programming: Efficiently Learning Action Maps in Multivariate Real-valued State-spaces. In Proceedings of the 8th International Conference of Machine Learning, Evanston, IL, USA, 1 June 1991; pp. 333–337. [CrossRef]
16. Zahmatkesh, M.; Emami, S.A.; Banazadeh, A.; Castaldi, P. Robust Attitude Control of an Agile Aircraft Using Improved Q-Learning. *Actuators* **2022**, *11*, 374. [CrossRef]
17. Sutton, R.S. First Results with Dyna, an Integrated Architecture for Learning, Planning and Reacting. In *Neural Networks for Control*; Miller, W.T., Sutton, R.S., Werbos, P.J., Eds.; The MIT Press: Cambridge, MA, USA, 1991. [CrossRef]
18. Tewari, A. *Atmospheric and Space Flight Dynamics*; Modeling and Simulation in Science, Engineering and Technology; Birkhäuser Boston: Boston, MA, USA, 2007; ISBN 978-0-81764-437-6. [CrossRef]
19. Martínez-Perez, I.; Garcia-Rodriguez, R.; Vega-Navarrete, M.A.; Ramos-Velasco, L.E. Sliding-mode based Thrust Vector Control for Aircrafts. In Proceedings of the 12th International Micro Air Vehicle Conference, Puebla, Mexico, 16–20 November 2021; pp. 137–143.

Article

Design and Determination of Aerodynamic Coefficients of a Tail-Sitter Aircraft by Means of CFD Numerical Simulation

Emmanuel Alejandro Islas-Narvaez [1], Jean Fulbert Ituna-Yudonago [1,*], Luis Enrique Ramos-Velasco [1], Mario Alejandro Vega-Navarrete [1] and Octavio Garcia-Salazar [2]

1 Department of Aerospace Engineering, Metropolitan Polytechnic University of Hidalgo, Boulevard Acceso a Tolcayuca 1009, Ex Hacienda San Javier, Tolcayuca 43860, Hidalgo, Mexico; 203220018@upmh.edu.mx (E.A.I.-N.); lramos@upmh.edu.mx (L.E.R.-V.); mvega@upmh.edu.mx (M.A.V.-N.)

2 Aerospace Engineering Research and Innovation Center, Faculty of Mechanical and Electrical Engineering, Autonomous University of Nuevo Leon, San Nicolás de los Garza 66455, Nuevo Leon, Mexico; octavio.garciasl@uanl.edu.mx

* Correspondence: jituna@upmh.edu.mx; Tel.: +52-464-112-02-81

Abstract: Vertical take-off and landing (VTOL) aircraft have become important aerial vehicles for various sectors, such as security, health, and commercial sectors. These vehicles are capable of operating in different flight modes, allowing for the covering of most flight requirements in most environments. A tail-sitter aircraft is a type of VTOL vehicle that has the ability to take off and land vertically on it elevators (its tail) or on some rigid support element that extends behind the trailing edge. Most of the tail-sitter aircraft are designed with a fixed-wing adaptation rather than having their own design. The design of the tail-sitter carried out in this work had the particularity of not being an adaptation of a quad-rotor system in a commercial swept-wing aircraft, but, rather, was made from its own geometry in a twin-rotor configuration. The design was performed using ANSYS SpaceClaim CAD software, and a numerical analysis of the performance was carried out in ANSYS Fluent CFD software. The numerical results were satisfactorily validated with empirical correlations for the calculation of the polar curve, and the performance of the proposed tail-sitter was satisfactory compared to those found in the literature. The results of velocity and pressure contours were obtained for various angles of attack. The force and moment coefficients obtained showed trends similar to those reported in the literature.

Keywords: flight dynamics identification; autonomous vehicles; model reduction; dynamic emulation; CFD simulation

check for **updates**

Citation: Islas-Narvaez, E.A.; Ituna-Yudonago, J.F.; Ramos-Velasco, L.E.; Vega-Navarrete, M.A.; Garcia-Salazar, O. Design and Determination of Aerodynamic Coefficients of a Tail-Sitter Aircraft by Means of CFD Numerical Simulation. *Machines* **2023**, *11*, 17. https://doi.org/10.3390/machines11010017

Academic Editor: Dan Zhang

Received: 13 November 2022
Revised: 6 December 2022
Accepted: 16 December 2022
Published: 23 December 2022

1. Introduction

The development of autonomous and unmanned aerial vehicles has been a key approach adopted by the aerospace industry in recent years. These new vehicles incorporate flight capabilities that allow them to operate in different scenarios and fulfill multiple operational purposes. Much of the qualities that these aircraft possess are attributed to their designs, which involve defining the flight mission, selecting the vehicle configuration, studying the known theory of similar aircraft, and relating the physical properties of each design to the results obtained using mathematical and computational tools.

Aircraft design processes can vary depending on the type of aircraft or part to be designed. There are some studies in the literature in which the design processes are different from the previous ones. Chung et al. [1] performed the design, fabrication, and flight testing of a flying wing UAV electric thruster. The design process consisted of defining the performance requirements, stall speed, maximum speed, rate of turn, cruise altitude, absolute ceiling, and radius. The wing loading and associated power loading were derived based on defined performance requirements. The aerodynamic and stability design of the aircraft began with the given wing reference area. Then, the shape and configuration of the aircraft were determined. XFLR5 and DATCOM software were used to build the 3D model to estimate the zero-lift drag coefficient and induced drag factor of the aircraft.The proposed flying wing UAV was made from composite materials. Asim [2] proposed the different stages to follow when designing an aircraft. The stages consist of defining the design requirement, the mission that the aircraft must perform, the load to be supported, the configuration of the main components of the aircraft, and the avionics, stability, and control systems. However, the paper does not present any particular design or analysis. Espinal-Rojas et al. [3] proposed design and construction processes for a prototype of an unmanned aerial vehicle equipped with artificial vision to search for people, which include: the design of a structure with the correct morphology of the drone, which allows for taking in-flight images; the design of a communication system that allows these images to be sent to a ground base; and, finally, the design and implementation of a control and artificial vision system for the correct flight and the identification of the existence of people on the ground. Chu et al. [4] carried out a review on the design, modeling, and control of morphing aircraft. Within this work, they proposed three main design steps: configuration design, dynamic modeling, and flight control. Rahman et al. [5] conducted a design and performance analysis of unmanned aerial vehicles delivering aid in remote areas. In this study, they proposed a design procedure consisting of: the definition of the mission requirement, the explanation of the design, the detailed design of the vehicle components, the configuration of the navigation system, and the final design. Similar to the previous authors, Khan et al. [6] proposed a design procedure that includes the following steps: the preliminary design, which includes the selection of the aerodynamic profile of the wing and empennage and the sizing and design of components; the determination of stability characteristics; and the detailed design. Likewise, Kontogiannis et al. [7] proposed four main design stages: conceptual design, preliminary design, aerodynamic analysis and optimization, and final design.

CAD modeling is the most important stage in aircraft design as it streamlines the design process and improves the visualization of sub-assemblies, parts, and the final product. It also enables an easier and more robust design documentation, including the geometry, dimensions, and bills of materials. This stage is complemented by FEA and CFD analysis, which allow for simulating engineering designs made with CAD to assess their characteristics, properties, feasibility, and profitability. Tyan et al. [8] performed a multidisciplinary design optimization of a tailless unmanned combat aerial vehicle (UCAV) using global variable fidelity aerodynamic analysis. The UAV design was developed using the CAD, Dassault CATIA. Three different geometric models (baseline, low-fidelity, GVFM) were designed and numerically simulated by CFD ANSYS Fluent to determine the properties of lift, drag, polar curve, and pressure distribution on the wing. Harasani [9] carried out the design, construction, and testing of an unmanned aerial vehicle. The analysis of the aerodynamic properties was performed using the commercial design software TORNADO. For stress analysis, ANSYS FEA was used, and an EXCEL spreadsheet was used for the performance analysis and stability derivatives. Iqbal and Sullivan [10] presented an integrated approach to a conceptual UAV design, which includes: the design stage with CAD software, structural analysis with FEA software, and aerodynamic analysis with CFD software. Sobester and Keane [11] described, illustrated, and demonstrated a conceptual UAV design system through a specific design case study. It was shown that commercial CAD tools can be integrated into the design process from the conceptual level, where, as parametric geometry engines, they can play the important role of providing the models required by the various lines of multidisciplinary analysis. This allows for the use of CFD and FEA solutions in the early design stages. Flynn [12] demonstrated techniques that can lead to optimized designs by using 3D solid modeling CAD software and then performing elementary CFD simulations and simple stress FEA analysis. In addition, he demonstrated how computer-optimized designs can be produced economically using state-of-the-art rapid prototyping and rapid production machines based on additive manufacturing techniques. Chowdhur et al. [13] carried out an aircraft design through the wing subsystems and a VTOL multi-rotor subsystem, involving the design of wing modules, fuselage, engines, stabilizers, and the battery. Then, they optimized the efficiency parameters L/D, autonomy, weight, and cost, but they carried out an aerodynamic analysis for different conditions—polar curve or performance analysis—with respect to similar designs. Katon and Kuntjoro [14] designed a new/original modular UAV platform in which different combinations of modules were assembled to create a fixed-wing UAV and a quadrotor UAV. A 3D CAD modeling was carried out to conceive the basic configuration of the modules and the fixing mechanisms between modules. Three-dimensional CAD modeling was also used to illustrate the optimized module and assembly configurations. The use of the CATIA parametric CAD system in aircraft aerodynamic design was investigated and demonstrated by Ronzheimer [15]. It was shown that a parametric CAD system can act as a geometry generator, producing a clean geometry input for CFD simulations. MohamedZain et al. [16] designed and simulated a small-sized unmanned aerial vehicle (UAV) using 3DEXPERIENCE software. The design process of the frame parts involved many methods to ensure that the parts can meet the requirements. Todorov [17] performed a structural and modal analysis of a wing box structure using numerical simulation. The research was based on the finite element analysis of the aircraft wing. The first six natural frequencies and mode shapes were obtained. Benaouali and Stanisław [18] presented a new procedure for the aircraft wing structure design process. The originality of this procedure lies in the complete automation of the design process, the complexity in the considered case of the wing structure, and the applicability to different types of wings through parametric modeling. The design process involved both geometric modeling and structural analysis through the integration of two commercial CAD and CAE tools. The aerodynamic and stability characteristics of a fixed-wing MPU RX-4, a flying wing UAV, were studied by Bliamis et al. [19]. The preliminary design phase was performed and the aerodynamic performance, as well as the stability and control behavior, were evaluated using semi-empirical correlations, which were specifically modified for winged drones light flywheels, and the CFD tool.

Among the properties that must be determined in the analysis and optimization phase of an aircraft design, the most important are the aerodynamic properties (lift and drag force, lift coefficient, drag coefficient, rolling moment coefficient, pitching moment coefficient, yawing moment coefficient, angle of attack, polar curve proportionality constant, speed for maximum lift–drag ratio, stall speed, and Oswald coefficient). These properties allow for validating the flight conditions suggested in the preliminary design stage. This analysis is performed using CFD simulation. The simulation of aerodynamic elements by means of CFD, in the aeronautical industry, represents a fundamental pillar in the development of new technologies and optimization of existing models. Menter et al. [20] determined the polar curve of an aircraft by numerical simulation. The results were successfully validated with experimental data. Şumnu et al. [21] demonstrated the effects of shape optimization on the missile performance at supersonic speeds. The aerodynamic coefficients of drag and lift under different Mach numbers and different angles of attack were investigated numerically by means of CFD ANSYS Fluent software. Lao and Wong [22] carried out an aerodynamic study through CFD simulation that allowed for knowing the behavior of the aerodynamic properties (drag, lift, and moment coefficients) of an aircraft under different angles of attack. Combining the results from the lift and drag coefficients, it could be summarized that the lift–drag relationship is greatly improved by the ground effect. Salazar and Lopez [23] carried out an aerodynamic study of a wing in two and three dimensions by means of CFD numerical simulation and reported the lift and drag coefficients for various angles of attack. They did not obtain results of aerodynamic moments or perform variations in the yaw angle. Manikantissar and Geete [24] presented a conceptual design of an aircraft and performed CAD drawing and CFD simulation with ANSYS CFX to determine the distribution of pressures and lift and drag forces for multiple angles of attack and geometric taper values. They did not determine any properties related to the aerodynamic performance. Lammers [25] carried out the modeling of a commercial aircraft. CFD simulation was performed to determine the pressure, velocity, density, and temperature fields of the air around the airplane, without an engine. Lift and drag forces were also calculated. The results were validated by means of experimental data. Gu et al. [26] performed CFD numerical modeling and simulations on a commercial aircraft to obtain the polar curve at different Mach numbers and with different numerical models. Kosík [27] performed the modeling and CFD simulation of a twin-engine aircraft to determine pressure contours, streamlines, and drag and lift coefficients versus the angle of attack by comparing ANSYS Fluent and OpenFOAM results with experimental data.

Apart from the aerospace field, CFD simulation also allows for predicting the aerodynamic behavior of land vehicles and wind turbines. In land vehicles, the most frequent cases are those of numerical studies focused on competition vehicles equipped with spoilers and other elements that allow them to improve their grip on the ground. Cravero and Marsano [28] used Ansys CFX code to investigate the ground effect on an open-wheel race car. This work focused on demonstrating the reliability of a RANS model to study the flow around the unprotected rotating wheels of a racing car, where the interaction with the multi-element inverted wing is very noticeable, including the entire front of the car, of an actual F1 model from the year 2000. Ravelli and Savini [29] modeled an F1 car and carried out CFD simulation with open source software to determine pressure contours and streamlines. Wang et al. [30] carried out a CFD simulation of an F1 car to obtain the aerodynamic forces and, later, the results were compared with wind tunnel tests data for different wind velocities. In wind turbines, the aerodynamic flow around the turbine blade is complex in nature and difficult to analyze through experiments. By using CFD, different aerodynamic properties of a blade can be easily analyzed. Fernandez-Gamiz et al. [31] used CFD simulation to analyze the aerodynamic behavior of Gurney flaps and microtabs in a wind turbine. Different CFD simulations were made to compute the lift-to-drag ratio for several angles of attack. In addition, the CFD simulations allowed for the sizing of the passive flow control devices based on the airfoil's aerodynamic performance. Aziz et al. [32] performed a numerical and experimental investigation of the drag and lift forces at low Reynolds numbers and at different angles of attack to optimize the design of a turbine blade. bCFD simulation was mainly applied to analyze the factors that are not possible to visualize in real time, the factors on which the drag coefficient is dependent, and how it should be minimized. Ung et al. [33] investigated the aerodynamic performance of a vertical axis wind turbine with an endplate design. CFD simulation was carried out using the sliding mesh method and the k-ω SST turbulence model on a two-bladed NACA0018 VAWT. The aerodynamic performance of a VAWT with offset, symmetric V, asymmetric, and triangular endplates was analyzed and compared against the baseline turbine.

Throughout the works reported above, it can be pointed out that most of the works related to the design of tail-sitter aircraft adapted a fixed-wing instead of making their own design, and that most of them only showed basic aerodynamics properties and did not take into account the contribution of aerodynamics moments or the slip angle variation. In addition, there are few publications where the design and simulation methodology is reported in a complete way. The present work develops: (a) a geometric design; (b) an aerodynamic analysis of a tail-sitter UAV through the use of CAD and CFD computational tools; (c) a performance evaluation of the final design through some efficiency parameters that give an idea of the quality of the design.

2. Material and Method

2.1. Tail-Sitter Design

The proposed tail-sitter design was based on a moderately swept trapezoidal wing. It is a twin-engine prototype capable of taking off and landing on rigid supports that are attached to the wing and extend below the trailing edge [34–37]. The different design steps are shown in Figure 1.

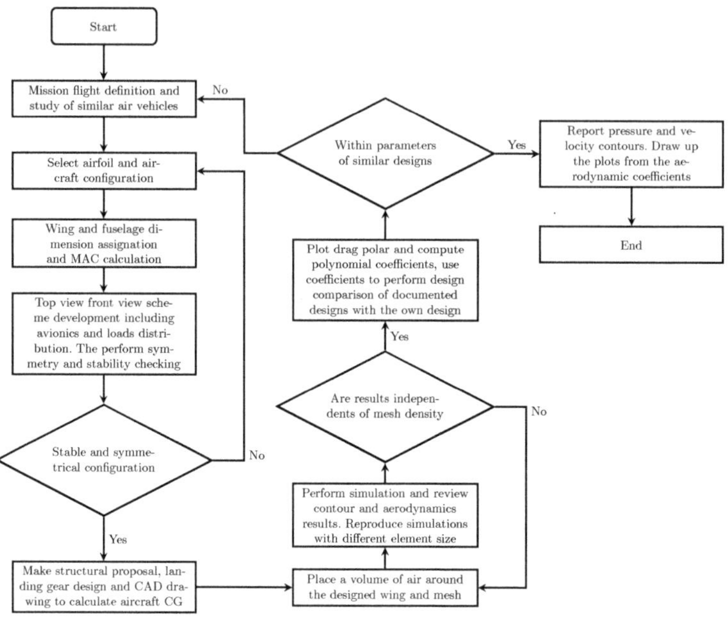

Figure 1. Flow chart of the wing design procedure.

2.1.1. Tail-Sitter Wing Sizing

This section consists of determining the sizes of the wing. In order to achieve this, an aerodynamic profile based on the MH60 profile was proposed [38], which is a profile frequently used in the design of composite wing or delta wing aircraft. For the central section of the wing that corresponds to the fuselage, re-scaling was applied to the original profile, with the purpose of having a greater internal space that allows for distributing the loads of the components installed inside the aircraft. Equations (1)–(4) were used to find the wing sizes shown in Table 1.

$$b_s = (L_l + L_{sc}) - \frac{b}{2} \tag{1}$$

$$\Lambda_r = \arctan\left(\frac{L_{ed}}{b - F_w}\right) \tag{2}$$

$$L_{ed} = C_{RW} - C_{TW} + T_{ed} \tag{3}$$

$$\Lambda_p = \arctan\left(\frac{T_{ed}}{b - F_w}\right) \tag{4}$$

In order to add some directional stability to the design, sweep angles were added to the wing. The first sweep angle was formed between the root and tip leading edges $\Lambda_{ba} = 7.34$; in the same way, the second sweep angle was formed between the trailing edges of the wing $\Lambda_{bs} = 16.00$.

The elevators are the mechanical elements responsible for the rotational movements that the tail-sitter can carry out. These elements were proposed with the dimensions described in Table 1. These mechanical elements were selected according to their dynamic effectiveness, which, in contrast to other documented designs [37], are similar in the moment coefficients that are produced. In the same way, the proportion of the elevator chord with respect to the mean wing chord was taken into account. The current proposal contemplates 20%, which is slightly below the most frequent proportion of 40%.

Table 1. Dimensions of tail-sitter wing.

Wing Description (Symbol)	Value [Units]
Fuselage width (F_w)	0.16 [m]
Half wing length (b_s)	0.76 [m]
Taper (λ)	0.76
Root chord (C_{Rw})	0.25 [m]
Tip chord (C_{Tw})	0.19 [m]
Dihedral angle (Γ)	3 [°]
Elevator description (Symbol)	
Elevator length (b_e)	0.35 [m]
Elevator root chord (C_{Re})	0.044 [m]
Elevator tip chord (C_{Te})	0.044 [m]
Reference chord (C_{ref})	0.3908 [m]
Fuselage description (Symbol)	
Center section length (L_{sc})	0.35 [m]
Side section length (L_l)	0.044 [m]
Offset length (L_D)	0.044 [m]
Fuselage length (L_{Fu})	0.3908 [m]
Motor support description (Symbol)	
Motor base (m_b)	0.0798 [m]
Airfoil part (a_{part})	0.162 [m]
Superior segment (S_{seg})	0.068
Elevon angle (δ_{elv})	71 [°]
Inferior segment (i_{seg})	0.046 [m]
Superior separation (s_{sep})	0.068 [m]
Inferior separation (i_{sep})	0.090 [m]
Thickness (t_s)	0.015 [m]

Figure 2a presents the elementary scheme of the wing with the dimensions of the root chord, the tip chord, half wing length, regressive sweep angle, and progressive sweep angle whose values are previously reported in Table 1. Figure 2b shows the detailed scheme of the wing, including the elevator dimensions. Values of these dimensions are previously reported in Table 1.

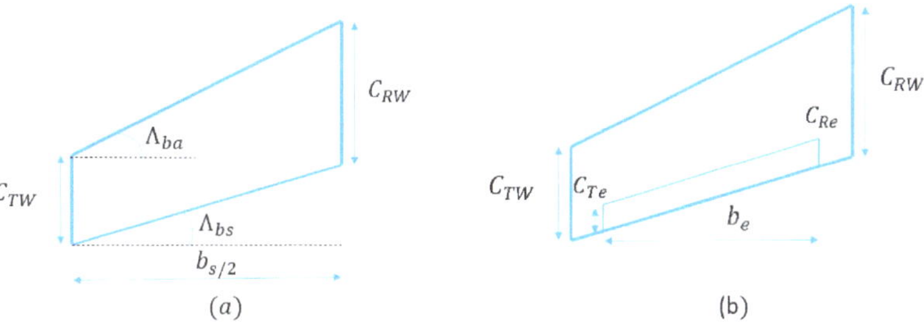

Figure 2. Wing schemes: (**a**) Elementary scheme, (**b**) Detailed scheme.

Once the dimensions of the wing are known, the mean aerodynamic chord can be determined. The mean chord is a parameter of great importance in the study of aerodynamic properties and in the placement and distribution of loads when carrying out weight and balance tasks. The mean aerodynamic chord can be calculated using Equation (5) [39].

$$MAC = \frac{2}{S} \int_0^{\frac{b}{2}} c(y)\,dy \qquad (5)$$

where S is the wing area, $d(y)$ is the function that describes the chord change with respect to the wing station, and $c(y)$ is the half wing chord that varies linearly from root to tip. The half wing chord can be calculated by the following Formula (6)

$$c(y) = C_{Rw} - \frac{2(C_{Rw} - C_{Tw})y}{b} \tag{6}$$

Carrying out the analytical development and simplifications of Equation (5) and substituting each quantity for its value yields $MAC = 0.2214$ [m]. This value mainly serves as a reference point for the placement of loads on board the tail-sitter.

2.1.2. Fuselage Sizing

In this section, the fuselage sizing is carried out from the middle of the aircraft. The corresponding scheme is shown in Figure 3a, and the dimensions are detailed in Table 1.

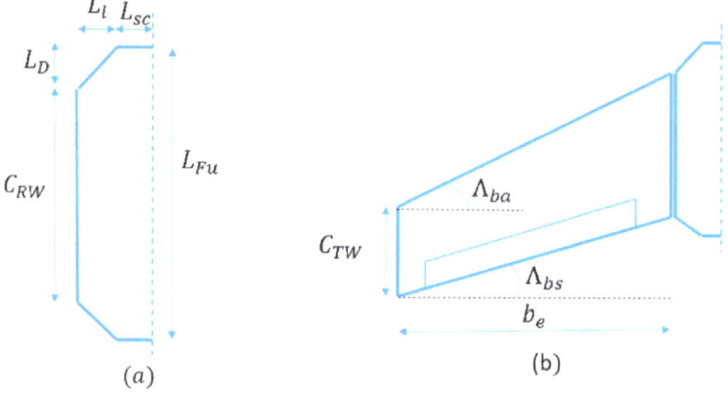

Figure 3. Fuselage scheme: (**a**) Half fuselage isolated, (**b**) Fusselaje and wing assembly.

The fuselage of the tail-sitter involves a longer central segment where avionics elements can be placed more freely. The design is based on an octagonal shape in which the two vertical sides have a height equivalent to the root chord of the semi-wings in order to facilitate the assembly of both elements as can be seen in Figure 3b.

2.1.3. Global Tail-Sitter Scheme

The global tail-sitter scheme is first presented in the front view in Figure 4 to visualize the dihedral angle (Γ) of the wings. This angle is responsible for modifying certain properties of lateral stability by adding a certain degree of robustness to the roll angle against disturbances or gusts of wind. Because greater angles can significantly displace the center of gravity, causing instability in the vertical flight, for the present prototype, three degrees of the dihedral angle were considered, as previously described in Table 1. The second global scheme of the tail-sitter is presented in top view in Figure 5 to visualize the sweep angle (Λ) of the wings. The distribution of tail-sitter components (wing, fuselage, and elevator) is symmetrical to the centerline where the center of gravity (CG) is located. The colored points represent instruments and hardware locations in the tail-sitter.

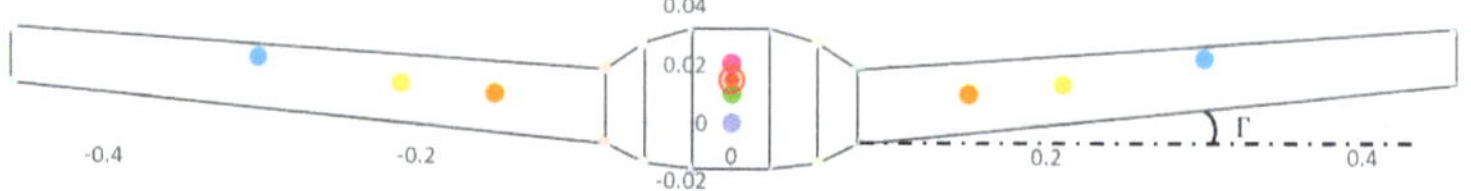

Figure 4. Front view of global tail-sitter scheme.

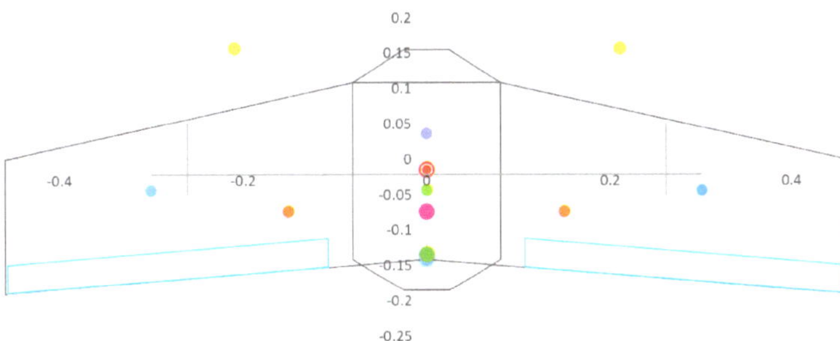

Figure 5. Top view of global tail-sitter scheme.

2.1.4. Structural Design of Tail-Sitter

The structural design of the tail-sitter components was based on the different models developed in the literature [40–42]. To carry out this structural design, firstly, two wing spars were proposed, whose positions are shown in Figure 6. The front spar is located at 13 percent of the middle chord from the leading edge, and the center spar at 58 percent. Each of these spars is made of balsa wood and has a square cross section with rounded edges. The side of the square measures 11 [mm] and the rounding at each vertex is 2.77 [mm].

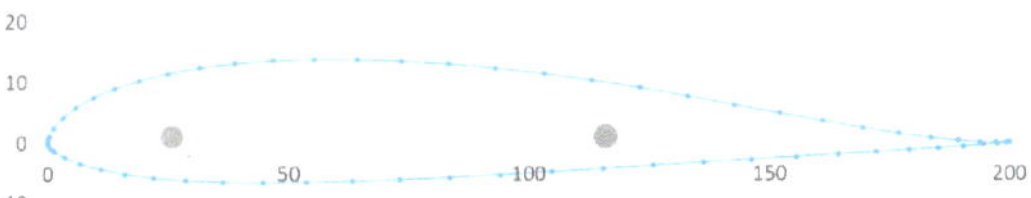

Figure 6. Side view of spar location.

The second step consisted of sizing the ribs and the motor support. Eight ribs made of balsa wood with a thickness of 3 [mm] were proposed in each half-wing, as well as a motor support (in each half-wing) made of square-shaped ABS polymer with a side of 27.5 [mm] and a total length of 358 [mm]. The bottom of each motor support has two diverging branches separated by an angle of 71°, between which, the elevator is mounted. The end of the support has a straight shape that is 114 [mm] long, with the same cross section as the trunk, which serves as landing gear. The motor support scheme is shown in Figure 7 and the dimensions and magnitudes of each part of the motor support are previously described in Table 1.

The spacing between ribs can be divided into two parts: the first distributes the first four ribs equidistantly from the root to the motor support, resulting in one rib every 12% of the half span. The second part joins four ribs from the motor support to the wingtip, resulting in one rib every 17% of the half span. Details of the spacing of each rib are presented in Table 2.

Table 2. Rib spacing.

Position [mm]	Chord [mm]	1st Spar [mm]	2nd Spar [mm]	Comments
0	250.00	32.50	145.00	
25	246.05	31.98	142.71	Internal elevator
72.5	238.55	31.01	138.36	
120	231.05	30.03	134.01	Motor support
185	220.07	28.70	128.05	
250	210.52	27.36	122.10	
315	200.02	26.03	116.15	
380	190.00	24.70	110.20	External elevator

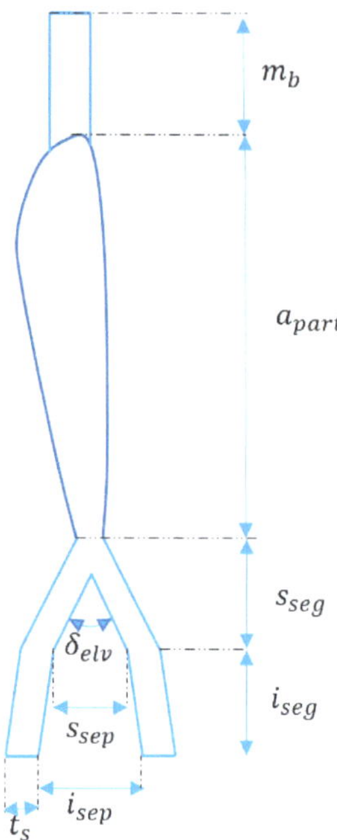

Figure 7. Motor support scheme.

The third step was to dimension the fuselage structure. The structure is made up of two ribs of balsa wood with a thickness of 3 [mm] and a cord of 340 [mm], both separated by 52 [mm]. These ribs are installed on two spars made of balsa wood, located in the same positions as those of the wings, with dimensions identical to the previous ones.

The fourth step involved drawing each component in CAD using ANSY Spaceclaim software. The assembly drawing of all of the structural components is shown in Figure 8.

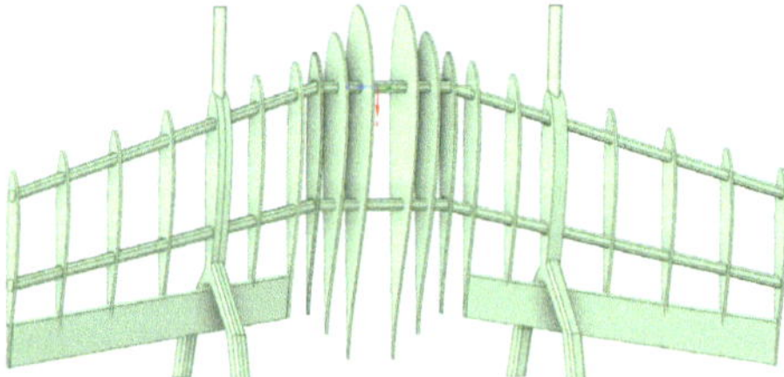

Figure 8. Detailed drawing of the tail-sitter structural arrangement.

The fifth step consisted of designing and locating the boxes that must house the various electronic components of the aircraft, such as Pixhawk flight computer, Raspberry pi main processing computer, battery, GPS antenna, electric motors, electronic speed controllers, servos, and ground supports. Each one of these boxes has the same dimensions as the physical component in terms of volume and assigned density, which allows it to match the mass to the real component. The dimensions and material of each component are described in Table 3. Regarding the location of each component, components such as flight computers, the battery, and the GPS antenna were placed above the longitudinal axis in a position close to the trailing edge because, in this way, a symmetry of loads can be guaranteed and the center of gravity can be positioned behind the center of pressures. The electric motors were placed at the tip of the ground support because, in this position, the propellers have free movement and the elevators can take advantage of the propwash generated by each propeller. The speed controllers usually have short cables that connect to each electric motor; therefore, these devices were placed outside the fuselage, especially in front of each elevator symmetrically from each other. The servos were also placed outside the fuselage, ahead of each elevator at a distance where the wing's convexity allowed them to protect the servos. Their location was approximately at the center of the elevator. The supports were placed at the same distance from the center of the wing, where each electric motor was located. The set of components can be seen in Figure 9.

Table 3. Dimensional characteristics of the components.

Name	Material	Length [mm]	Width [mm]	Depth [mm]	Mass [kg]
Pixhawk	Plastic and semiconductors	50	15.5	81.5	0.038
Rasp berry pi	Plastic and semiconductors	88	58	19.5	0.046
Battery	Lithium polymer	105	33	24	0.08
Antenna GPS	Ceramics and semiconductors	54	54	15	0.03
Electric motor 1	Metal	27.5	27.5	30	0.033
Electric motor 2	Metal	27.5	27.5	30	0.033
ESC 1	Semiconductors	55	28	7	0.01
ESC 2	Semiconductors	55	28	7	0.01
Servo 1	Plastic	40	20	38	0.022
Servo 2	Plastic	40	20	38	0.022
Ground support 1	ABS polymer	40	20	38	0.17
Ground support 2	ABS polymer	40	20	38	0.17

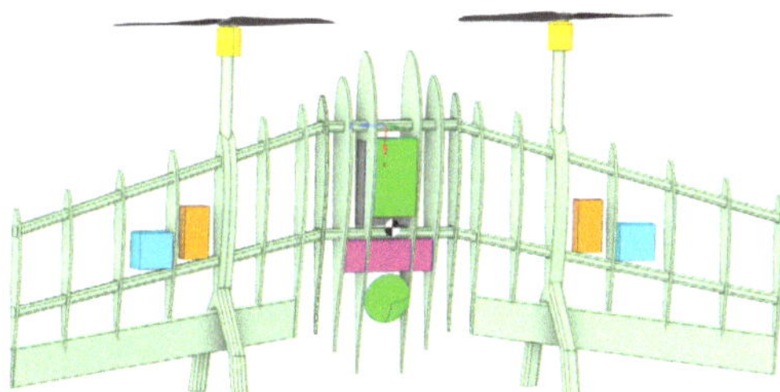

Figure 9. Detailed drawing of the tail-sitter structural arrangement and components.

The sixth step was to choose the material for the skin of the aircraft and its coating. A shrink-wrap film known as "monokote" was chosen, which is 1 [mm] thick and has a weigh per unit area of 0.061 [kg/m^2]. This material covers the entire external surface of wings and fuselage as can be seen in different views in Figure 10.

The last step consisted of calculating the center of gravity of the prototype designed. The calculation was based on the formulas proposed by Sandonis [43].

$$x_{CG} = \frac{\sum_i m_i x_i}{M} \tag{7}$$

$$y_{CG} = \frac{\sum_i m_i y_i}{M} \tag{8}$$

$$z_{CG} = \frac{\sum_i m_i z_i}{M} \tag{9}$$

where m_i is the individual mass of each component, M is the total mass of the components, (x_i, y_i, z_i) is the center of gravity of each component whose reference point is the origin of the aircraft's nose coordinates, and (x_{CG}, y_{CG}, z_{CG}) is the position of the center of gravity. The mass of each component was obtained from the manufacturers' data sheet. For the motor support, the mass was determined from the CAD Space Claim. The value of the mass of each component is previously described in Table 3. Thus, the total mass of the tail-sitter was 0.942 [kg], which corresponds to the sum of the masses of all of the components. The (x_i, y_i, z_i) of each component was determined according to the location of each one.

These three parameters allowed us to calculate the coordinates of the center of gravity (x_{CG}, y_{CG}, z_{CG}), whose values in [m] were, respectively, $[0.178, 7 \times 10^{-7}$, and $0.0152]$.

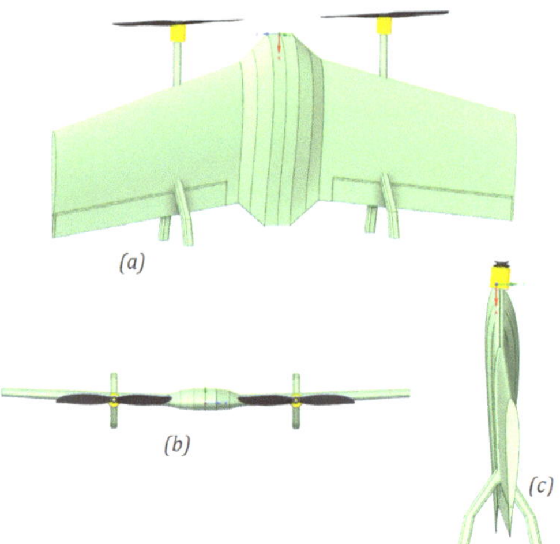

Figure 10. Virtual prototype of the tail-sitter: (**a**) Top view. (**b**) Front view. (**c**) Side view.

2.2. Determination of Aerodynamic Coefficients by Means of CFD

2.2.1. Mathematical Model for CFD Simulation of Tail-Sitter

The mathematical model used to develop the numerical solution was carried out by means of a 3D model of the equations of mass, momentum, and turbulence in steady state.

The Spalart Allmaras model was chosen to analyze the turbulent behavior of the airflow around the tail-sitter. This model is commonly used in aerodynamic problems and is very suitable for predicting pressure gradients within the boundary layer [22]. The y^+ corresponding to this model is very low ($y^+ < 5$), which allows it to accurately predict aerodynamic properties close to the boundary layer compared to 2-equation models such as $k - \epsilon$, whose y^+ is higher ($y^+ < 30$) [44–46].

The following considerations were taken into account for the development of the mathematical model:

- The simulation takes place in an isothermal environment.
- Air is the working fluid and is assumed to be viscous, incompressible, and behave like an ideal gas.
- The gravity force acts in a direction normal to the intrados of the wing.
- The density and viscosity of the fluid remain constant.

The governing equations were formulated compactly as follows:
Mass conservation equation:

$$\nabla(\rho v) = 0 \tag{10}$$

Momentum conservation equation:

$$\nabla \cdot (\rho v T) = \nabla \left(\frac{\lambda}{C_p} \nabla T \right) + \Psi \tag{11}$$

Turbulent viscosity equation for Spalart Allmaras model formulated by [47]:

$$u_j \frac{\partial \tilde{v}}{\partial x_j} = C_{b1}[1 - f_{t2}]\tilde{S}\tilde{v} + \frac{1}{\sigma}\left\{ \nabla \cdot [(v + \hat{v})\nabla\tilde{v}] + C_{b2}|\nabla\tilde{v}|^2 \right\}$$

$$- \left[C_{w1}f_w - \frac{C_{b1}}{k^2}f_{t2} \right] \left(\frac{\tilde{n}u}{d} \right)^2 + f_{t1}\Delta U^2 \tag{12}$$

where $\nu = \frac{\mu}{\rho}$ is the molecular kinematic viscosity and *rho* is the density, $\nabla \cdot [(\nu + \tilde{\nu})\nabla\tilde{\nu}]$ is the classical diffusion operator, σ is the Prandtl turbulence number, $C_{b2}|\nabla\tilde{\nu}|^2$ is a non-conservative term, and $C_{b1}[1 - f_{t2}]\tilde{S}\tilde{\nu}$ is the turbulence destruction term in which

$$\tilde{S} \equiv S + \frac{\nu}{k^2 d^2} f_{v2} \tag{13}$$

where d is the distance between the nearest surface and S is defined as

$$S = \sqrt{2\Omega_{ij}\Omega_{ij}} \tag{14}$$

where Ω_{ij} is the rotation tensor given by

$$\Omega_{ij} = \frac{1}{2}\left(\frac{\partial u_i}{\partial x_j} - \frac{\partial u_j}{\partial x_i}\right) \tag{15}$$

$$f_{v2} = 1 - \frac{\chi}{1 + \chi f_{v1}} \tag{16}$$

$$f_{t1} = C_{t1} g_t exp\left(-C_{t2}\frac{\omega_t^2}{\Delta U^2}\left[d^2 + g_t^2 d_t^2\right]\right) \tag{17}$$

The turbulence destruction term is contemplated by

$$\left[C_{w1} f_w - \frac{C_{b1}}{k^2} f_{t2}\right]\left(\frac{\tilde{\nu}}{d}\right)^2 \tag{18}$$

where

$$f_w = g\left[\frac{1 + C_{w3}^6}{g^6 + +C_{w3}^6}\right] \tag{19}$$

$$g = r + C_{w2}(r^6 - r) \tag{20}$$

$$r \equiv \frac{\nu}{\tilde{S} k^2 d^2} \tag{21}$$

$$f_{t2} = C_{t3} exp(C_{t4}\chi^2) \tag{22}$$

The constants included in the turbulence model are described in Table 4.

Table 4. Constants included in the turbulence model.

Constant	Value
σ	2/3
C_{b1}	0.1355
C_{b2}	0.622
k	0.41
C_{w1}	$C_{b1}/k^2 + (1 + C_{b2})/\sigma$
C_{w2}	0.3
C_{w3}	2
C_{v1}	7.1
C_{t1}	1
C_{t2}	2
C_{t3}	1.1
C_{t4}	2

2.2.2. Geometric Model

The geometric model for the numerical simulation as shown in Figure 11a is composed of two domains: the tail-sitter domain and the air enclosure domain. This model was carried out using the CAD ANSYS DesignModeler, in which, the domain of the tail-sitter was exported from the prototype built in CAD ANSYS SpaceClaim and the air enclosure domain was drawn based on [22,48]; the dimensions are described in Table 5. The computational domain shown in Figure 11b surrounded the tail-sitter aircraft, which has the same dimensions as the design stage. However, motors and motor support were not considered since the fluid dynamic analysis focuses on the aerodynamic forces.

Table 5. Air enclosure dimensions.

Description	Value [Units]
Length	830 [mm]
Width	300 [mm]
Height	60 [mm]

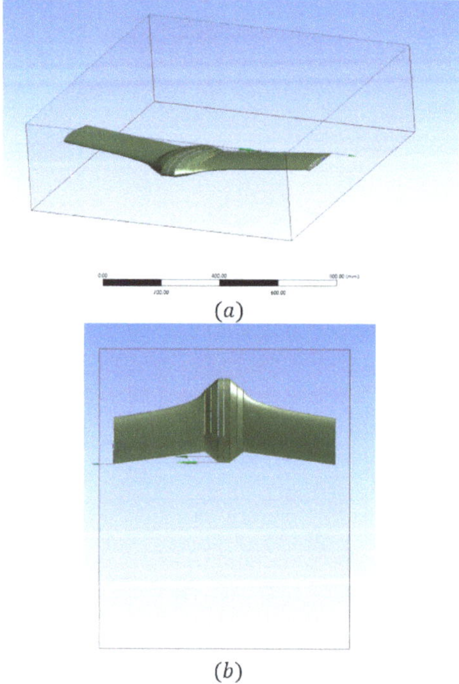

(*a*)

(*b*)

Figure 11. Computational domains for CFD simulation: (**a**) Isometric view, (**b**) Top view.

2.2.3. Boundary Conditions

The walls of the air enclosure domain, as shown in Figure 11, were the boundaries for the numerical simulation. The front wall was considered as the air flow inlet boundary, wose condition was a velocity inlet with a magnitude of 16 [m/s] (value used as a condition to design the tail-sitter). The rear wall was considered as the outlet boundary, with a gauge pressure outlet condition of 0 [Pa] (since the absolute pressure is equal to the atmospheric pressure). The other walls (upper, lower, right, and left sides) were considered as symmetry boundary conditions in a similar way to that of [22,48]. This allowed the air velocity to be maintained above zero in the area close to the walls to avoid the hydrodynamic boundary layer effect.

Unlike the wall boundary condition, the symmetry boundary condition allows for simulations with reduced air domains. The use of this symmetry condition allowed us to eliminate the boundary layer effect due to the reduction in the air velocity close to the wall [22,49]. Boundary conditions are resumed in Table 6.

Table 6. Boundary conditions.

Location	Boundary Type	Value [Units]
Front face	Velocity inlet	16 [m/s]
Rear face	Gauge pressure outlet	0 [Pa]
Upper face	Symmetry	-
Lower face	Symmetry	-
Right face	Symmetry	-
Left face	Symmetry	-

2.2.4. Mesh Independence Analysis

The mesh independence study was carried out with the purpose of finding the mesh size limit at the point where the predictions obtained from the simulation are independent at greater mesh size reductions [50]. This analysis was preceded by the meshing process of the air enclosure domain. This choice is justified by the fact that the goal pursued by the CFD simulation is to determine the aerodynamic properties throughout the external surface of the tail-sitter. However, it is not necessary to consider the tail-sitter domain in the mesh. To perform this process, an unstructured mesh was made in order to reduce the number of nodes and, consequently, the computational time of the analysis. The meshing was carried out using the ANSYS Meshing software included in the ANSYS Workbench platform [51]. The meshed computational domain is shown in Figure 12. It is possible to see a global view of the mesh domain, as well as a region close to the aircraft where polygons conform a fine mesh in the zone that makes contact between the air domain and aircraft skin. The mesh refinement was made using body sizing with capture curvature and capture proximity in order to reduce the size of the mesh close to the aircraft skin. The distance of the first border node was 0.10949 [mm]. This value was checked with free software [52] using properties such as the density ρ, viscosity μ, mean aerodynamic chord MAC, y plus y^+, and air velocity U_∞. The distance calculated by means of this software was 0.116074268 [mm], which is close to that of the generated mesh. In the zoom box, it is possible to see the size ratio between the grid near the wing and the grid at the bottom boundary.

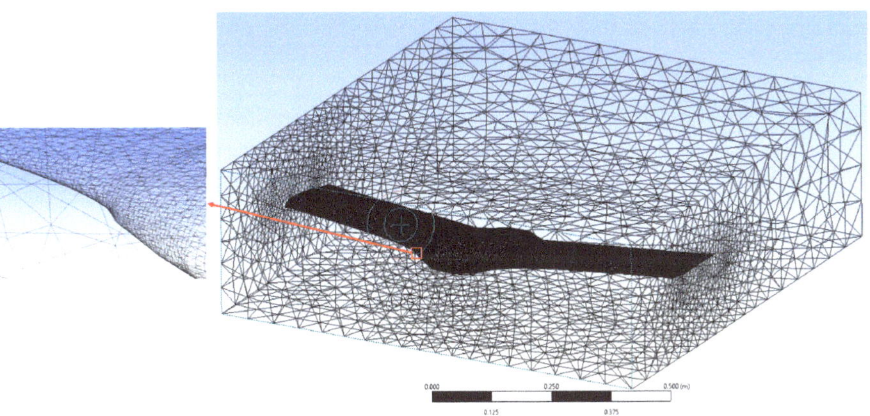

Figure 12. Mesh domain.

The mesh independence analysis consisted of determining the behavior of the yaw moment coefficient (C_l) with respect to the number of elements of the mesh. The angle of attack of zero degrees was selected to place the computational domain in a cruise flight position as can be seen in Figure 12. To perform this analysis, four different mesh sizes were selected. The characteristics of each mesh are described in Table 7

Table 7. Statistics in the study of mesh independence.

Characteristics	Mesh Sizes			
	M1	**M2**	**M3**	**M4**
Nodes	59,909	66,701	74,868	84,543
Elements	328,718	350,992	403,549	455,174
Mesh metric Skewness				
Minimum	2.138×10^{-4}	2.293×10^{-4}	2.5457×10^{-4}	3.340×10^{-4}
Maximum	0.7415	0.7979	0.9251	0.9795
Average	0.2322	0.2328	0.2338	0.2331
Standard deviation	0.1219	0.1225	0.1237	0.12156

Each mesh was simulated in a steady state using CFD ANSYS Fluent from the Workbench platform. The flow and turbulence models, as well as the boundary conditions described in the previous section, were used. Regarding the pressure–velocity coupling algorithm, the coupled segregation algorithm was used, since it is the most appropriate algorithm to predict the behavior of pressure and velocity on the wing surface of aircraft [25]. For the discretization scheme of momentum and turbulent viscosity equations, second-order upwind was used for more precise results and, furthermore, upwind is the most appropriate scheme for discretizing the turbulent flow model [25]. Regarding the residuals, 10^{-4} was used for continuity and velocity in the x, y, and z direction and 10^{-6} for the turbulent kinetic viscosity (nut).

The results of this analysis are presented in Figure 13. It can be seen that the yaw moment coefficient is independent of the mesh size when the element number exceeds 403,549. Therefore, it was decided to use the mesh of 455,174 elements in the present study in order to obtain more precise results.

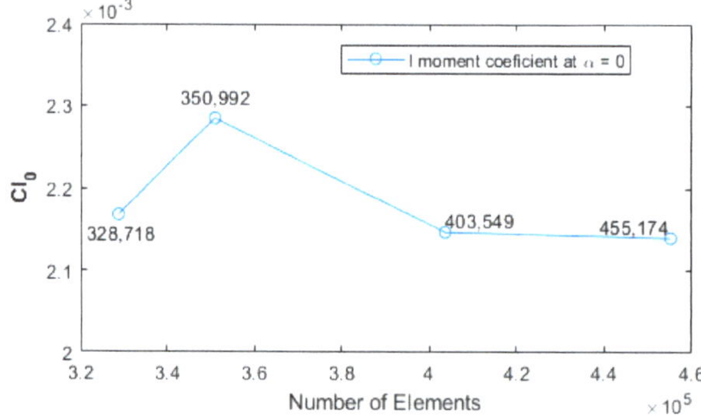

Figure 13. Mesh independence of moment coefficient C_l.

3. Results and Discussions

3.1. Validation of Tail-Sitter CFD Simulation Results

The polar curve is an algebraic representation that describes the value of the drag coefficient as a quadratic function of lift and is an invaluable resource when evaluating the efficiency of an aircraft under particular conditions [53]. This curve was used to validate the results of the numerical simulation carried out in this work. Figure 14 compares the polar tail-sitter curve produced by the results of the CFD simulation with that produced by an empirical correlation [54].

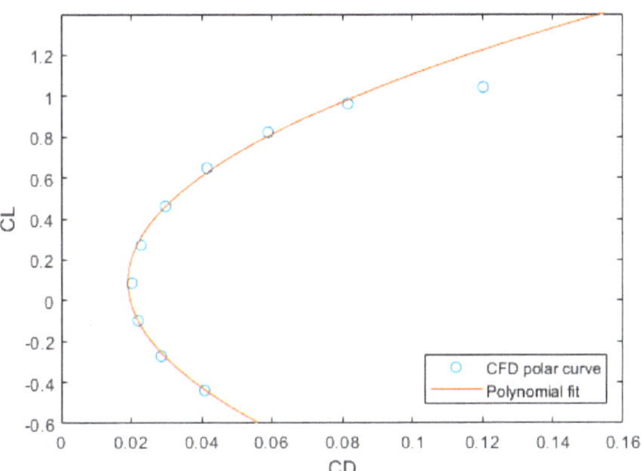

Figure 14. Tail-sitter polar curve: CFD vs. polynomial fit.

A similarity between the polar curve generated by the empirical correlation and that generated by numerical simulation was observed. Figure 15a,b describe the error percentages between the CFD simulation results and the polynomial fit of the lift and drag coefficients, respectively. A good approximation was observed between the two results. The maximum error for the C_L is around 10%, whereas, for the drag coefficient, the error reaches 25% when the C_D produced by the CFD simulation is equal to 0.12, but the average error is 10%, which is within a suitable range.

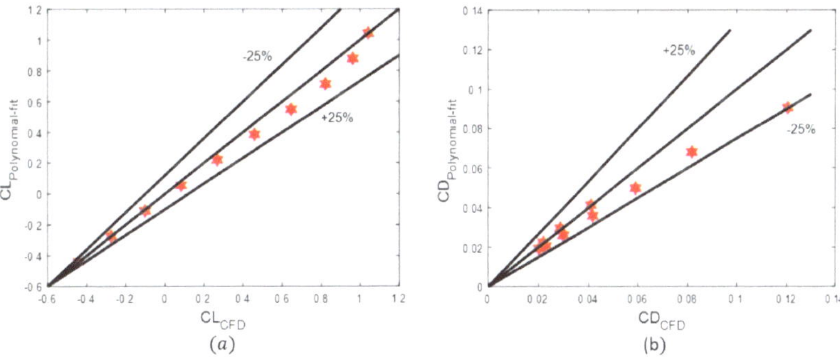

Figure 15. Percentage of errors between CFD and polynomial fit: (**a**) C_L and (**b**) C_D.

Therefore, the numerical model developed in the present study can predict the aerodynamic behavior of the designed tail-sitter.

3.2. Aerodynamic Properties of the Tail-Sitter

The main aerodynamic properties in this study are the coefficients of forces and moments. However, to better understand the effects of aerodynamic forces on the tail-sitter at different angles of attack (α), the pressure contours on the tail-sitter surface, as well as the airspeed distribution around the airfoil, were reviewed. The pressure contours on the extrados zones and intrados zones are shown in Figures 16 and 17, respectively. Positive pressures were mainly obtained on the extrados and negative pressures on the intrados when $\alpha = -4$. This behavior is normal because, when the aircraft is tilted down, the airflow is intercepted by the upper surface (extrados), causing greater pressure. The pressure decreases progressively when the angle of attack approaches zero, and when the angle of attack begins to increase as shown by the pressure contours when $\alpha = 4$ and $\alpha = 8$, the pressure decreases on the extrados and increases on the intrados, allowing the aircraft to lift. The wing reaches a minimum pressure of -133.5 [Pa] at the beginning of the lower surface and a maximum pressure of 142 [Pa] at the leading edge, which is clearly acceptable for the zone where each pressure is located.

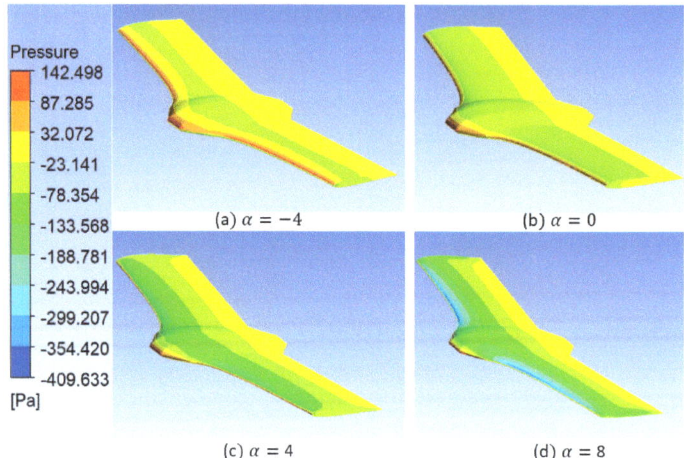

Figure 16. Pressure distribution on the tail-sitter surface at different angles of attack at upper surface.

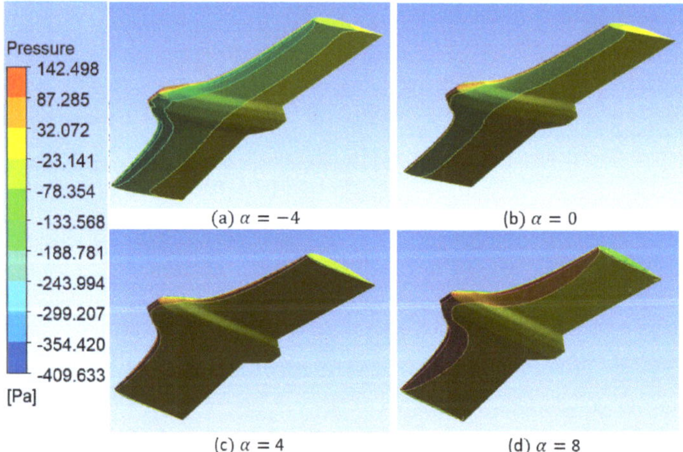

Figure 17. Pressure distribution on the tail-sitter surface at different angles of attack at lower surface.

The inverse of the pressure behaviors around the tail-sitter area is observed in Figure 18a–c for the velocity distributions. When the aircraft is tilted downwards, the airflow is intercepted by the upper surface (extrados), causing a decrease in the air velocity and, consequently, an increase on the lower surface. The air velocity remains almost in balance at approximately 18 [m/s] on both sides when the angle of attack approaches zero, and when the angle of attack starts to increase (see the velocity contours when $\alpha = 4$ and $\alpha = 8$), the air velocity decreases on the intrados by up to 10 [m/s] and increases on the extrados by up to 23.3 [m/s], allowing for a better lift of the aircraft. In addition, the air velocity decreases progressively along the mid-span from root to tip. This avoids a greater wake effect at the tip.

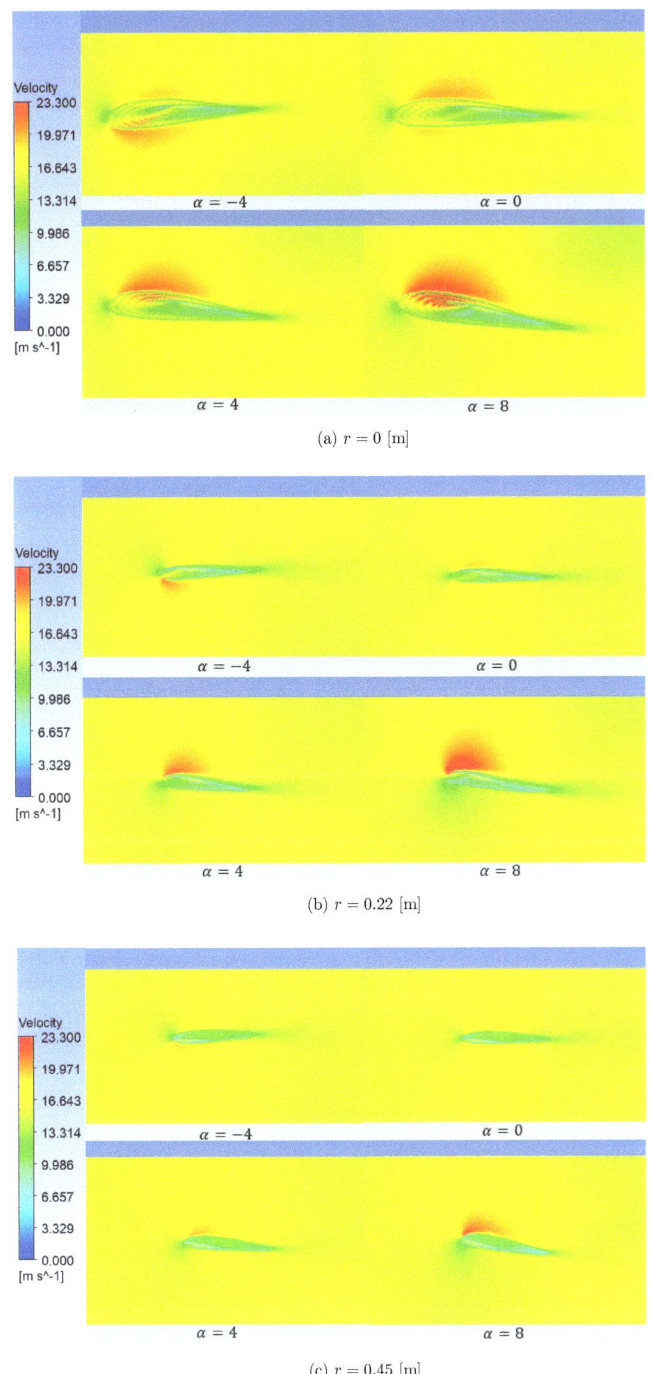

(a) $r = 0$ [m]

(b) $r = 0.22$ [m]

(c) $r = 0.45$ [m]

Figure 18. Velocity contours at different angles of attack: (**a**) local radius $r = 0$ [m], (**b**) local radius $r = 0.22$ [m] and (**c**) local radius $r = 0.45$ [m] .

In general, the pressure and velocity contours show reasonable behavior from an aerodynamic point of view. This once again confirms the good performance of the tail-sitter design proposed in this study.

The effects of the changes in the angle of attack and angle of slip (β) on the force coefficients and moment coefficients are analyzed below.

Figure 19a shows the behavior of the lift coefficient against the variations in both angles. The lift coefficient increases linearly from -0.4 to 1 as the angle of attack increases from $-6°$ to $10°$. The trend is slightly changed when the angle of attack exceeds $10°$. The variations in the angle β do not provide significant effects on the lift coefficient since the proposed tail-sitter has a swept trapezoidal wing. This behavior matches those reported in the literature for the trapezoidal wings [1,19,55,56].

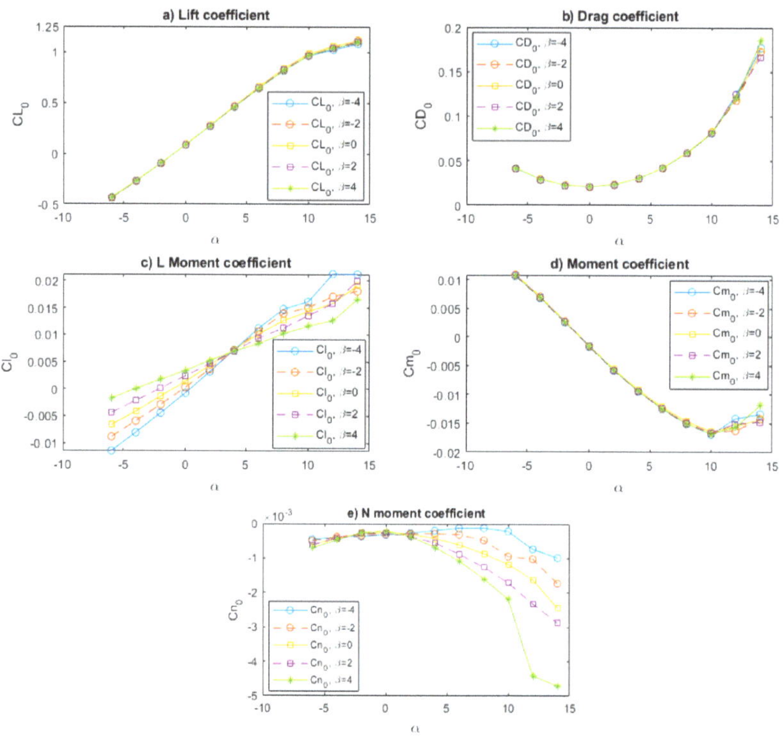

Figure 19. Effects of changes in angle of attack and angle of slip on the force coefficients and moment coefficients: (**a**) lift coefficient, (**b**) drag coefficient, (**c**) L moment coefficient, (**d**) M moment coefficient, (**e**) N moment coefficient.

Figure 19b shows the behavior of the drag coefficient against changes in the angles α and β. The drag coefficient does not increase linearly like the lift coefficient. It has a tendency of a curve of second degree, as is usual in the behavior of the drag coefficient of trapezoidal wings when the angle of attack varies; see Refs. [57,58]. Similar to the lift coefficient, variations in the angle β do not provide a significant change in the drag coefficient. The drag coefficient reaches its minimum value of 0.02017 when the angle of attack is $0°$ and reaches 0.17 when $\alpha = 14$. This magnitude is within the adequate range for trapezoidal wings as reported in the literature [1,19,55,56].

Figure 19c describes the effects of changes in the angles α and β on the moment coefficient around the X axis (responsible for the roll movement). A linear increase in the L moment coefficient is observed as the angle of attack α increases from $-0.4°$ to

14°. In addition, increasing the angle β modifies the slope of each line. This behavior is adequate since the modification of the slip angle causes the displacement of the center of pressure of the aircraft in a lateral direction, producing a torque that contributes to the roll movement [59].

The effects of changes in the angles α and β on the moment coefficient around the Y axis (responsible for the pitch movement) are described in Figure 19d. The moment coefficient Cm_0 decreases from 0.012 to -0.017 when the angle α increases from $-4°$ to $10°$. This behavior is due to changes in the magnitude of the center of pressure and its position in the longitudinal axis, which results in an increase in the longitudinal axis torque that contributes to the pitch motion.

Figure 19e shows the behavior of the moment coefficient around the Z axis (responsible for the yaw movement) against the changes in the angles α and β. A decrease in Cn_0 was observed when the angle α exceeds $2°$. Furthermore, the increase in the angle β causes an increase in Cn_0 when the angle α exceeds $2°$. This is because the modification of the angle β causes the air to hit the leading edge of one half wing more directly than the other. For this reason, the center of pressure moves laterally, producing the torque caused by the aerodynamic moment.

For aircraft with a symmetrical wing, the moments Cl_0 and Cn_0 are usually zero when $\beta = 0$ as explained by [60] in the topics of flight mechanics. This behavior is adequately demonstrated in Figure 19c,e.

3.3. Performance Evaluation of the Final Design

The application of CFD numerical simulation techniques to new aircraft designs allows for the knowledge of not only their aerodynamic characteristics, but also their performance. In Table 8, the performance parameter values of the integrated and trapezoidal wing design of four aircraft, documented in the literature, are compared with those of the design carried out in the present study. The results show that the performance parameter values of the design performed in the present study are within the very acceptable range. The metrics that allow for a performance comparison with other designs are presented in Table 9.

Table 8. Comparison of documented designs with our own design.

Parameter	[19]	[55]	[56]	[1]	Own Design
W/S	63.765 [N/m^2]	-	21.88 [N/m^2]	23.544 [N/m^2]	44.004 [N/m^2]
C_{Lmax}	1.235	0.97	1.09	1.26	1.12
V_s	10.3435 [m/s]	-	6.4363 [m/s]	6.209 [m/s]	9.004 [m/s]
V_{E^*}	16.864 [m/s]	-	9.359 [m/s]	10.7863 [m/s]	13.099 [m/s]
C_{D0}	0.023	0.007	0.0195	0.009	0.0207
K	0.1108	0.0893	0.0734	0.0516	0.0739
e	0.546	0.6393	0.86	0.82	0.92

Table 9. Dimensional characteristics of the components.

Metrics (Symbol)	Formula	Description
W/S	$\left(\dfrac{W}{S}\right) = \frac{1}{2}\rho_\infty V_s^2 C_{LMax}$	Wing loading required at the stall speed [1]
V_s	$V_s = \sqrt{\dfrac{2}{\rho_\infty C_{LMax}}\left(\dfrac{W}{S}\right)}$	Stall speed [1]
V_{E^*}	$V_{E^*} = \sqrt{\dfrac{2}{\rho_\infty}\left(\dfrac{W}{S}\right)}\sqrt{\dfrac{K}{C_{D0}}}$	Speed for the maximum lift-to-drag ratio [1]
K	$K = \dfrac{1}{\pi AR e}$	Induced drag factor

4. Conclusions

The present work was based on performing the geometric design, determining the aerodynamic coefficients, and analyzing the performance of a tail-sitter UAV using CAD and CFD computational tools. The design methodology mainly consisted of proposing and adjusting the dimensions and properties of the wing according to wing designs reported in various research works.

To carry out the structural design, an arrangement consisting of the distribution of ribs, spars, and loads from the instrumentation of the vehicle was proposed. The structural components and the instrumentation component boxes were drawn in 3D using ANSYS SpaceClaim CAD software to obtain the virtual prototype of the proposed tail-sitter.

For the CFD numerical simulation, a computational domain of air was drawn that encloses the entire part of the geometric model of the tail-sitter. The domain was meshed in an unstructured way to minimize the number of nodes and elements. A mesh independence analysis was performed based on the yaw moment coefficient. The results show that the yaw moment coefficient is independent of the mesh size when the number of elements exceeds 403,549. Therefore, it was decided to use the mesh of 455,174 elements to obtain more precise results. Based on this mesh, numerical simulations were performed in ANSYS Fluent CFD software to determine the aerodynamic coefficients. The numerical results were satisfactorily validated with empirical correlations for the calculation of the polar curve and comparison of the performance of the proposed tail-sitter with those found in the literature.

Satisfactory results of velocity and pressure contours were obtained for various angles of attack. The results of force and moment coefficients showed trends similar to those reported in the literature. Based on these results, it was concluded that the methodology proposed in this present work is feasible in the design and determination of the aerodynamic coefficients for the tail-sitter.

The results of the performance evaluation of the final design showed that the values of the performance parameters of the proposed tail-sitter design are within the very acceptable range.

As future work, it is proposed to carry out a vertical flight stability analysis of tail-sitter trough CFD simulation in order to identify the vertical flight dynamics for the controller design. In addition, the data of aerodynamic coefficients obtained in this work will be used to carry out the dynamic model of the tail-sitter, with the purpose of designing laws of control that allow for controlling the horizontal and vertical flight phases of the vehicle.

Author Contributions: Conceptualization, L.E.R.-V. and M.A.V.-N.; Methodology, E.A.I.-N., J.F.I.-Y. and L.E.R.-V.; Software, E.A.I.-N., J.F.I.-Y.; Validation, E.A.I.-N., J.F.I.-Y. and L.E.R.-V.; Formal Analysis, E.A.I.-N., J.F.I.-Y. and L.E.R.-V.; Investigation, E.A.I.-N. and O.G.-S.; Writing—Original Draft Preparation, E.A.I.-N., J.F.I.-Y. and L.E.R.-V. ; Writing—Review & Editing, E.A.I.-N., O.G.-S.; Supervision, L.E.R.-V. and M.A.V.-N.; Project Administration, L.E.R.-V. All authors have read and agreed to the published version of the manuscript..

Funding: This research received no external funding.

Acknowledgments: Emmanuel A. Islas-Narvaez thanks Conacyt for the support provided during his master's studies at the Universidad Politécnica Metropolitana de Hidalgo. He also thanks Octavio Garcia-Salazar of the Aerospace Engineering Research and Innovation Center, Faculty of Mechanical and Electrical Engineering, UANL for the support received during the research stay.

Conflicts of Interest: The authors declare no conflict of interest.

Nomenclature

AR	Aspect ratio
C_{D0}	Zero-lift drag coefficient
C_{Lmax}	Maximum lift coefficient
e	Oswald coefficient
K	Induced drag factor
ρ_∞	Air density
S	Wing surface
V_{E*}	Speed for the maximum lift-to-drag ratio
V_s	Stall speed
W	Aircraft weight

References

1. Chung, P.H.; Ma, D.M.; Shiau, J.K. Design, Manufacturing, and Flight Testing of an Experimental Flying Wing UAV. *Appl. Sci.* **2019**, *9*, 3043. [CrossRef]
2. Asim, M.; Khan, A.A. Designing and Development of Unmanned Aerial Vehicle. In Proceedings of the International Conference of Aeronautical Sciences, Toronto, ON, Canada, 8–13 September 2002; Volume 1, pp. 134.1–134.9.
3. Rojas, A.E.; Espinal, A.A.; Ramos, L.; Aux, J.H.E. Design and Construction of a Prototype of an Unmanned Aerial Vehicle Equipped with Artificial Vision for the Search of People. *Sist. Telemática* **2017**, *15*, 9–26. [CrossRef]
4. Chu, L.; Li, Q.; Gu, F.; Du, X.; He, Y.; Deng, Y. Design, Modeling, and Control of Morphing Aircraft: A Review. *Chin. J. Aeronaut.* **2022**, *35*, 220–246. [CrossRef]
5. Ur Rahman, R.; Nayeem, M.H.K.; Hossain, M.; Roman, M.; Rabby, H.F. Design and Performance Analysis of Unmanned Aerial Vehicle (UAV) to Deliver Aid to the Remote Area. In Proceedings of the 4th International Conference on Mechanical, Industrial and Materials Engineering, Atlanta, GA, USA, 25–27 October 2017.
6. Khan, M.I.; Salam, M.A.; Afsar, M.R.; Huda, M.N.; Mahmud, T. Design, Fabrication & Performance Analysis of an Unmanned Aerial Vehicle. In Proceedings of the 11th International Conference on Mechanical Engineering, Dhaka, Bangladesh, 18–20 December 2016; Volume 1754.
7. Kontogiannis, S.G.; Ekaterinaris, J.A. Design, Performance Evaluation and Optimization of a UAV. *AIP Conf. Proc.* **2016**, *29*, 339–350.
8. Tyan, M.; Nguyen, N.V.; Lee, J.W. A Tailless UAV Multidisciplinary Design Optimization Using Global Variable Fidelity Modeling. *Int. J. Aeronaut. Space Sci.* **2017**, *18*, 662–674. [CrossRef]
9. Harasani, W.I. Design, Build and Test an Unmanned Air Vehicle. *JKAU: Eng. Sci.* **2010**, *18*, 662–674. [CrossRef]
10. Iqbal, L.; Sullivan, J. Application of an Integrated Approach to the UAV Conceptual Design. In Proceedings of the 46th AIAA Aerospace Sciences Meeting and Exhibit, Reno, Nevada, 7–10 January 2008.
11. Sobester, A.; Keane, A. Multidisciplinary Design Optimization of UAV Airframes. In Proceedings of the Structures, Structural Dynamics, and Materials Conference, Sao Paulo, Brazil, 1–4 May 2006; Volume 9, pp. 160–169.
12. Flynn, E.P. Low-cost Approaches to UAV Design Using Advanced Manufacturing Techniques. In Proceedings of the 3rd IEEE Integrated STEM Education Conference, Princeton, NJ, USA, 9 March 2013.
13. Chowdhury, S.; Maldonado, V.; Patel, R. Conceptual Design of a Multi-ability Reconfigurable Unmanned Aerial Vehicle (UAV) Through a Synergy of 3D CAD and Modular Platform Planning. In Proceedings of the 15th AIAA/ISSMO Multidisciplinary Analysis and Optimization Conference, Atlanta, GA, USA, 16–20 June 2014.
14. Katon, M.; Kuntjoro, W. Static Structural Analysis of Blended Wing Body II-E2 Unmanned Aerial Vehicle. *J. Appl. Environ. Biol. Sci.* **2017**, *7*, 91–98.
15. Ronzheimer, A. CAD in Aerodynamic Aircraft Design. In Proceedings of the Deutscher Luft- und Raumfahrtkongress, Müchen, Germany, 5–7 September 2017.
16. MohamedZain, A.O.; Chua, H.; Yap, K.; Uthayasurian, P.; Jiehan, T. Novel Drone Design Using an Optimization Software with 3D Model, Simulation, and Fabrication in Drone Systems Research. *Drones* **2022**, *6*, 97. [CrossRef]
17. Todorov, M. 3-D CAD Modeling and Modal Analysis of Light Aircraft Wing Using Solid Works and ANSYS softwares. *Mach. Technol. Mater.* **2014**, *8*, 51–53.
18. Benaouali, A.; Kachel, S. An Automated CAD/CAE Integration System for the Parametric Design of Aircraft Wing Structures. *J. Theor. Appl. Mech.* **2014**, *55*, 447–459. [CrossRef]
19. Bliamis, C.; Zacharakis, I.; Kaparos, P.; Yakinthos, K. Aerodynamic and Stability Analysis of a VTOL Flying Wing UAV. *IOP Conf. Ser. Mater. Sci. Eng.* **2021**, *1024*, 1–8. [CrossRef]
20. Menter, F.R.; Galpin, P.F.; Esch, T.; Kuntz, M.; Berner, C. CFD Simulations of Aerodynamic Flows with a Pressure-Based Method. In Proceedings of the 24th International Congress of the Aeronautical Sciences, Yokohama, Japan, 29 August–3 September 2004; Volume 24, pp. 1–11.
21. Şumnu, A.; Güzelbey, İ.H.; Öğücü, O. Aerodynamic Shape Optimization of a Missile Using a Multiobjective Genetic Algorithm. *Int. J. Aerosp. Eng.* **2020**, *2020*, 1528435. [CrossRef]
22. Lao, C.T.; Wong, E.T.T. CFD Simulation of a Wing-in-Ground-Effect UAV. *IOP Conf. Ser. Mater. Sci. Eng.* **2018**, *370*, 1–8. [CrossRef]
23. Salazar-Jiménez, G.; López-Aguilar, H.A.; Gómez, J.A.; Chazaro-Zaharias, A.; Duarte-Moller, A.; Pérez-Hernández, A. Blended wing' CFD Analysis: Aerodynamic Coefficients. *Int. J. Math. Comput. Simul.* **2018**, *12*, 33–43.
24. Manikantissar; Geete, A. CFD Analysis of Conceptual Aircraft Body. *Int. Res. J. Eng. Technol.* **2017**, *4*, 217–222.
25. Lammers, K. Aerodynamic CFD Analysis on Experimental Airplane. Master's Thesis, Mechanical Engineering 'Fluid Dynamics', University of Twente, Enschede, The Netherlands; RMIT University, Melbourne, VIC, Australia, 2015; pp. 8–30.
26. Gu, X.; Ciampa, P.D.; Nagel, B. An Automated CFD Analysis Workflow in Overall Aircraft Design Applications. *CEAS Aeronaut. J.* **2018**, *8*, 3–13. [CrossRef]
27. Kosík, A. The CFD Simulation of the Flow Around the Aircraft Using OPENFOAM and ANSA. In Proceedings of the 5th ANSA & µETA International Conference, Thessaloniki, Greece, 5–7 June 2013.
28. Cravero, C.; Marsano, D. Computational Investigation of the Aerodynamics of a Wheel Installed on a Race Car with a Multi-Element Front Wing. *Fluids* **2022**, *7*, 182. [CrossRef]

29. Ravelli, U.; Savini, M. Aerodynamic Simulation of a 2017 F1 Car with Open-Source CFD Code. *J. Traffic Transp. Eng.* **2018**, *6*, 155–163.

30. Wang, J.; Li, H.; Liu, Y.; Liu, T.; Gao, H. Aerodynamic Research of a Racing Car Based on Wind Tunnel Test and Computational Fluid Dynamics. In Proceedings of the 4th International Conference on Mechatronics and Mechanical Engineering (ICMME 2017), Kuala Lumpur, Malaysia, 28–30 November 2017; p. 153.

31. Fernandez-Gamiz, U.; Gomez-Mármol, M.; Chacón-Rebollo, T. Computational Modeling of Gurney Flaps and Microtabs by POD Method. *Energies* **2018**, *11*, 2091. [CrossRef]

32. Aziz, S.; Khan, A.; Shah, I.; Khan, T.A.; Ali, Y.; Sohail, M.U.; Rashid, B.; Jung, D.W. Computational Fluid Dynamics and Experimental Analysis of a Wind Turbine Blade's Frontal Section with and without Arrays of Dimpled Structures. *Energies* **2022**, *15*, 7108. [CrossRef]

33. Ung, S.; Chong, W.; Mat, S.; Ng, J.; Kok, Y.; Wong, K. Investigation into the Aerodynamic Performance of a Vertical Axis Wind Turbine with Endplate Design. *Energies* **2022**, *15*, 6925. [CrossRef]

34. Li, B.; Zhou, W.; Sun, J.; Wen, C.-Y.; Chen, C.-K. Development of Model Predictive Controller for a tail-sitter VTOL UAV in Hover Flight. *Sensors* **2018**, *18*, 2859. [CrossRef] [PubMed]

35. Zhou, J.; Lyu, X.; Li, Z.; Shen, H. A Unified Control Method for Quadrotor tail-sitter UAVs in all Flightmodes: Hover, Transition, and Level Flight. In Proceedings of the International Conference on Intelligent Robots and Systems (IROS), Vancouver, BC, Canada, 24–28 September 2017.

36. Jin, W.; Bifeng S.; Liguang W.; Wei. L_1 Adaptive Dynamic Inversion Controller for an X-wing tail-sitter MAV in Hover Flight. *Procedia Eng.* **2015**, *99*, 969–974. [CrossRef]

37. Yang, Y.; Zhu, J.; Zhang, X.; Wang. Active Disturbance Rejection Control of a Flying-Wing tail-sitter in Hover Flight. In Proceedings of the 2018 IEEE/RSJ International Conference on Intelligent Robots and Systems (IROS), Madrid, Spain, 1–5 October 2018.

38. Martin Hepperle. 2015. MH 60. (mh60-il). Airfoiltools.com. Available online: http://airfoiltools.com/airfoil/details?airfoil=mh60-il (accessed on 5 October 2022).

39. Abbott, I.H. *Theory of Wing Sections*; Dover Publications, Inc.: New York, NY, USA, 1959.

40. Raja Sekar, K.; Ramesh, M.; Naveen, R.; Prasath, M. S.; Vigneshmoorthy, D. Aerodynamic Design and Structural Optimization of a Wing for an UnmannedAerial Vehicle (UAV). *IOP Conf. Ser. Mater. Sci. Eng.* **2020**, *764*, 1–10. [CrossRef]

41. Makthal, S.; Dinesh, S.; Tabish, M. Design and Analysis of Amphibious Flying Wing UAV. *Int. J. Eng. Res. Technol. (IJERT)* **2020**, *9*, 135–148.

42. Huang, H. Optimal Design of a Flying-Wing Aircraft Inner Wing Structure Configuration. Ph.D. Thesis, Cranfield University, Cranfield, UK, 2012.

43. Ohanian, H.C.; Markert, J.T. Physics for Engineers and Scientists. *Mc Graw Hill* **2020**, *1*, 313–321.

44. Elcner, J.; Lizal, F.; Jicha, M. Comparison of Turbulent Models in the Case of a Constricted Tube. *Exp. Fluid Mech. 2016* **2017**, *143*, 02020. [CrossRef]

45. Razzaghi, M.J.P.; Sani, S.M.R.; Masoumi, Y.; Xu, C. Comparison and Modication of Various Turbulence Models for Analysis of Fluid Microjet Injection into the Boundary Layer over a Flat Surface. *Res. Sq.* **2022**, 1–25. [CrossRef]

46. Versteeg, H.M.W. *An Introduction to Computational Fluid Dynamics, The Finite Volume Method*, 2nd ed.; Pearson Education: Glasgow, UK, 2010.

47. Spalart, P.; Allmaras, S. A One-Equation Turbulence Model for Aerodynamic Flows. In Proceedings of the 30th Aerospace Sciences Meeting and Exhibit, American Institute of Aeronautics and Astronautics, Reno, NV, USA, 6–9 January 1992; Volume 1, pp. 5–21.

48. Valencia, E.; Hidalgo, V. Innovative Propulsion Systems and CFD Simulation for Fixed Wings UAVs. In *Chapter in the Book: Aerial Robots—Aerodynamics, Control and Applications*; Editorial Intechopen: London, UK, 2017.

49. Khan, S.; Malik, M.S.; Junaid, M.; Jafry, A.T. To investigate and compare the wing planform's effect on the aerodynamic parameters of aircraft wings using computational fluid dynamics (CFD). *Res. Sq.* **2022**, 1–17. Preprint (Version 1).

50. Ardila-Marín, J.G.; Hincapié-Zuluaga, D.A.; Sierra-del-Rïo, J.A. Independencia de Malla en Tubos Torionados para Intercambio de Calor: Caso de Estudio. *Rev. Fac. Cienc.* **2016**, *5*, 124–140. [CrossRef]

51. Ansys Inc. Southpointe. *ANSYS Meshing User's Guide*; ANSYS, Inc.: Troy , MI, USA, 2013.

52. CADENCE:22. Compute Grid Spacing for a Given Y+. (s. f.). Cadence. Available online: https://www.cadence.com/koKR/home/tools/systemanalysis/computatonalfluiddynamics (accessed on 2 October 2022).

53. Anderson, J.D., Jr. *Aircraft Performance and Design*, 1st ed.; Tata Mcgraw-Hill Edition: New York, NY, USA, 2010.

54. Vassberg, J.C.; Tinoco, E.N.; Mani, M.; Levy, D.; Zickuhr, T.; Mavriplis, D.J.; Wahls, R.A.; Morrisonk, J.H. Comparison of NTF Experimental Data with CFD Predictions from the Third AIAA CFD Drag Prediction Workshop. In Proceedings of the 26th AIAA Applied Aerodynamics Conference, Honolulu, HI, USA, 18–21 August 2008; Volume 6, pp. 1–25.

55. Na, D.; You, N.; Jung, S.N. Comprehensive Aeromechanics Predictions on Air and Structural Loads of HART I Rotor. *Int. J. Aeronaut. Space Sci.* **2017**, *18*, 165–173. [CrossRef]

56. Hamada, A.; Sultan, A.; Abdelrahman, M. Design, Build and Fly a Flying Wing. *Athens J. Technol. Eng.* **2018**, *5*, 223–250. [CrossRef]

57. Betancourth, N.J.P.; Villamarin, J.E.P.; Rios, J.J.V.; Bravo-Mosquera, P.D.; Cerón-Muñoz, H.D. Design and Manufacture of a Solar-Powered Unmanned Aerial Vehicle for Civilian Surveillance Missions. *J. Aerosp. Technol. Manag.* **2016**, *8*, 385–396. [CrossRef]

58. Anderson, J.D., Jr. *Introduction to Flight*, 8th ed.; McGraw-Hill Series in Aeronautical and Aerospace Engineering; McGraw-Hill: New York, NYm USA, 2014.
59. Chang, A.K.K.; Sheen, Do.; Jo, Yo.; Shim, H.J. CFD Analysis of the Sides Lip Angle Effect Around a BWB type Configuration. *Int. J. Aerosp. Eng.* **2019**, *2019*, 4959265.
60. Stevens, B.L.; Lewis, F.L. *Aircraft Control and Simulation*, 2nd ed.; John Wiley and Sons, Inc.: New York, NY, USA, 2003.

Article

Performance Evaluation of an H-VTOL Aircraft with Distributed Electric Propulsion and Ducted-Fans Using MIL Simulation

Juan Manuel Bustamante Alarcon, José Leonel Sánchez Marmolejo, Luis Héctor Manjarrez Muñoz, Eduardo Steed Espinoza Quesada, Antonio Osorio Cordero and Luis Rodolfo García Carrillo

[1] Center for Research and Advanced Studies of the National Polytechnic Institute, UMI-LAFMIA, Av. Instituto Politécnico Nacional 2508, San Pedro Zacatenco, Mexico City 07360, Mexico; juan.bustamantea@cinvestav.mx (J.M.B.A.); jose.sanchezm@cinvestav.mx (J.L.S.M.); luis.manjarrez@cinvestav.mx (L.H.M.M.); eduardo.espinoza@cinvestav.mx (E.S.E.Q.); aosorio@cinvestav.mx (A.O.C.)
[2] National Council of Humanities Science and Technology, Av. Insurgentes Sur, No 1582, Mexico City 03940, Mexico
[3] Klipsch School of Electrical and Computer Engineering, New Mexico State University, Las Cruces, NM 88003-8001, USA
*   Correspondence: luisillo@nmsu.edu; Tel.: +1-(575)-646-3173

Abstract: This paper deals with the problem of increasing the energy efficiency of a hybrid vertical take-off and landing aircraft. To this end, an innovative aerial vehicle was developed, featuring a distributed electrical propulsion system with ducted-fan rotors. To compare and analyze the effectiveness of the proposed propulsion system, two configurations with a different number of ducted-fan rotors were examined: a four-rotor configuration and a six-rotor configuration. The mathematical model of the four-rotor configuration was derived using the Newton–Euler formalism, allowing the design and implementation of a control strategy for conducting model-in-the-loop simulations. These simulations enabled the evaluation and analysis of the performance of the proposed propulsion system, where the numerical results demonstrated the functionality of both designs and showed that, during the multirotor flight, the configuration with six rotors increased its energy efficiency by up to 11%, providing higher vertical lift with the same power consumption. This was achieved by distributing its weight among a higher number of engines. The incorporation of two additional ducted fans increased the weight and the drag of the six-rotor configuration, resulting in a low augmentation in power consumption of 1%. Finally, this caused a decrease in airspeed by up to 4% during the cruise speed phase.

Keywords: distributed electric propulsion; VTOL; model-in-the-loop simulation; X-plane; ducted-fan; unmanned aircraft system

Citation: Juan Manuel Bustamante Alarcon, José Leonel Sánchez Marmolejo, Luis Héctor Manjarrez Muñoz, Eduardo Steed Espinoza Quesada, Antonio Osorio Cordero and Luis Rodolfo García Carrillo Performance Evaluation of an H-VTOL Aircraft with Distributed Electric Propulsion and Ducted-Fans Using MIL Simulation. *Machines* **2023**, *11*, 852. https://doi.org/10.3390/machines11090852

Academic Editors: Octavio Garcia-Salazar, Anand Sanchez-Orta and Aldo Jonathan Muñoz-Vazquez

Received: 15 July 2023
Revised: 14 August 2023
Accepted: 23 August 2023
Published: 25 August 2023

1. Introduction

The development of an aircraft that combines the advantages of both fixed-wing and rotorcraft, to overcome their weakness, has prompted the creation of a wide variety of new aircraft designs, which have been classified as hybrid VTOLs (H-VTOLs). This type of aircraft offers several advantages over the traditional ones, since its VTOL capabilities eliminate the need for runways, enabling it to operate in confined spaces or remote areas, and providing enhanced versatility [1,2]. Zong et al. [3] mentioned that the primary focus of developing this type of aircraft revolves around urban air mobility applications, given the potential to effectively and safely handle urban transportation needs. Furthermore, prominent organizations such as NASA and Uber are considering how the utilization of H-VTOLs could bring revolutionary changes in industries such as transportation, search and rescue, package delivery, and security patrols.

Despite their advantages, H-VTOLs also face challenges that require further research and development, such as the need for additional power to conduct vertical take off, compared with the power required for the fixed-wing phase. This increased power demand can affect the overall efficiency and endurance of the vehicle, limiting its flight time or its payload capacity. In this sense, the key points that must be taken into account in the design of these vehicles include the addition of dead weight; the requirement for higher speeds, which might not be achieved if the same propulsion system is used for both flying phases; the fact that VTOL efficiency means greater lift and less required power; the fact that cruising efficiency means greater lift and less drag; and, finally, the need for safer aircraft, seeking to reduce the number of accidents, at least compared with the number of accidents with conventional aircraft [4].

Nowadays, there exist many configurations of H-VTOLs, which are mainly classified as compound aircraft, tilt-rotor, tilt-wing, tail-sitter, lift-fan, and vectored-thrust vehicles [5]. These types of configurations are clearly exemplified by the Jump 20 [6], a compound aircraft with a conventional plane configuration that includes a dedicated quadrotor system for multirotor flight; the Quantum Systems Tron F90 [7], a tilting-rotor aircraft, similar to the Jump 20 UAV with the addition of a tilting mechanism that uses the same propulsion system for vertical and cruise flight; the JAXA's QTWUAV AKITSU tilt-wing [8], which consists of tandem tilting wings and propellers mounted on the leading edge of each UAV wing; and the WintraOne [9], a tail-sitter UAV with two propellers that allow taking-off vertically and a mechanism to tilt the whole airframe forward for cruise flight. All of these aircraft address the issue of the extra power consumed during multirotor flight by reducing the dead weight through using the same propulsion system for all phases of flight, with the exception of the compound aircraft whose main advantage lies in separating each propulsion system, in order to optimize them for each phase of flight.

Alternatively, in order to increase the VTOL efficiency of the H-VTOl aircraft, the distributed electric propulsion (DEP) technique integrates the propulsion system from multiple small-size engines and propellers instead of using a large motor with an equally large propeller. In this technique, propulsion units are distributed on the frame to stabilize the attitude of the aircraft, which features a higher propulsion efficiency [10–12]. The effectiveness of this technology has been demonstrated by Kim et al. [13], where the authors employed DEP on a NASA demonstration aircraft, a conventional take-off and landing vehicle (CTOL) called X-57. This aircraft was reconfigured with a much smaller wing than the reference aircraft Tecman P2006T. Such a smaller wing was achieved thanks to the lift provided by 12 small electric propellers along the leading edge of the wing during the take-off and landing phases of flight [14].

Similarly, the Joby S2 concept incorporates DEP technology, positioning its engines throughout the aircraft, without increasing mechanical complexity and weight. This aircraft incorporates twelve motors with twelve fixed-pitch rotors. During aircraft mode, the induced velocity generated by the rotors, which become propellers, provides the vehicle with a lift-enhancing effect, similar to that produced by hyper-sustainable mechanisms, as demonstrated by Stoll et al. [15]. The Greased Lightning GL-10 vehicle, a joint effort of the Advanced Aircraft Company and NASA is an aircraft that incorporates 10 rotors for multirotor flight, which subsequently change to propellers through a tilting system, to propel the vehicle as a fixed wing [16]. The Langley Aerodrome 8 is a tandem tilt-wing electric VTOL configuration that has two tilt wings, eight electric motors, and several control surfaces, for a total of twenty independent control actuators. This aircraft is utilized for testing and evaluation of new technologies for H-VTOL vehicles, such as the examination of new developments in DEP, advanced flight control systems, and complex aerodynamics [17].

1.1. Related Work

In 2016, the Lilium company conducted a flight test of its all-electric two-passenger VTOL vehicle. This aircraft consists of thirty-six ducted fans mounted along its 10 m wingspan. When taking-off, the engines are pointed downwards, to produce lift, and then the engines rotate to the horizontal position to produce thrust for the aircraft mode [18]. Similarly, the Lightning Strike VTOL X-Plane, an aircraft that began as a project by DARPA, features twenty-four electric ducted fans distributed across the vehicle's wingspan and canard surface, whose functionality was validated through conducting flight tests with a scaled-down version [19]. In 2018, Zhao and Zhou [20] presented a lift-propulsion VTOL concept, which adopted the idea of the supporting body incorporating the fuselage as a lift element and the wing section. The drive unit composed two main ducted fans at the wingtips and another two at the rear of the fuselage, to generate thrust in airplane mode. They presented the aircraft design, the prototype construction, and vertical flight tests. In [21], Hoeveler et al. developed a fan-in-wing VTOL using two ducted fans in the wing. In their design, the authors incorporated a step for each ducted-fan outlet at the bottom of the wing, in order to obtain a better performance during airplane mode. This vehicle has a tilting ducted fan on the front, used to perform the transition in flight. In their research work, the geometry of the duct with the added step was analyzed and compared against a wing with a standard duct and a wing without ducts. The lift/drag relation for the three studied cases showed an increase of up to 66% when the step at the bottom of the wing was used. However, the performance with respect to the ductless wing characteristics was considerably improved.

In this research work, two H-VTOL concepts are presented: the XEVTOL-2FNW and XEVTOL-4FNW. The two experimental vehicles consist of a flying wing aircraft with a tilting system incorporated at the front for the flight transition and one or two rotors embedded in the wing, according to the 2FNW or 4FNW configuration. These concepts deal with the extra power demanded during the VTOL phase using a tilt-rotor configuration, which reduces the dead weight through using the cruise propulsion as a lift system during multirotor flight. The DEP technique was employed, together with the incorporation of ducted-fans to increase the efficiency of the propulsion system during the VTOL phase, and we performed a comparison of the 2FNW and 4FNW concepts, making use of model-in-the-loop (MIL) simulations. The aim of these simulations was to evaluate the performance of the aircraft when the required thrust is distributed throughout a different number of ducted-fans and, simultaneously, to evaluate their effects on the aerodynamics of the aircraft.

1.2. Main Contributions

The main contributions of this work can be summarized as follows:

- A mathematical model of two unconventional H-VTOLs was derived, to obtain the corresponding flight controllers;
- An evaluation is presented regarding the advantages and drawbacks of using 4 or 6 ducted fans during the cruising phase by conducting a study of the performance of the two platforms based on MIL simulations using X-Plane flight simulator and Matlab.

The remainder of this paper is organized as follows: Section 2 presents the XEVTOL-2FNW and XEVTOL-4FNW concepts, proposed to reduce energy consumption and to increase the lift produced by the rotor during multirotor flight and incorporating DEP technology, as well as a description of the dynamic model of the XEVTOL-2FNW using the Newton–Euler formulation, considering the forces generated by the rotors, as well as the aerodynamic forces and moments generated by the aircraft wing. Section 3 presents a 3D model of the two aircraft concepts developed on Plane Maker and the flight controllers implemented for the model-in-the-loop simulations using the X-Plane flight simulator and Matlab-Simulink. Section 4 shows the simulation results of the two vehicles, and the energy parameters of the aircraft are compared to evaluate the advantage of the DEP and the drawbacks of increasing the number of ducted fans. Finally, concluding remarks about this research are provided in Section 5.

2. Mathematical Model of the XEVTOL-2FNW

The XEVTOL-2FNW and XEVTOL-4FNW aircraft concepts are based on DEP technology, which states that it is more efficient to have several light small engines, distributed along a vehicle than having a single powerful but heavy engine. This implies that the engines operate in a more efficient region, where they generate more lift with a lower energy consumption (g/W). In addition, ducted fans were utilized for thrust generation in multirotor flight, to improve the energy efficiency of each rotor and provide an additional layer of safety since the blades are enclosed within the housing provided by the ducts, reducing the risk of injury or damage to surrounding objects or people in close proximity to the aircraft [22–24]. The design of both aircraft concepts was based on the commercial UAV Mapper V1.8, manufactured by TUFFWING [25], a flying wing with a wingspan of 120 cm.

Based on the above considerations, the proposed XEVTOL concepts can be defined as a flying wing aircraft with a dedicated lift system of ducted fans in the wing, with a tilting dual-ducted fan on the front of the aircraft for the transition phase (see Figure 1). The XEVTOL-2FNW concept incorporates two ducted fans embedded in the wing and a tilting dual-ducted fan in the front of the vehicle. The tilting ducted fans allow carrying out the transition from the multirotor to the fixed-wing mode and vice versa. In the fixed-wing phase, two elevons, acting as the elevator and the aileron, are the control surfaces. For the XEVTOL-4FNW, two ducted fans were added at the rear of the vehicle, which required increasing the distance of the tilting dual-ducted fan from the fuselage, in order to counterbalance the torque of the rear ducted fans.

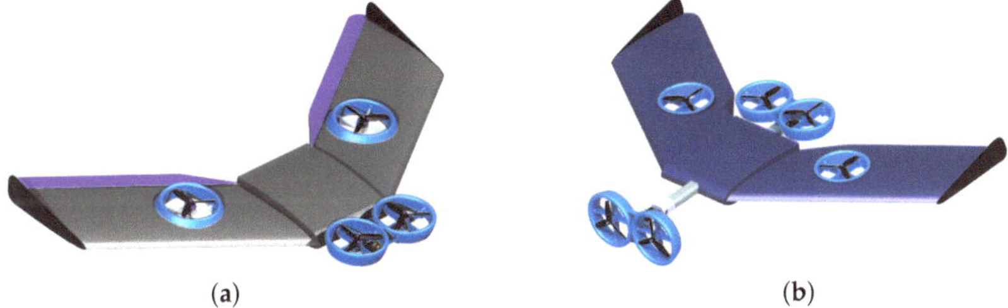

(a)   (b)

Figure 1. Aircraft concepts based on DEP technology. (**a**) XEVTOL-2FNW concept in VTOL mode. (**b**) XEVTOL-4FNW concept in fixed-wing mode.

Furthermore, the addition of two rear ducted fans to the XEVTOL-4FNW configuration primarily altered the center of gravity (c.g.), the MTOW of the platform and, to a minor degree, the aerodynamic center (a.c.), given that the latter is more influenced by the characteristics of the wings. The c.g. of the XEVTOL-4FNW is located at 40% of the fuselage chord, while for the XEVTOL-4FNW it is located at 67% of the fuselage chord. The empty weight for the XEVTOL-2FNW is 2.1 kg, while for the XEVTOL-4FNW configurations it is 2.7 kg. Figure 2 depicts the location of the c.g and the a.c. for both platforms.

MIL simulations were utilized to evaluate the effects of incorporating the additional ducted fans on the vehicle's performance during the VTOL phase. The XEVTOL-2FNW and XEVTOL-4FNW concepts were introduced to assess factors such as energy efficiency, power consumption, and lift and drag forces. In this sense, the development of the mathematical model is described below, in order to obtain the corresponding flight controllers to be implemented in the MIL simulations.

The mathematical model of XEVTOL-2FNW was derived using the Newton–Euler formalism. To this end, in addition to the forces generated by the rotors, the contribution of the aerodynamic forces and the moments generated by the aircraft wing was considered.

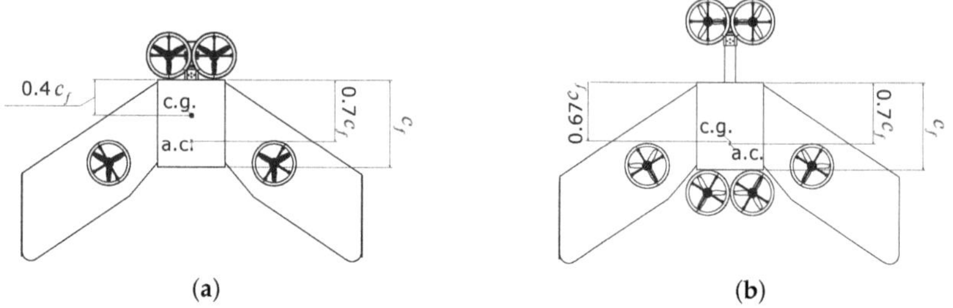

(a)   (b)

Figure 2. Location of the c.g. and the a.c. of the XEVTOL concepts. (**a**) The 2FNW and (**b**) the 4FNW. Where c_f is the fuselage chord of the vehicles.

2.1. Dynamics

Consider the XEVTOL-2FNW aerial platform as a rigid body with six degrees of freedom with a body frame $\mathbb{B} = [x_{\mathbb{B}},\ y_{\mathbb{B}},\ z_{\mathbb{B}}] \in \mathbb{R}^3$ located at its c.g. and an inertial frame $\mathbb{I} = [x_{\mathbb{I}},\ y_{\mathbb{I}},\ z_{\mathbb{I}}] \in \mathbb{R}^3$ fixed in the earth as shown in Figure 3.

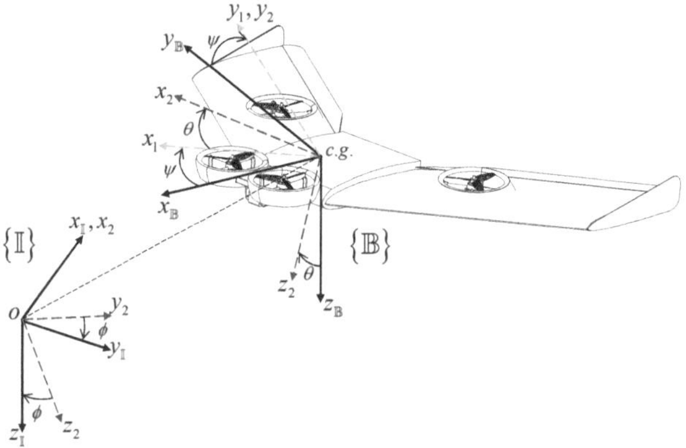

Figure 3. Reference frames and their rotations, to describe the transformation of coordinates of the XEVTOL-2FNW.

The mathematical model of the aircraft was derived using the Newton–Euler formalism [26] as:

$$\begin{bmatrix} m\mathbf{I} & \mathbf{0} \\ \mathbf{0} & \mathcal{I}_{\mathbb{B}} \end{bmatrix} \begin{bmatrix} \dot{V}_{\mathbb{B}} \\ \dot{\Omega}_{\mathbb{B}} \end{bmatrix} + \begin{bmatrix} \Omega_{\mathbb{B}} \times mV_{\mathbb{B}} \\ \Omega_{\mathbb{B}} \times \mathcal{I}_{\mathbb{B}}\Omega_{\mathbb{B}} \end{bmatrix} = \begin{bmatrix} F_T \\ M_T \end{bmatrix} \tag{1}$$

where m is the mass of the aircraft, and $\mathbf{I} \in \mathbb{R}^{3\times3}$ and $\mathbf{0} \in \mathbb{R}^{3\times3}$ are the identity matrix and the zero matrix, respectively. The linear velocity in the body frame is denoted by $V_{\mathbb{B}} = [U,\ V,\ W]^T \in \mathbb{R}^3$, where $U,\ V,\ W$ represent the linear velocity about the $x_{\mathbb{B}},\ y_{\mathbb{B}},\ z_{\mathbb{B}}$ axes, respectively, and whose derivative with respect to time is denoted as $\dot{V}_{\mathbb{B}}$. The angular velocity in the body frame is given as $\Omega_{\mathbb{B}} = [p,\ q,\ r]^T \in \mathbb{R}^3$, where $p, q,$ and r represent the rate of change in the pitch, roll, and yaw angle expressed in the body frame, respectively. The derivative with respect to time of this angular velocity is denoted as $\dot{\Omega}_{\mathbb{B}}$. $F_T \in \mathbb{R}^3$ and $M_T \in \mathbb{R}^3$ represent the forces and moments on the body frame acting on the c.g. of the aircraft. The matrix $\mathcal{I}_{\mathbb{B}}$ is the inertia tensor that contains the vehicle's moments of inertia. Since the $x_{\mathbb{B}}$-$z_{\mathbb{B}}$ plane is symmetric, $I_{xy} = I_{yx} = 0$ and $I_{xz} = I_{zx}$ [27], and the inertia tensor is defined as

$$\mathcal{I}_{\mathbb{B}} = \begin{bmatrix} I_{xx} & 0 & I_{xz} \\ 0 & I_{yy} & 0 \\ I_{xz} & 0 & I_{zz} \end{bmatrix} \tag{2}$$

Equation (1) is defined in the body frame $\{\mathbb{B}\}$ and it has to be transformed to the inertial frame $\{\mathbb{I}\}$, in order to obtain the mathematical model in terms of the inertial frame. This is done by performing axis rotation, first rotating about the $z_{\mathbb{B}}$ axis a yaw angle (ψ). Then, rotating about the y_1 axis a pitch angle (θ). Finally, rotating about the x_2 axis a theta angle (ϕ). Obtaining the $R_{\mathbb{B}\to\mathbb{I}}$ matrix, which is an orthogonal rotation matrix leading from the body frame to the inertial frame [28], is given as:

$$R_{\mathbb{B}\to\mathbb{I}} = \begin{bmatrix} c_\psi c_\theta & c_\psi s_\theta s_\phi - s_\psi c_\phi & c_\psi s_\theta c_\phi + s_\psi s_\phi \\ s_\psi c_\theta & s_\psi s_\theta s_\phi + c_\psi c_\phi & s_\psi s_\theta c_\phi - c_\psi s_\phi \\ -s_\theta & c_\theta s_\phi & c_\theta c_\phi \end{bmatrix} \tag{3}$$

where $c_* = cos(*)$ and $s_* = sin(*)$

While the matrix $W_{\mathbb{B}\to\mathbb{I}}$ gives us the equation for the rotational kinematics as a function of the angular velocities [28], which is given by

$$W_{\mathbb{B}\to\mathbb{I}} = \begin{bmatrix} 1 & s_\phi s_\theta / c_\theta & c_\phi s_\theta / c_\theta \\ 0 & c_\phi & -s_\phi \\ 0 & s_\phi / c_\theta & c_\phi / c_\theta \end{bmatrix} \tag{4}$$

Then, by grouping Equations (3) and (4), the transformation that permits the transfer of linear and angular velocities and accelerations from the body frame to the inertial frame is obtained. Equation (5) presents the transformation of linear and angular acceleration from the body to the inertial frame:

$$\begin{bmatrix} \ddot{P}_{\mathbb{I}} \\ \ddot{\Theta}_{\mathbb{I}} \end{bmatrix} = \begin{bmatrix} R_{\mathbb{B}\to\mathbb{I}} & 0 \\ 0 & W_{\mathbb{B}\to\mathbb{I}} \end{bmatrix} \begin{bmatrix} \dot{V}_{\mathbb{B}} \\ \dot{\Omega}_{\mathbb{B}} \end{bmatrix} \tag{5}$$

The linear acceleration vector in the inertial frame is defined as $\ddot{P}_{\mathbb{I}} = [\ddot{x},\, \ddot{y},\, \ddot{z}]^T$, while the angular acceleration vector in the inertial frame is $\ddot{\Theta}_{\mathbb{I}} = [\ddot{\phi},\, \ddot{\theta},\, \ddot{\psi}]^T$.

2.2. Forces

Figure 4 depicts the forces (red arrows) and moments (blue arrows) in the body frame, the c.g. of the body on the diagram, and the a.c. and the origin of the inertial frame $\{o\}$. The total force on the vehicle given in Equation (6) is expressed as F_T, and this is given by the vector sum of 3 types of forces: the propulsion forces F_p due to propellers/rotors, the aerodynamic forces F_a due to the wing, and the gravitational force F_W.

$$F_T = \begin{bmatrix} X \\ Y \\ Z \end{bmatrix} = F_p + F_a + F_W \tag{6}$$

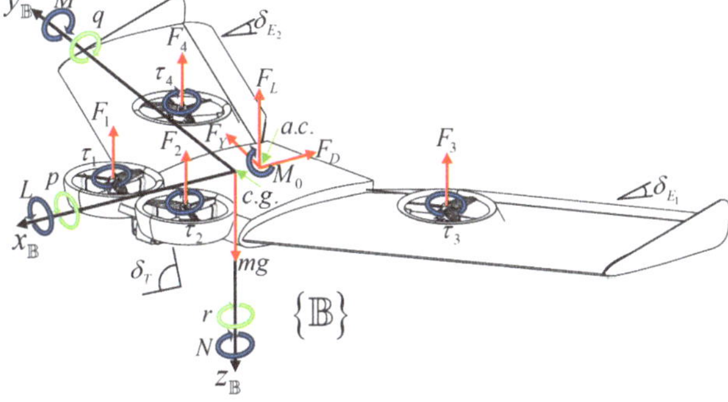

Figure 4. The forces and moments within the body frame, which is fixed at the c.g. of the aircraft.

2.2.1. Propulsion Forces

The propulsion forces generated by the vehicle are denoted by Equation (7). Where $F_i = k\omega_i^2$, $i = 1, 2, 3, 4$ define the thrust forces of the prop/rotors, k is a proportional constant, and ω is the angular velocity of the motor. The thrust forces due to the front ducted fans (F_1 and F_2) have a contribution to the body frame $x_\mathbb{B}$ or $y_\mathbb{B}$, depending of the tilting angle δ_T [26]. It can be observed that the ducted fans F_3 and F_4 only make a contribution in the vertical axis $z_\mathbb{B}$.

$$F_p = \begin{bmatrix} F_{px} \\ F_{py} \\ F_{pz} \end{bmatrix} = \begin{bmatrix} s_{\delta_T} & s_{\delta_T} & 0 & 0 \\ 0 & 0 & 0 & 0 \\ -c_{\delta_T} & -c_{\delta_T} & -1 & -1 \end{bmatrix} \begin{bmatrix} F_1 \\ F_2 \\ F_3 \\ F_4 \end{bmatrix} \tag{7}$$

Note that $\delta_T = 0°$ for the multirotor mode and $\delta_T = 90°$ for the fixed-wing mode.

2.2.2. Aerodynamic Forces of the Wing

The aerodynamic forces are mainly due to the interaction of the wing with the wind, generating the lift force F_L, the drag force F_D, and the lateral force F_Y, as defined in Equation (8):

$$F_D = \frac{1}{2}\rho_\infty V_\infty^2 S C_{Dw}$$

$$F_Y = \frac{1}{2}\rho_\infty V_\infty^2 S C_{Yw} \tag{8}$$

$$F_L = \frac{1}{2}\rho_\infty V_\infty^2 S C_{Lw}$$

The rotation matrix $R_{\mathcal{W}\to\mathbb{B}}$ in Equation (9) allows these aerodynamic forces to be transformed from the wind reference frame to the body frame [29], by performing rotations of the wind frame $\{\mathcal{W}\}$ about the $z_{\mathcal{W}}$ with a slip angle β and a rotation of the stability frame $\{\mathcal{S}\}$ about the $y_{\mathcal{S}}$ and incidence angle α, as shown in Figure 5.

$$F_a = \begin{bmatrix} F_{ax} \\ F_{ay} \\ F_{az} \end{bmatrix} = \underbrace{\begin{bmatrix} -c_\alpha c_\beta & -c_\alpha s_\beta & s_\alpha \\ s_\beta & c_\beta & 0 \\ -c_\beta s_\alpha & -s_\alpha s_\beta & -c_\alpha \end{bmatrix}}_{R_{\mathcal{W}\to\mathbb{B}}} \begin{bmatrix} F_D \\ F_Y \\ F_L \end{bmatrix} \tag{9}$$

which can be written as

$$F_a = \begin{bmatrix} -c_\alpha(c_\beta F_D + s_\beta F_y) + s_\alpha F_L \\ s_\beta F_D + c_\beta F_y \\ -s_\alpha(c_\beta F_D + s_\beta F_y) + c_\alpha F_L \end{bmatrix} \tag{10}$$

where ρ_∞ is the wing density; V_∞ is the relative wind speed, S is the wing surface; $C_{Dw} = C_{D_0 w} + C_{D_L w}$ is the wing drag coefficient; $C_{D_0 w}$ is the wing drag coefficient at zero lift; $C_{D_L w}$ is the drag coefficient due to wing lift; C_{Y_w} is the lateral force coefficient; $C_{Lw} = a_0 + a_1\alpha + a_2\delta_E + a_3\beta_\eta$ is the lift coefficient of the wing; a_0, a_1, and a_2, a_3 are the aerodynamic coefficients; δ_E is the angle of deflection of the elevon; and β_η is the elevon trim [30,31].

2.2.3. Gravitational Force

Since, in Equation (6), the force of gravity is defined in the inertial frame $\{\mathbb{I}\}$, a rotation is performed as a function of the Euler angles θ and ϕ, in order to transfer this force to the body frame $\{\mathbb{B}\}$ using Equation (11), where $g = 9.81$ m/s^2 is the gravitational acceleration.

$$F_W = \begin{bmatrix} F_{Wx} \\ F_{Wy} \\ F_{Wz} \end{bmatrix} = mg \begin{bmatrix} -s_\theta \\ c_\theta s_\phi \\ c_\theta c_\phi \end{bmatrix} \tag{11}$$

2.3. Moments Acting on the Aircraft

The moments acting on the platform are due to the propulsion systems M_p, the gyroscopic effect M_{gyro}, and the aerodynamics moments M_a, whose total sum is denoted as:

$$M_T = \begin{bmatrix} L \\ M \\ N \end{bmatrix} = M_p + M_{gyro} + M_a \tag{12}$$

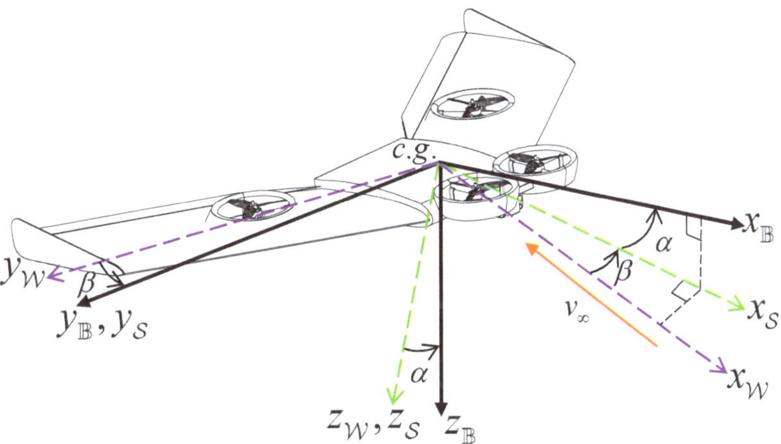

Figure 5. Definition of the wind frame, stability frame, and the aerodynamic angles.

2.3.1. Propulsion Moments

The moments generated by the propulsion system are given by two phenomena. The first is due to the force generated by the motors because of their position with respect to the c.g. and are denoted as M_{L_p}. The second results from the torque produced by the motors during their rotation and is denoted as $M\tau_P$. Combining these two expressions, the moments due to the propulsion system are given as

$$M_p = \begin{bmatrix} d_{ty}k(\omega_2^2 - \omega_1^2) + d_{Dy}k(\omega_3^2 - \omega_4^2) \\ c_{\delta_T}d_{tx}k(\omega_1^2 + \omega_2^2) - d_{Dx}k(\omega_3^2 + \omega_4^2) \\ 0 \end{bmatrix} + \begin{bmatrix} s_{\delta_T}\lambda k(\omega_1^2 - \omega_2^2) \\ 0 \\ \lambda k(\omega_4^2 - \omega_3^2) + c_{\delta_T}\lambda k(\omega_2^2 - \omega_1^2) \end{bmatrix} \tag{13}$$

Figure 6 shows the body diagrams used for the analysis of the moments about the three axes of the vehicle ($x_\mathbb{B}$, $y_\mathbb{B}$, $z_\mathbb{B}$), which are given by:

$$M_p = \begin{bmatrix} M_{px} \\ M_{py} \\ M_{pz} \end{bmatrix} = M_{L_p} + M\tau_p \tag{14}$$

where $\lambda k\omega_i^2 = \tau_i$, $i = 1, 2, 3, 4$ is the torque generated by the motors and λ is a proportional constant used to relate the force generated by the motors with the torque they produce [26].

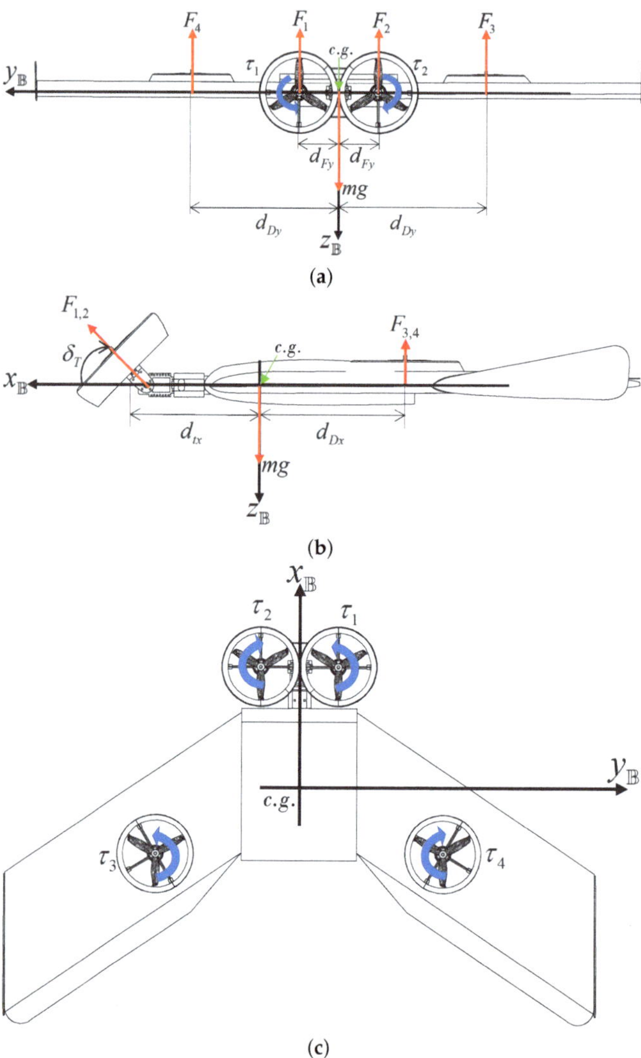

Figure 6. Free body diagrams of the XEVTOL-2FNW platform: (**a**) Frontal view, plane $y_\mathbb{B}$-$z_\mathbb{B}$ (moments around $x_\mathbb{B}$). (**b**) Lateral view, plane $x_\mathbb{B}$-$z_\mathbb{B}$ (moments around $y_\mathbb{B}$). (**c**) Top View, plane $x_\mathbb{B}$-$y_\mathbb{B}$ (moments around $z_\mathbb{B}$).

2.3.2. Gyroscopic Moments

The gyroscopic moments of the propellers are given in Equation (15). In the first term of the right hand side of Equation (15), i.e., inside the vector product between the angular speed of the body and each motor, a rotation of each motor is modeled as a function of the tilting angle δ_T, which applies only to motors 1 and 2, since they change their gyroscopic moment contribution depending on their orientation, due to the variation of the δ_T angle, while motors 3 and 4 remain fixed to the body:

$$M_{gyro} = \begin{bmatrix} M_{gyro_x} \\ M_{gyro_y} \\ M_{gyro_z} \end{bmatrix} = \sum_{i=1}^{2} J_{p_i}(\Omega_\mathbb{B} \times \omega_i [s_{\delta_T},\ 0,\ c_{\delta_T}]^T) + \sum_{i=3}^{4} J_{p_i}(\Omega_\mathbb{B} \times \omega_i) \tag{15}$$

$$M_{gyro} = \begin{bmatrix} q\{c_{\delta_T}(w_1 J_{p1} - w_2 J_{p2}) - w_3 J_{p3} + w_4 J_{p4}\} \\ p\{c_{\delta_T}(w_1 J_{p1} - w_2 J_{p2}) + w3 J_{p3} - w_4 J_{p4}\} + r s_{\delta_T}(w_2 J_{p2} - w_1 J_{p1}) \\ q s_{\delta_T}(w_1 J_{p1} - w_2 J_{p2}) \end{bmatrix} \tag{16}$$

where J_{p_i}, $i = 1, 2, 3, 4$ is the moment of inertia of the propellers about the axes of rotation of the engines [32].

2.3.3. Aerodynamic Moments

For delta wing aircraft, the moments are generated exclusively by the aerodynamic effects of the wing, since they lack the stabilizing surfaces present in conventional aircraft. This contribution of aerodynamic moments is linked to the wing coefficients. The aerodynamic moments of the aircraft are expressed by the Equation (17), derived from the diagram shown in Figure 7. These moments result from the aerodynamic forces F_a, acting at a distance l_{wx} from the a.c. to the c.g.

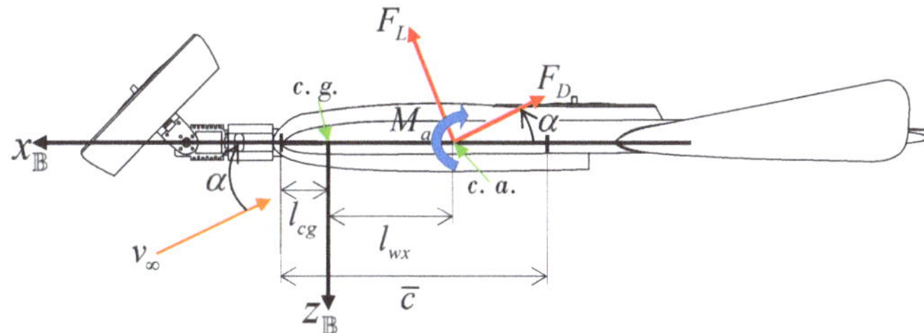

Figure 7. Free body diagram of the XEVTOL-2FNW vehicle in the plane $x_{\mathbb{B}}$-$z_{\mathbb{B}}$ with the aerodynamic forces and moments illustrated in red and blue arrows, respectively.

As shown in Figure 7, the aerodynamic moments that act on the vehicle are determined by the aerodynamic forces and the distances to the c.g. (l_{cg}) and to the c.a. (l_{wx}) from the mean aerodynamic chord ($\overline{c}$) of the wing, and they are given as follows:

$$M_a = \begin{bmatrix} M_{ax} \\ M_{ay} \\ M_{az} \end{bmatrix} = [r_a \times F_a] + \begin{bmatrix} l_a \\ m_a \\ n_a \end{bmatrix} \tag{17}$$

$$M_a = \begin{bmatrix} 0 \\ -l_{wx}[s_\alpha(c_\beta F_D + s_\beta F_y) - c_\alpha F_L] \\ -l_{wx}[s_\beta F_D + c_\beta F_y] \end{bmatrix} + \begin{bmatrix} 1/2\rho_\infty V_\infty^2 SbC_l \\ 1/2\rho_\infty V_\infty^2 S\overline{c}C_m \\ 1/2\rho_\infty V_\infty^2 SbC_n \end{bmatrix} \tag{18}$$

where $r_a = [-l_{wx}, \ 0, \ 0]^T$ is the position vector of the a.c. in the body frame, b is the wingspan, $\overline{c}$ is the mean aerodynamic chord, and the vector $[l_a, \ m_a, \ n_a]^T$ represents the aerodynamic moments as a function of the dimensionless aerodynamic coefficients C_l, C_m and C_n [33].

2.4. Equations of Motion

From the Newton–Euler formulation, it can be observed that the equations of motion of the XEVTOL-2FNW for the multirotor, transition, and fixed-wing modes are given as

$$\begin{bmatrix} \ddot{x} \\ \ddot{y} \\ \ddot{z} \end{bmatrix} = \begin{bmatrix} c_\psi c_\theta & c_\psi s_\theta s_\phi - s_\psi c_\psi & c_\psi s_\theta c_\phi + s_\psi s_\phi \\ s_\psi c_\theta & s_\psi s_\theta s_\phi + c_\psi c_\phi & s_\psi s_\theta c_\phi - c_\psi s_\phi \\ -s_\theta & c_\theta s_\phi & c_\theta c_\phi \end{bmatrix} \begin{bmatrix} \dot{U} \\ \dot{V} \\ \dot{W} \end{bmatrix} \tag{19}$$

$$\begin{bmatrix} \ddot{\phi} \\ \ddot{\theta} \\ \ddot{\psi} \end{bmatrix} = \begin{bmatrix} 1 & s_\phi s_\theta/c_\theta & c_\phi s_\theta/c_\theta \\ 0 & c_\phi & -s_\phi \\ 0 & s_\phi/c_\theta & c_\phi/c_\theta \end{bmatrix} \begin{bmatrix} \dot{p} \\ \dot{q} \\ \dot{r} \end{bmatrix} \tag{20}$$

$$\begin{bmatrix} \dot{U} \\ \dot{V} \\ \dot{W} \end{bmatrix} = \begin{bmatrix} Vr - Wq \\ Wp - Ur \\ Uq - Vp \end{bmatrix} + \frac{1}{m}\begin{bmatrix} X \\ Y \\ Z \end{bmatrix} \tag{21}$$

$$\begin{bmatrix} \dot{p} \\ \dot{q} \\ \dot{r} \end{bmatrix} = \begin{bmatrix} \dfrac{1}{I_{xx}I_{zz} - I_{xz}^2}\left\{ \left[(I_{yy} - I_{zz})I_{zz} - I_{xz}^2\right]qr + I_{xz}(I_{xx} - I_{yy} + I_{zz})pq + I_{xz}N + I_{zz}L \right\} \\[2ex] \dfrac{1}{I_{yy}}\left[(I_{zz} - I_{xx})rp - I_{xz}(p^2 - r^2) + M\right] \\[2ex] \dfrac{1}{I_{xx}I_{zz} - I_{xz}^2}\left\{ \left[I_{xz}^2 + I_{xx}(I_{xx} - I_{yy})\right]pq + \left[I_{xz}(I_{yy} - I_{zz} - I_{xx})\right]qr + I_{xx}N + I_{xz}L \right\} \end{bmatrix} \tag{22}$$

Equations (19) and (20) represent the linear and angular velocities of the vehicle with respect to the inertial frame, while Equations (21) and (22) represent the linear and angular velocities referring to the body axes.

Since the XEVTOL-4FNW configuration is similar to the one developed in this section, the derivation of its mathematical model is omitted for brevity. The main difference between the two configurations is in the forces and moments acting on the vehicle, due to the addition of another two ducted fans in the XEVTOL-4FNW configuration. In this sense, only the corresponding terms of forces and moments for the propulsion system will be modified.

Multirotor Flight Mode Simplification

During multirotor flight, the tilting engines are in the vertical position $\delta_T = 0°$, with an incidence angle $\alpha = 0°$ and a slip angle $\beta = 0°$. The aerodynamic forces and moments are neglected, i.e., $F_a = 0$ and $M_a = 0$. For hovering flight conditions, the vehicle's roll, pitch, and yaw angles have small variations, so the angular velocities can be considered as $[p, q, r]^T \approx [0, 0, 0]^T$. Since $[p, q, r]^T \approx [0, 0, 0]^T$ and the inertia of the propellers is small, the gyroscopic effects of the propellers are neglected. Considering the body axes coincide with the main axes of the vehicle, the inertia product I_{xz} is zero [29]. Applying these considerations to Equations (19)–(22), the dynamic model for multirotor flight yields

$$\ddot{x} = -u_T(c_\psi s_\theta c_\phi + s_\psi s_\phi)/m \tag{23}$$

$$\ddot{y} = -u_T(s_\psi s_\theta c_\phi - c_\psi s_\phi)/m \tag{24}$$

$$\ddot{z} = g - u_T c_\theta c_\phi/m \tag{25}$$

$$\ddot{\phi} = M_\phi/I_{xx} + M_\theta s_\phi t_\theta/I_{yy} + M_\psi c_\phi t_\theta/I_{zz} \tag{26}$$

$$\ddot{\theta} = M_\theta c_\phi/I_{yy} - M_\psi s_\phi/I_{zz} \tag{27}$$

$$\ddot{\psi} = M_\theta s_\phi/I_{yy}c_\theta + M_\psi c_\phi/I_{zz}c_\theta \tag{28}$$

where $u_T = F_1 + F_2 + F_3 + F_4$ is control force and $M_\phi = d_{Dy}(F_2 + F_3 - F_4 - F_1)$, $M_\theta = d_{tx}(F_1 + F_2) - d_{Dx}(F_3 + F_4)$, and $M_\psi = (\tau_4 - \tau_3) + (\tau_2 - \tau_1)$ are the control moments.

3. Model-in-the-Loop Simulations in X-PLANE

Testing control algorithms in a physical aircraft is an expensive and time-consuming process that requires many repetitive tests during the development phase. In addition, current UAV regulations are too restrictive, since experimental testing is a dangerous task. Moreover, a possible failure of the aircraft sensors could cause the estimates made by the vehicle to be inaccurate. This does not allow correct evaluation of the control algorithms and other systems during flight tests. Therefore, a common alternative for flight validation is to develop flight tests using virtual environments by performing model-in-the-loop (MIL) simulations [34].

MIL simulations refer to a type of simulation where a mathematical or computational model is incorporated into a closed-loop system. The model interacts with other components of the system, such as sensors, actuators, and control algorithms, to simulate the behavior of the overall system. In MIL simulations, the model is typically a simplified representation of the real system, often developed using mathematical equations or computer-based algorithms. The purpose of using a model is to simulate the system's behavior under various conditions, test different control strategies, or evaluate the performance of specific components [35].

In this sense, MIL simulations were employed to evaluate and compare the performance of the two XEVTOL concepts. The position and attitude of the XEVTOL concepts were obtained using the software X-Plane, which is a flight simulator developed by Laminar Research with realistic flight physics. This allowed us to verify, in this case, the XEVTOL concepts, providing pivotal feedback for the design phase and for the design of control strategies, and obtaining simulation results of the aircraft performance close to a flight in real conditions. To this end, X-Plane utilizes blade element momentum theory (BEMT) to analyze the geometric configuration of an aircraft and computes its flight behavior. This approach involves subdividing the aircraft into smaller elements and continuously calculating the forces acting on each element. These forces are then translated into accelerations, which are integrated over time to derive the aircraft's velocities and positions, to describe the dynamics of the aircraft motion in the simulation [36].

Thong [37] tested the accuracy of X-Plane by means of a series of simulations with a model of a F-15E. They concluded that the simulator was accurate in the sense that the input parameters were captured as close to reality as possible. Moreover, the information obtained through the simulations was compared with that obtained both analytically and from the aircraft flight manuals. Similarly, Fayyaz [38] developed simulations with the Pilatus PC-9 modeled in X-Plane 10 with the least number of discrepancies with respect to the real aircraft. The simulation results were compared with theoretical and real data, showing that the performance and stability characteristics of the aircraft remained within a range of ten percent compared to the theoretical/actual values, with the exception of lateral stability.

In our simulation environment, the scheme used to conduct MIL simulations used X-Plane and Matlab-Simulink. In this case, the X-Plane program was in charge of generating the dynamic model of the vehicle from a 3D model developed in Plane Maker software. The position and orientation of the vehicle was sent to Simulink through the UDP protocol. Once in Simulink, this information was used to compute the corresponding flight controllers and, finally, the obtained control signals were sent to the vehicle engines in X-Plane for the operation of the vehicle. Figure 8 schematizes this simulation process.

The development of the 3D model of the XEVTOL concepts in Plane Maker, as well as the control scheme used for the development of the MIL simulations, are described in Section 3.1.

3.1. Plane Maker 3D Model

Plane Maker is the software part of X-Plane that allows users to create and edit their customized aircraft, based on existing models or entirely new designs. By setting different physical attributes such as the weight, wingspan, control surfaces, propulsion system, and wing profile, among others, users can define the characteristics of their custom aircraft. Once the model has been developed and imported into X-Plane, the software estimates and simulates how the aircraft would perform in real conditions [39].

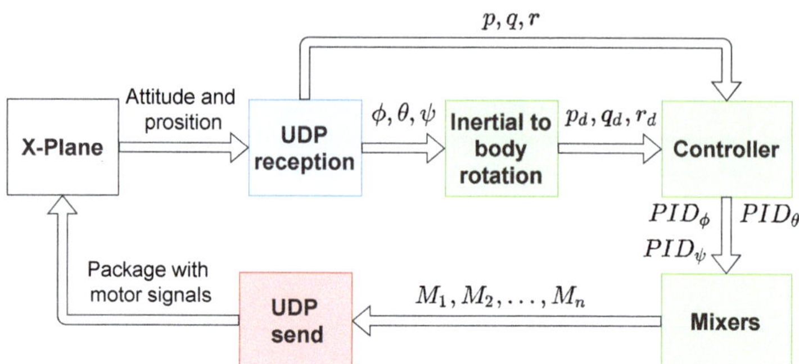

Figure 8. Block diagram describing the MIL simulation process. The reception block is represented in blue, the control blocks developed in Simulink in green, the transmission of the control signals for the motors in red, and the X-Plane simulator in white.

The design methodology used to obtain the 3D model of the aircraft in plane maker is presented in the diagram in Figure 9. This corresponds to the design of the XEVTOL-2FNW concept; however, the same methodology applies for the design of the XEVTOL-4FNW version. The main difference is in the definition of the propulsion system, as two additional engines needed to be incorporated for the XEVTOL-4FNW version.

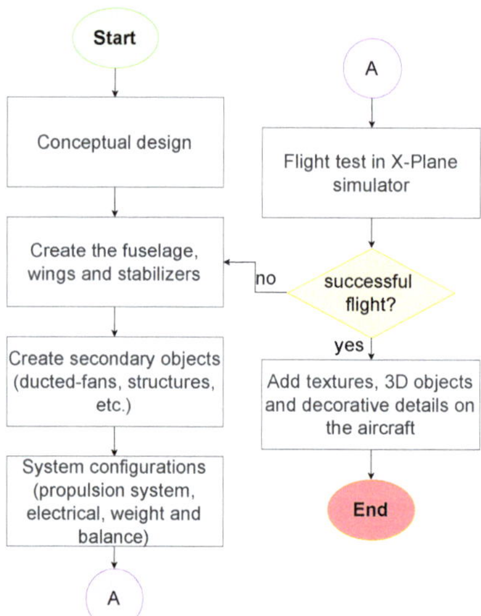

Figure 9. Plane maker design methodology. The design process started with the conceptual design that helps to create the components of the aircraft in the software. Then, flight tests are conducted, in order to evaluate the performance of the aircraft, and adjustments to the geometry are made if necessary.

The geometry of the vehicle's wing was defined in the following way: the central part of the vehicle that serves as the fuselage is part of the aircraft's wing, so it was modeled like a wing section, considering this section as a rectangular wing. The right semi-wing was considered in three sections, with two wing sections located at the tip and root, and a central part where the ducted-fan is located, as shown in Figure 10.

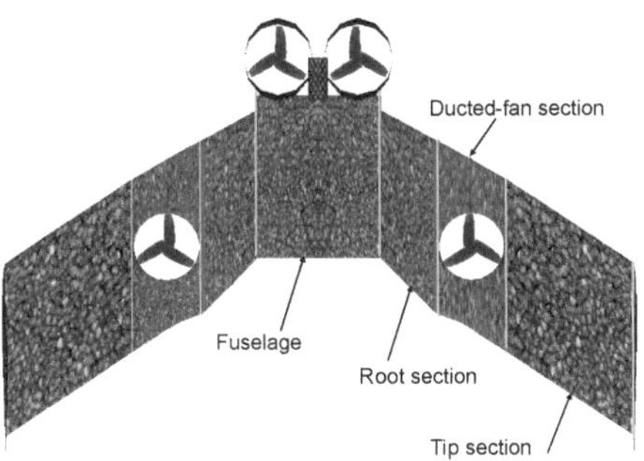

Figure 10. Top view of the 3D model created using Plane Maker for the XEVTOL-2FNW aircraft.

Next, the electrical system was configured in the Systems module within the Standard tab in Plane Maker. The electrical parameters to be used during the simulation were determined, including the source of the electrical energy, the buses through which the energy is distributed, and the inverters. A 59 Wh battery, a single channel, and no inverters were defined. The weight and balance of the aircraft were defined using the module "Weight & Balance" where the location of the c.g. on the longitudinal axis was defined, indicating a maximum and minimum value, and the c.g. on the vertical axis was also indicated; defining for the flying wing a c.g. at 40% and 67% of the fuselage chord, an empty weight of 2.1 kg and 2.7 kg for the XEVTOL-2FNW and XEVTOL-4FNW configurations, respectively. The 3D models of the XEVTOL concepts developed in Plane Maker are shown in Figure 11.

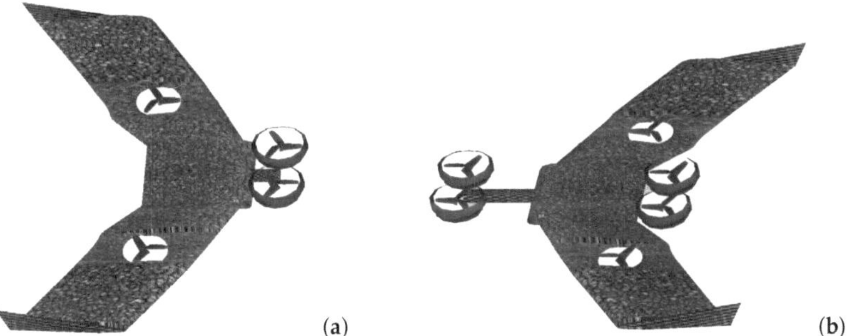

(a)   (b)

Figure 11. The 3D models of the H-VTOLs developed in Plane Maker for the MIL simulation in X-Plane: (**a**) XEVTOL-2FNW. (**b**) XEVTOL-4FNW.

3.2. Multirotor Flight Control

From the mathematical model for multirotor flight presented in Equations (23)–(28), the complete system was divided into four subsystems: the altitude, directional, lateral, and longitudinal subsystems. To control each subsystem, PID controllers were utilized, as described below.

3.2.1. Altitude Controller

The altitude dynamics are given by Equation (25) as:

$$\ddot{z} = g - \frac{1}{m}(u_T)c_\theta c_\phi \tag{29}$$

then, defining the altitude error $e_z = z - z_d$, where z_d is the desired altitude, the following computation can be performed:

$$\ddot{e}_z = \ddot{z} - \ddot{z}_d \tag{30}$$

Substituting Equation (29) into Equation (30), the following equation is obtained: $\ddot{e}_z = g - (1/m)u_T c_\theta c_\phi - \ddot{z}_d$. Since the control objective is to guarantee that $\ddot{e}_z = 0$, a PD controller is proposed, resulting in: $\ddot{e}_z = g - (1/m)u_T c_\theta c_\phi - \ddot{z}_d = -K_{pz}e_z - K_{dz}\dot{e}_z$. Solving for u_T, Equation (31) is obtained as

$$u_T = -(-\ddot{z}_d - K_{pz}e_z - K_{dz}\dot{e}_z - g)\left(\frac{m}{c_\theta c_\phi}\right) \tag{31}$$

and finally, substituting Equation (31) into Equation (29) and considering z_d as a constant, Equation (32) is obtained as

$$\ddot{z} = u_z \tag{32}$$

where $u_z = -K_{pz}e_z - K_{dz}\dot{e}_z$.

3.2.2. Directional Controller

The dynamics given by Equation (28) correspond to the directional dynamics represented by the yaw angle. By considering small angle approximations, i.e., $\psi \approx 0$, Equation (33) is obtained as

$$\ddot{\psi} = \frac{M_\theta \phi}{I_{yy}c_\theta} + \frac{M_\psi}{I_{zz}c_\theta} \tag{33}$$

and defining the error in yaw as $e_\psi = \psi - \psi_d$, where ψ_d is the desired yaw angle, yields

$$\ddot{e}_\psi = \ddot{\psi} - \ddot{\psi}_d \tag{34}$$

By substituting Equation (33) into Equation (34), the following equation is obtained: $\ddot{e}_\psi = \frac{M_\theta \phi}{I_{yy}c_\theta} + \frac{M_\psi}{I_{zz}c_\theta} - \ddot{\psi}_d$. Since the control objective is to guarantee that $\ddot{e}_z = 0$, a PD controller is proposed, resulting in $\frac{M_\theta \phi}{I_{yy}c_\theta} + \frac{M_\psi}{I_{zz}c_\theta} - \ddot{\psi}_d = -k_{p\psi}e_\psi - K_{d\psi}\dot{e}_\psi$. By solving for M_ψ, Equation (35) is obtained as

$$M_\psi = I_{zz}c_\theta\left(-\frac{M_\theta \phi}{I_{yy}c_\theta} - k_{p\psi}e_\psi - K_{d\psi}\dot{e}_\psi\right) \tag{35}$$

and, substituting Equation (35) into Equation (33), Equation (36) is obtained as

$$\ddot{\psi} = u_\psi \tag{36}$$

where $u_\psi = -k_{p\psi}e_\psi - K_{d\psi}\dot{e}_\psi$.

3.2.3. Lateral Controller

The lateral subsystem is given in Equations (24) and (26). By considering small angle approximations $\phi \approx 0$, $\psi \approx 0$, $s_\phi \approx \phi$ and substituting u_T from Equation (31), the lateral sub-system is rewritten as in Equations (37) and (38) as

$$\ddot{y} = -\frac{\phi g}{c_\theta c_\phi} \tag{37}$$

$$\ddot{\phi} = \frac{M_\phi}{I_{xx}} \tag{38}$$

$$\ddot{e}_y = \ddot{y} - \ddot{y}_d \tag{39}$$

where the lateral error is defined as $e_y = y - y_d$, with y_d being the desired lateral position.

By substituting Equation (37) into Equation (39), the following equation is obtained $\ddot{e}_y = -\frac{\phi g}{c_\theta c_\phi} - \ddot{y}_d$. Since the control objective is to guarantee that $\ddot{y} = 0$, a PD controller is proposed, resulting in $-\dfrac{\phi g}{c_\theta c_\phi} - \ddot{y}_d = -k_{py}e_y - K_{dy}\dot{e}_y$. Solving for ϕ yields that the desired angle to stabilize the lateral dynamics is given as in Equation (40):

$$\phi_d = -(-k_{py}e_y - K_{dy}\dot{e}_y)\frac{c_\theta c_\phi}{g} \tag{40}$$

and substituting Equation (40) into Equation (37), Equation (41) is obtained:

$$\ddot{y} = \phi \tag{41}$$

Defining the error in roll as $e_\phi = \phi - \phi_d$ and substituting Equation (38) into Equation (42), given as

$$\ddot{e}_\phi = \ddot{\phi} - \ddot{\phi}_d \tag{42}$$

the following equation is obtained: $\ddot{e}_\phi = \dfrac{M_\phi}{I_{xx}} - \ddot{\phi}_d$. To guarantee the convergence in the roll angle, a PD controller is proposed, and by solving for M_ϕ, Equation (43) is obtained as

$$M_\phi = I_{xx}(-k_{p\phi}e_\phi - K_{d\phi}\dot{e}_\phi + \ddot{\phi}_d) \tag{43}$$

and substituting Equation (43) into Equation (38) yields

$$\ddot{\phi} = u_\phi \tag{44}$$

where $u_\phi = -k_{p\phi}e_\phi - K_{d\phi}\dot{e}_\phi$.

3.2.4. Longitudinal Controller

The longitudinal subsystem is given by Equations (23) and (27). Considering small angle approximations $\phi \approx 0$, $\psi \approx 0$, and by substituting u_T from Equation (31), the longitudinal subsystem is rewritten in Equations (45) and (46) as

$$\ddot{x} = -g \tan \theta \tag{45}$$

$$\ddot{\theta} = \frac{M_\theta}{I_{yy}} \tag{46}$$

where $\tan \theta$ is the trigonometric function tangent. Defining the longitudinal error as $e_x = x - x_d$, where x_d is the desired longitudinal position, yields

$$\ddot{e}_x = \ddot{x} - \ddot{x}_d \tag{47}$$

Following the same methodology used for the lateral controller, the desired pitch angle is obtained in Equation (48) as

$$\theta_d = \tan^{-1}\left(-\frac{-k_{px}e_x - K_{dx}\dot{e}_x + \ddot{x}_d}{g}\right) \tag{48}$$

and M_θ yields

$$M_\theta = I_{yy}(\ddot{\theta}_d - k_{p\theta}e_\theta - K_{d\theta}\dot{e}_\theta) \tag{49}$$

By substituting Equation (49) into the dynamic of $\ddot{\theta}$ given by Equation (46), the following equation is obtained:

$$\ddot{x} = \theta \tag{50}$$

$$\ddot{\theta} = u_\theta \tag{51}$$

where $u_\theta = -k_{p\theta}e_\theta - K_{d\theta}\dot{e}$.

Finally, the multirotor flight controllers for the translational dynamics are given as:

$$\ddot{y} = -k_{py}e_y - k_{dy}\dot{e}_y \tag{52}$$

$$\ddot{x} = -k_{px}e_x - k_{dx}\dot{e}_x \tag{53}$$

$$\ddot{z} = -k_{pz}e_z - k_{dz}\dot{e}_z \tag{54}$$

while for the rotational dynamics, this is given by

$$\ddot{\phi} = -k_{p\phi}e_\phi - k_{d\phi}\dot{e}_\phi \tag{55}$$

$$\ddot{\theta} = -k_{p\theta}e_\theta - k_{d\theta}\dot{e}_\theta \tag{56}$$

$$\ddot{\psi} = -k_{p\psi}e_\psi - k_{d\psi}\dot{e}_\psi \tag{57}$$

From the control law of the roll dynamics given in Equation (55), it can be observed that $\ddot{\phi} \to 0$ when $e_\phi \to 0$; thus, solving the right hand side of Equation (55) for $\dot{\phi}$, the expression $\dot{\phi} = -\dfrac{k_{p\phi}}{k_{d\phi}}e_\phi + \dot{\phi}_d$ is obtained. Since $\ddot{\phi} \to 0$, the following is obtained:

$$\dot{\phi}_d = \underbrace{\frac{k_{p\phi}}{k_{d\phi}}}_{k_\phi} e_\phi \tag{58}$$

In the same way, Equations (59) and (60) are obtained for the pitch and yaw angles:

$$\dot{\theta}_d = \underbrace{\frac{k_{p\theta}}{k_{d\theta}}}_{k_\theta} e_\theta \tag{59}$$

$$\dot{\psi}_d = \underbrace{\frac{k_{p\psi}}{k_{d\psi}}}_{k_\psi} e_\psi \tag{60}$$

where k_ϕ, k_θ, and k_ψ are positive constants relating the orientation error with the desired angular velocity. To relate the desired angular velocities of the inertial frame to the desired angular velocities in the body frame, the transformation matrix $W_{\mathbb{I}\to\mathbb{B}}$ is utilized. The angular velocity transformation using this matrix is given as

$$\Omega_d = \begin{bmatrix} p_d \\ q_d \\ r_d \end{bmatrix} = \underbrace{\begin{bmatrix} 1 & 0 & -s_\theta \\ 0 & c_\phi & s_\phi c_\theta \\ 0 & -s_\phi & c_\phi c_\theta \end{bmatrix}}_{W_{\mathbb{I}\to\mathbb{B}}} \begin{bmatrix} \dot{\phi}_d \\ \dot{\theta}_d \\ \dot{\psi}_d \end{bmatrix} \tag{61}$$

3.2.5. Control Laws for the Angular Velocities in the Body Frame

In order to guarantee that the desired angles given in Equations (58)–(60) can be achieved, the following PID controllers were employed in the body frame:

$$\ddot{u}_p = kp_p e_p + kd_p \dot{e}_p + kI_p \int e_p \, dt \tag{62}$$

$$\ddot{u}_q = kp_q e_q + kd_q \dot{e}_q + kI_q \int e_q \, dt \tag{63}$$

$$\ddot{u}_r = kp_r e_r + kd_r \dot{e}_r + kI_r \int e_r \, dt \tag{64}$$

where kp_i, kd_i, and kI_i, $i = p, q, r$ are positive constants defined heuristically for tuning the corresponding control law.

Finally, Figure 12 shows a block diagram of the control laws for multirotor flight described above.

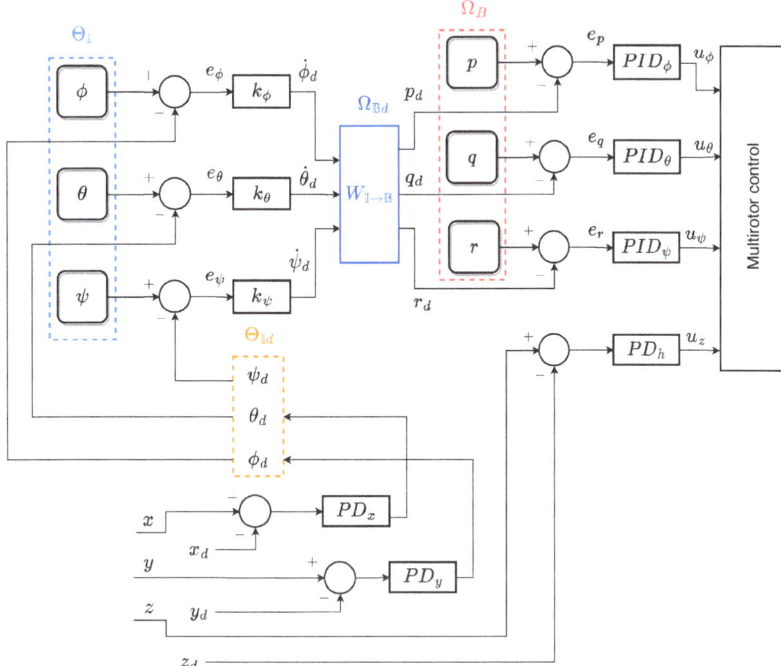

Figure 12. Block diagram of the control laws used to stabilize the vehicle in orientation and altitude by using K-PID controllers during the multirotor flight mode.

3.3. Control Schemes for Fixed-Wing and Transition Modes

The control scheme used for the flight phase in fixed-wing mode is shown in Figure 13. The controllers used during this flight phase were adapted in the PX4 firmware of the Pixhawk autopilot, since the proposed aircraft were not defined in the available configurations of the firmware [40]. With this control strategy, the dynamics of the roll and pitch angles were stabilized, as well as the aircraft altitude. Note that the desired roll angle is equal to zero, while the desired pitch depends on the altitude error.

The multirotor and fixed-wing flight mode controllers are activated depending on the desired flight mode. They operate separately or together during the transition. The diagram in Figure 14 depicts a simplified control diagram of the transition mode. It can be see that, depending on the selected flight mode, a customized mixer is used to send the control signals to the engines and control surfaces of the aircraft in X-Plane.

At the beginning of the flight transition, the aircraft tilts the frontal ducted fans to gain airspeed, while gradually activating the fixed-wing mode controller at the same time that the multirotor controller begins to lose authority, until it is completely deactivated and the vehicle is fully governed by the fixed-wing mode controller.

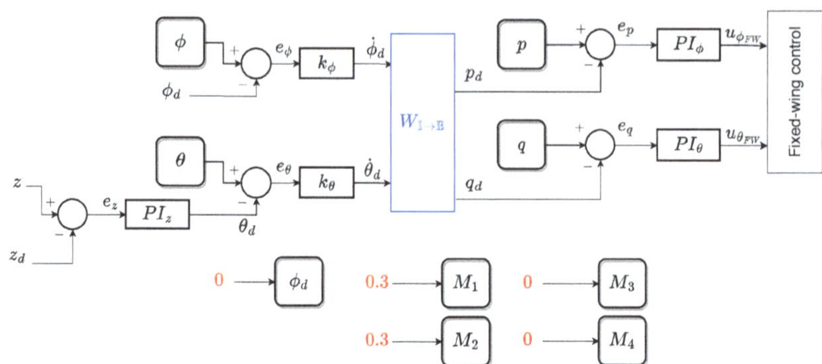

Figure 13. Block diagrams of the control laws used to stabilize the vehicle in orientation and altitude during fixed-wing mode.

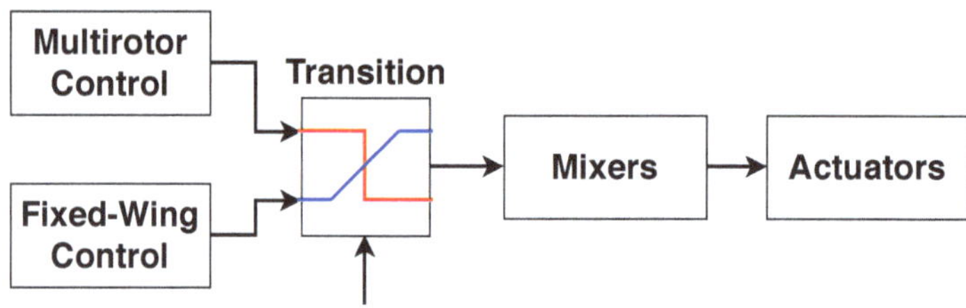

Figure 14. Transition diagram, which illustrates the selection of the flight mode to send the control law to the mixers and finally the control signals to the actuators.

3.4. Mixers

In order to determine the control signals for each of the engines, the signals obtained from the PID controls are mixed, taking into account the location of the engines with respect to the c.g. of the aircraft, in order to consider the effects that they have on the vehicle dynamics.

The corresponding signal for each engine configuration of the vehicles are defined from the geometry of the aircraft depicted in Figure 6. Based on such a geometry, the mixers for each motor M_i of both configurations are given in Table 1.

Table 1. Mixer configurations for both platforms.

XVTOL-2FNW	XVTOL-4FNW
$M_1 = u_z - d_{Fy}\, u_\phi + d_{tx}\, u_\theta - u_\psi$	$M_1 = u_z - l_3\, u_\phi + l_1\, u_\theta + u_\psi$
$M_2 = u_z + d_{Fy}\, u_\phi + d_{tx}\, u_\theta + u_\psi$	$M_2 = u_z + l_3\, u_\phi + l_1\, u_\theta - u_\psi$
$M_3 = u_z - d_{Dy}\, u_\phi - d_{Dx}\, u_\theta - u_\psi$	$M_3 = u_z - l_4\, u_\phi - l_2\, u_\theta - u_\psi$
$M_4 = u_z + d_{Dy}\, u_\phi - d_{Dx}\, u_\theta + u_\psi$	$M_4 = u_z + l_4\, u_\phi - l_2\, u_\theta + u_\psi$
	$M_5 = u_z - l_5\, u_\phi - l_6\, u_\theta + u_\psi$
	$M_6 = u_z + l_5\, u_\phi - l_6\, U_\theta - u_\psi$

where $d_{tx} = 0.2194$ m, $d_{Dx} = 0.1584$ m, $d_{Fy} = 0.0762$ m, $d_{Dy} = 0.29$ m, $l_1 = 0.4053$ m, $l_2 = 0.0822$ m, $l_3 = 0.07$ m, $l_4 = 0.29$ m, $l_5 = 0.79$ m and $l_6 = 0.17$ m are the distances from each motor to the c.g. on the plane x_B-y_B.

4. Simulation Results and Discussions

This section describes the conditions used in the MIL simulations utilizing X-Plane and Matlab-Simulink. The results obtained from these simulations were obtained wih the XEVTOL-2FNW and XEVTOL-4FNW configurations. For both configurations, a comparison was made regarding the energy efficiency and aerodynamic performance.

MIL flight simulations were carried out under the atmospheric conditions of Mexico City (see Table 2). The test consisted of the vehicles taking-off in multirotor flight, climbing until the aircraft reached a transition height defined at 30 m, and then starting the transition phase to plane mode, and once this was completed, maintaining a height of 30 m in straight and level flight. This mission is illustrated in Figure 15. The same scenario was used to perform a direct comparison between the two VTOL UAV concepts with embedded ducts. By using the same test environment, a fair and accurate evaluation of the performance of both configurations under identical conditions was ensured.

Table 2. Atmospheric conditions for the simulations.

Magnitude	Value	Unit
Mexico City elevation	2250	m
Air density	0.9814	kg/m^3
Dynamic viscosity	1.77×10^{-5}	kg/m s
Temperature	15	$^\circ$C
Time (GMT-6)	12	h
Wind	0	kt

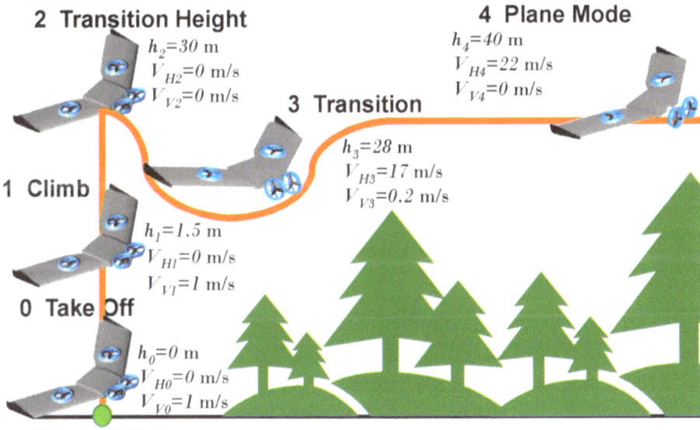

Figure 15. Mission profile. The aircraft performs a vertical take-off from the home point, ascends and transitions to fixed-wing mode. Where V_V is the climb speed, V_H is the airspeed, and h is the height of the vehicle.

The MIL flight tests were conducted using two computers connected in parallel, one running X-Plane 11.5 and the other one Matlab-Simulink 2022a. The flight tests were carried out with the minimum graphical characteristics, keeping the number of frames per second during the simulation within 30 frames per second (fps), where an interval between 25–35 fps is considered ideal for simulation [41]. The specifications of the computers used for the development of these simulations were as follows: Laptop ASUS ROG GL752VW with processor Intel Core I7-6700HQ 2.60 GHz, 16 GB RAM, graphics NVIDIA Geforce GTX 960M and Windows 10 Home, and Xtreme PC Gamer AMD Radeon Vega Renoir Ryzen 7 5700G 3.8 GHZ, RAM 16GB, and Windows 10 Pro.

Figures 16–19 show the simulation results for height, attitude (pitch and roll), transition signals, energy parameters, and aerodynamic performance using X-Plane and Simulink after simulating each of the vehicles using MIL under the same conditions.

It can be seen from the height plot of the Figure 16 that the vehicle ascended to a height of 30 m in multirotor flight at a climb velocity of 1 m/s, remaining at this height for 5 s. Then, it began the transition to fixed-wing mode 35 s after taking-off. First, it rotated forward the tilting mechanism in a step-wise manner, while gradually increasing the thrust of its two front engines until it reached 30% of the throttle, in order to gain airspeed. At 40 s, the multirotor mode controller action started to decrease, along with the dedicated ducted fans embedded inside the wing; turning them off avoids pushing air through the duct, eliminating the additional resistance caused by the spinning blades and reducing the total energy consumption and increasing the aircraft's endurance.

During the transition, the engines were controlled by two systems: the multirotor controller, and the fixed-wing controller. Due to this tilting action, the front engines experienced a decrease in performance related to pitching dynamics, resulting in a noticeable nose-down pitch of the vehicle, as shown in Figure 16. This nose-down pitch during the transition, as long as it is small, has no influence on the roll dynamics. However, if its magnitude increases, it starts to affect the roll of the vehicle. The transition phase ends at 60 s, when the propulsion reaches its established throttle at 30%, the roll and pitch angles converge to zero degrees, while the vehicle flies level at a height of 30 m.

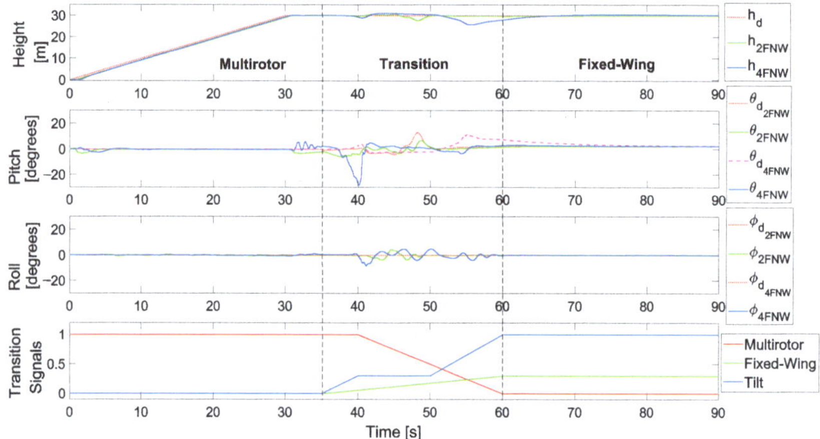

Figure 16. Flight simulation results for the two platforms. From top to bottom: altitude, orientation (pitch and roll), and the signals used to command the transition phase.

The results depicted in Figure 17 show three energy parameters obtained during the simulation of the two vehicles. The power graph shows the electrical power consumption of the propulsion system from second 0 to 35 during the multirotor phase. From second 35 to 40, the power consumption increased, since at this point the vehicle was transitioning from the multirotor mode to the fixed-wing mode and it required more energy to compensate for the loss of altitude. After reaching the mark at second 40, a reduction in power consumption became evident. This was attributed to the gradual transition from multicopter mode to fixed-wing mode. In this phase, the aircraft entered into a nose-down dive due to the decrease in thrust generated by the front ducted fans and the insufficient lift to sustain fixed-wing flight mode. At this point, the altitude dynamic was commanded by the control surfaces, while the throttle input was increased until the vehicle reached sufficient airspeed to exclusively use the fixed-wing controller.

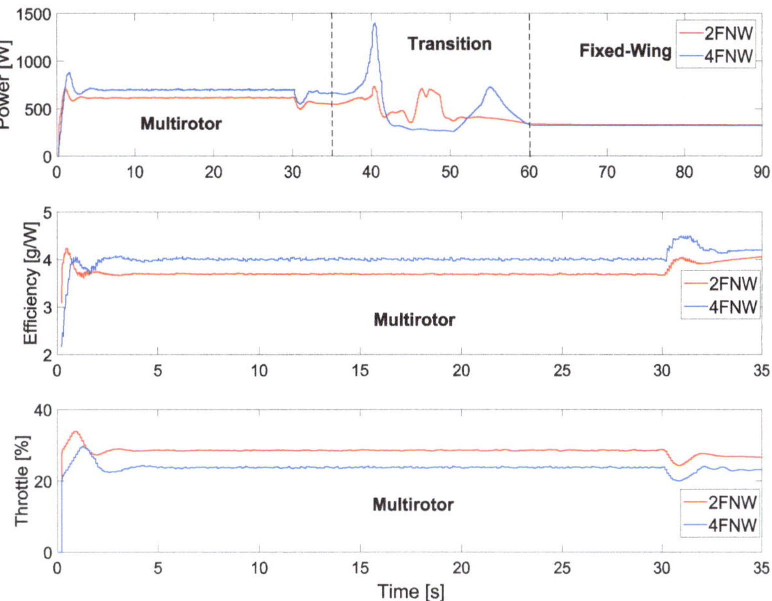

Figure 17. Energy parameters. From top to bottom: electrical power consumed by the propulsion systems, energy efficiency (propulsion/electrical power), and throttle.

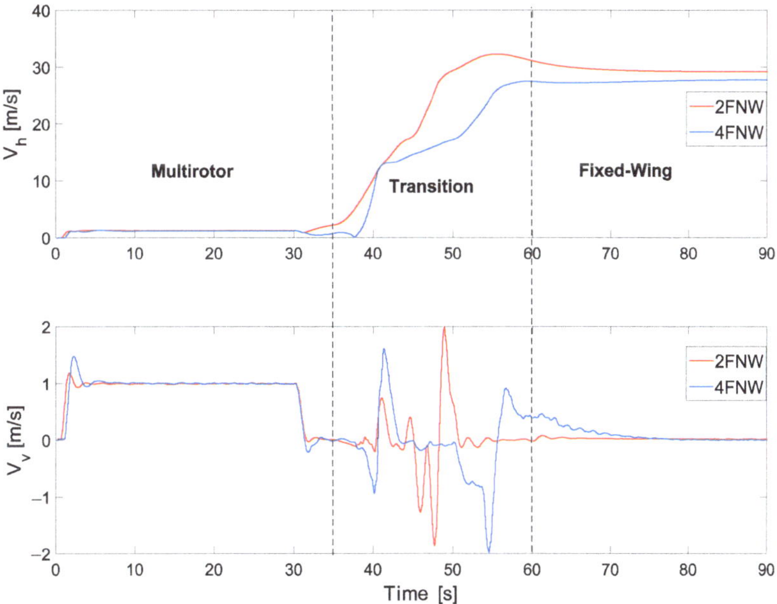

Figure 18. Airspeed and climb speed (V_h, V_v).

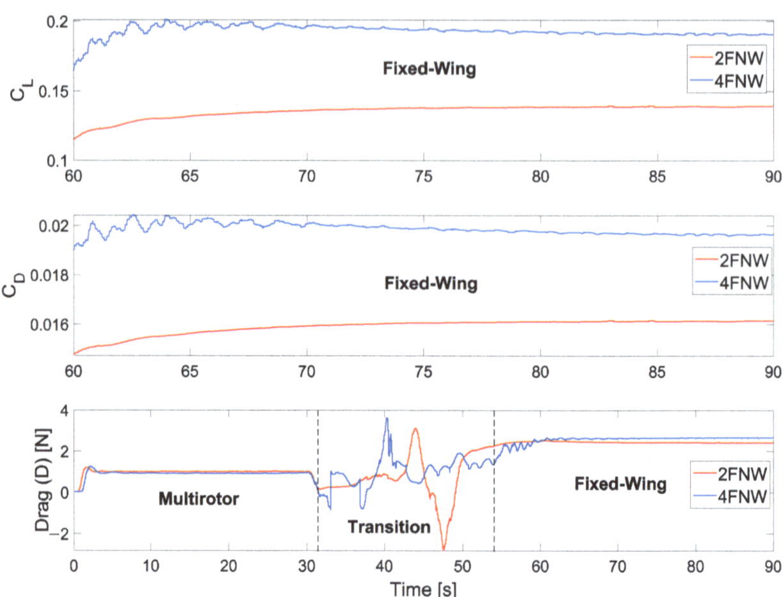

Figure 19. Aerodynamic Results. Lift coefficient (C_L), Drag coefficient (C_D), and Drag force (D).

In Figure 17, related to the obtained power results, the beginning and end of the transition phase are indicated by dashed lines at t = 35 s and t = 60 s, respectively. These lines represent the start of the tilting action and its conclusion. Once the transition was finalized, the tilting ducted fans only produced forward thrust, at t = 60 s. We can see that at second 60, there was a significant decrease of 100% in the power consumption during the fixed-wing phase. This reduction can be attributed to the aerodynamic lift generated by the wings, enabling efficient sustained flight. This contrasts with the multirotor mode, where the lift was primarily a result of rotors thrust, demanding a continuous power supply to maintain the vehicle flight.

Analyzing the multirotor phase, the 4FNW platform consumed additional power due to the inclusion of two extra ducted fans, in contrast to the 2FNW, which operated with a lower power demand; however, a closer examination of the energy efficiency graph reveals that the 4FNW reached an energy efficiency of 4 g/W, while for the 2FNW this was 3.6 g/W, proving that the 4FNW was the more energy efficient vehicle. This advantage stems from the distribution of the vehicle's weight among a greater number of ducted fans, enabling them to function at lower throttle percentages and resulting in a more efficient operation. Looking at the throttle graph, it can also be seen that for the case of vehicles 2FNW and 4FNW, increasing the number of engines lowered the percentage of throttle required.

The graph in Figure 18 shows the airspeed and climb speed. Both platforms were observed to ascend at a constant climb speed of 1 m/s during the multirotor flight. On the other hand, an increase in airspeed can be noticed as both aircraft reached their maximum speeds while flying fixed-wing, employing only 30% of throttle, and achieving an airspeed of 29 and 27 m/s for the 2FNW and 4FNW aircraft, respectively. The 2FNW vehicle, being lighter in weight compared to the 4FNW, achieved a higher airspeed.

Figure 19 depicts the aerodynamic results obtained during the MIL simulations, and the three graphs correspond to the lift coefficient (C_L), drag coefficient (C_D), and the drag force (D).

The coefficient results are presented for the fixed-wing flight mode. It can be observed that, at the start of this phase, there was an increase in C_L due to the vehicles needing to ascend to reach the target altitude. At the beginning of this phase, the vehicles were flying at low airspeed, and the method of compensating for this was raising the angle of attack. This adjustment led to an increase in C_L. However, as the airspeed continued to rise, the coefficient decreased until it stabilized at a constant value. This stabilization indicated that the vehicle had entered a cruising phase. During cruising, the 4FNW was the vehicle with the higher C_L, due to its greater weight compared to the 2FNW, necessitating increased lift.

On the other hand, the C_D exhibited a similar behavior to the C_L, increasing at the beginning due to induced drag, but then decreasing until it reached the minimum value during the cruise phase. This demonstrates the increase in drag for the 4FNW during this stage in comparison to the other aircraft. This can be attributed to the dirtier configuration resulting from the additional ducted fans.

The drag results during the multirotor flight presented a value of approximately 0.9 N, similarly for the two vehicles. Drag is encountered due to the fact that a vehicle generates a force that opposes its vertical displacement. This drag is influenced by factors such as the wing surface area and the rate of climb.

5. Conclusions and Future Work

This research addressed the challenge of improving the energy efficiency of H-VTOL aircraft. To deal with this issue, an innovative aerial vehicle design incorporating a DEP system in conjunction with ducted-fan rotors was proposed. Mathematical models for the XEVTOL-2FNW configuration were developed using the Newton–Euler formalism, enabling the design and implementation of a control strategy for conducting model-in-the-loop simulations in X-Plane. Two configurations were investigated, characterized by a different number of ducted-fan rotors: a four-rotor configuration and a six-rotor configuration. The developed MIL flight simulations enabled the evaluation and analysis of the performance of the proposed propulsion system, showing the following:

An increase of up to 10% in energy consumption for the 4FNW vehicle during the multirotor phase, attributed to the two extra ducted fans consuming more power compared with the 2FNW. However, the 4FNW proved to be 11% more energy efficient on a per rotor basis, as its propulsion system generated more thrust for the same power consumption.

A decrease in electrical power consumption on both platforms when switching from multirotor to the fixed-wing phase, reducing the power consumed by 100%. The power consumed by the 4FNW during the fixed wing phase, increased by only 1% compared with the 2FNW vehicle.

During the fixed-wing phase, the 2FNW vehicle achieved an up to 7.4% higher airspeed compared to the 2FNW vehicle during cruising, attributed to its lighter weight.

The aerodynamic results from the MIL simulation show that both platforms presented a drag force during vertical ascent due to the wing surface, which is in contrast to conventional multirotors, where this force is considered insignificant. Moreover, it was determined that during the fixed-wing flight in cruise condition, the 4FNW produced an increase of up to 46% in the lift coefficient, due to the fact that more lift was required because it is a heavier vehicle. With respect to the vehicle's drag, the 4FNW produced an increase of up to 19% in the drag coefficient when the vehicle was in cruising flight, due to the extra ducted fans.

In forthcoming research, an analysis of the propulsion system performance will include an analysis of the drag generated by the wing surface during vertical take off, as well as the optimization of the climbing speed during the VTOL phase, in order to reduce the energy consumption due to the extra drag force. Moreover, integrating an altitude controller, through utilizing the pitch angle in conjunction with the thrust to enhance the vehicle's altitude response during the transition and fixed-wing flight modes, will be considered in the future development of our project. Further certainty regarding the aerodynamic performance of these vehicles could potentially be obtained through additional analyses based on CFD and wind tunnel tests. The study of these XEVTOL concepts using software-in-the-loop and hardware-in-the-loop simulations could provide more information about the performance of these vehicles, to provide feedback for the design process. Similarly, conducting experimental flight tests to validate the numerical results obtained from the simulations will be considered. Finally, in order to improve the performance of the VTOL platforms in the transition phase, the possibility of using the embedded wing ducted fans to generate lift, compensating for the absence of airflow over the wings until the aircraft reaches the desired airspeed, will be evaluated.

Author Contributions: Conceptualization, J.M.B.A. and E.S.E.Q.; Methodology, J.M.B.A. and E.S.E.Q.; X-Plane and Simulink Simulations, J.M.B.A., J.L.S.M. and L.H.M.M.; Dynamics Modeling and Control, J.M.B.A. and E.S.E.Q.; Investigation, J.M.B.A. and E.S.E.Q.; Writing—Original Draft Preparation, J.M.B.A., E.S.E.Q. and L.R.G.C.; Writing—Review and Editing, J.M.B.A., J.L.S.M., L.H.M.M., E.S.E.Q., A.O.C. and L.R.G.C.; Supervision, E.S.E.Q. and A.O.C.; Project Administration, E.S.E.Q. All authors have read and agreed to the published version of the manuscript.

Funding: This work was partially supported by the Mexican National Council for Science and Technology through Project: Laboratorio Nacional en Vehículos Autónomos y Exoesqueletos.

Data Availability Statement: Data sharing is not applicable to this article.

Acknowledgments: Juan M. Bustamante-Alarcon thanks Conahcyt in recognition of the invaluable support provided throughout his PhD studies at the Center for Research and Advanced Studies of the National Polytechnic Institute Zacatenco, Mexico.

Conflicts of Interest: The authors declare no conflict of interest.

Nomenclature

a.c.	Aerodynamic center
c.g.	Center of gravity
$\mathbb{B}$	Body axes
C_D	Drag coefficient
c_f	Fuselage chord
C_L	Lift coefficient
F_T	Total external forces vector $[X,\ Y,\ Z]^T$
F_a	Aerodynamic forces vector
F_D	Drag force
F_L	Lift force
F_p	Propulsion forces vector
F_W	Gravity force vector
F_Y	Lateral force
g	Unit of mass, gram
h	Height
$\mathcal{I}_{\mathbb{B}}$	Aircraft moments of inertia about the axes $x_{\mathbb{B}}$, $y_{\mathbb{B}}$, $z_{\mathbb{B}}$
$\mathbb{I}$	Inertial Frame
J_p	Moments of inertia of propellers
k	Constant of proportionality of the propulsion force of the engines
m	Mass of the vehicle
M_a	Aerodymic moments
M_{gyro}	Gyroscopic moments
MIL	Model In the Loop
M_p	Propulsion moments
M_T	Total external moments vector, $[L,\ M,\ N]^T$
MTOW	Maximum Take-Off Weight
$\ddot{P}_{\mathbb{I}}$	Linear acceleration vector in the inertial framework, $[\ddot{x},\ \ddot{y},\ \ddot{z}]^T$
$R_{\mathbb{B}\to\mathbb{I}}$	Orthogonal rotation matrix of the body axes to the inertial frame
r_a	Aerodynamic vector position relative to body axes, $[-l_{wx},\ 0,\ 0]^T$
$R_{\mathcal{W}\to\mathbb{B}}$	Rotation matrix from the wind frame to the body frame
S	Wing area
UAV	Unmanned Aerial Vehicle
$V_{\mathbb{B}}$	Linear speed vector in the body axes, $[U,\ V,\ W]^T$

$\dot{V}_{\mathbb{B}}$	Linear acceleration vector in the body axes, $[\dot{U},\ \dot{V},\ \dot{W}]^T$
V_V	climb speed
V_H	airspeed
V_∞	Relative wind speed
W	Unit of electric power, Watt
$W_{\mathbb{B}\to\mathbb{I}}$	Orthogonal rotation matrix for rotational kinematics in function of angular velocities
β_η	Elevator angle adjustment(Trim)
δ_T	Tilting angle
δ_E	Elevon deflection angle
$\ddot{\Theta}_{\mathbb{I}}$	Angular acceleration vector in the inertial frame, $[\ddot{\phi},\ \ddot{\theta},\ \ddot{\psi}]^T$
θ	Pitch angle
ρ_∞	Wind density
ϕ	Roll angle
ψ	Yaw angle
$\Omega_{\mathbb{B}}$	Angular velocity in body axes, $p,\ q,\ r^T$
$\dot{\Omega}_{\mathbb{B}}$	Angular acceleration in body axes, $[\dot{p},\ \dot{q},\ \dot{r}]^T$
ω	Engine rotational speed

References

1. Dündar, Ö.; Bilici, M.; Ünler, T. Design and performance analyses of a fixed wing battery VTOL UAV. *Eng. Sci. Technol. Int. J.* **2020**, *23*, 1182–1193. [CrossRef]
2. Ducard, G.J.; Allenspach, M. Review of designs and flight control techniques of hybrid and convertible VTOL UAVs. *Aerosp. Sci. Technol.* **2021**, *118*, 107035. [CrossRef]
3. Zong, J.; Zhu, B.; Hou, Z.; Yang, X.; Zhai, J. Evaluation and Comparison of Hybrid Wing VTOL UAV with Four Different Electric Propulsion Systems. *Aerospace* **2021**, *8*, 256. [CrossRef]
4. Zhou, Y.; Zhao, H.; Liu, Y. An evaluative review of the VTOL technologies for unmanned and manned aerial vehicles. *Comput. Commun.* **2020**, *149*, 356–369. [CrossRef]
5. McCormick, B. *Aerodynamics of V/STOL Flight*, Dover ed.; Dover Publications: Mineola, NY, USA , 1999.
6. Naval Technology. JUMP 20 VTOL Unmanned Aerial Vehicle. 2016. Available online: https://www.naval-technology.com/projects/jump-20-vtol-unmanned-aerial-vehicle/ (accessed on 14 June 2023).
7. Quantum Systems Tron F90+ eVTOL Fixed-Wing PPK UAV. Available online: https://1updrones.com/product/quantum-systems-tron-f90-evtol-fixed-wing-ppk-uav/ (accessed on 14 June 2023).
8. Sato, M.; Muraoka, K. Flight Controller Design and Demonstration of Quad-Tilt-Wing Unmanned Aerial Vehicle. *J. Guid. Control Dyn.* **2015**, *38*, 1071–1082. [CrossRef]
9. Wingtra. WingtraOne Gen. Available online: https://wingtra.com/es/dron-mapeo-wingtraone/ (accessed on 14 June 2023).
10. Colucci, F. *Lift Where You Need It*; Vertiflite: Fairfax, VA, USA, 2016; pp. 26–30.
11. Sahoo, S.; Zhao, X.; Kyprianidis, K. Review of Concepts, Benefits, and Challenges for Future Electrical Propulsion-Based Aircraft. *Aerospace* **2020**, *7*, 44. [CrossRef]
12. Zhao, W.; Zhang, Y.; Peng, T.; Wu, J. The Impact of Distributed Propulsion on the Aerodynamic Characteristics of a Blended-Wing-Body Aircraft. *Aerospace* **2022**, *9*, 704. .:10.3390/aerospace9110704. [CrossRef]
13. Kim, H.; Berton, J.; Jones, S. Low Noise Cruise Efficient Short Take-Off and Landing Transport Vehicle Study. In Proceedings of the 6th AIAA Aviation Technology, Integration and Operations Conference (ATIO), Wichita, KS, USA, 25–27 September 2006. [CrossRef]
14. Borer, N.K.; Patterson, M.D.; Viken, J.V.; Moore, M.D.; Bevirt, J.; Stoll, A.M.; Gibson, A.R. Design and Performance of the NASA SCEPTOR Distributed Electric Propulsion Flight Demonstrator. In Proceedings of the 16th AIAA Aviation Technology, Integration and Operations Conference (ATIO), Washintong, DC, USA, 13–17 June 2016. [CrossRef]
15. Stoll, A.M.; Stilson, E.V.; Bevirt, J.; Pei, P.P. Conceptual Design of the Joby S2 Electric VTOL PAV. In Proceedings of the 14th AIAA Aviation Technology, Integration, and Operations Conference, Atlanta, GA, USA, 16–20 June 2014. [CrossRef]
16. Advanced Aircraft Company. Greased Lightning. 2015. Available online: https://advancedaircraftcompany.com/greased-lightning/ (accessed on 21 June 2023).
17. North, D.D.; Busan, R.C.; Howland, G. Design and Fabrication of the LA-8 Distributed Electric Propulsion VTOL Testbed. In Proceedings of the AIAA Scitech 2021 Forum, virtual event, 11–15, 19–21 January 2021. [CrossRef]
18. Lilium. Introducing the First Electric Vertical Take-Off and Landing Jet. 2016. Available online: https://lilium.com/jet (accessed on 21 June 2023).
19. DARPA. DARPA Completes Testing of Subscale Hybrid Electric VTOL X-Plane. 2017. Available online: https://www.darpa.mil/news-events/2017-04-04 (accessed on 14 June 2023).
20. Zhao, X.; Zhou, Z.; Zhu, X. Design of a Lift-propulsion VTOL UAV system. In Proceedings of the 2018 IEEE International Conference on Mechatronics and Automation (ICMA), Changchun, China, 5–8 August 2018; pp. 1908–1913. [CrossRef]

21. Hoeveler, B.; Wolf, C.C.; Raffel, M.; Janser, F. Aerodynamic Study on Efficiency Improvement of a Wing Embedded Lifting Fan Remaining Open in Cruise Flight. In Proceedings of the 2018 Applied Aerodynamics Conference, Atlanta, GA, USA, 25–29 June 2018. . [CrossRef]
22. Abrego, A.I.; Bulaga, R.W. Performance Study of a Ducted Fan System. In Proceedings of the Aerodynamics, Acoustics, and Test and Evaluation Technical Specialists Meeting, San Francisco, CA, USA, 23–25 January 2002;
23. Zhang, T.; Barakos, G. Review on ducted fans for compound rotorcraft. *Aeronaut. J.* **2020**, *124*, 941–974. [CrossRef]
24. Gudmundsson, S. *General Aviation Aircraft Design*, 2nd ed.; Butterworth-Heinemann: Cambridge, MA, USA, 2022; pp. 643–644.
25. Tuffwing. UAV Mapper User Guide V1.8 [Manufactured 2019]. Available online: http://www.tuffwing.com/support/UAV_Mapper_Quick_Start_Manual_1.8.html (accessed on 5 May 2023).
26. Nam, K.J.; Joung, J.; Har, D. Tri-Copter UAV with Individually Tilted Main Wings for Flight Maneuvers. *IEEE Access* **2020**, *8*, 46753–46772. [CrossRef]
27. Castillo, P.; Lozano, R.; Dzul, A.E. *Modelling and Control of Mini-Flying Machines*, 1st ed.; Springer: London, UK, 2005; pp. 145–146.
28. Rumit, K.; Alireza, N.; Manish, K.; Rajnikant, S.; Kelly, C.; Franck, C. Tilting-Rotor Quadcopter for Aggressive Flight Maneuvers Using Differential Flatness Based Flight Controller. In Proceedings of the ASME 2017 Dynamic Systems and Control Conference, Tysons, VA, USA, 11–13 October 2017. [CrossRef]
29. Etkin, B.; Reid, L.D. *Dynamics of Flight Stability and Controls*, 2nd ed.; John Wiley and Sons: Hoboken, NJ, USA, 1996.
30. Roskam, J.; Lan, C.E. *Airplane Aerodynamics and Performance*, 1st ed.; DARcorporation: Lawrence, KS, USA, 1997.
31. Cook, M.V. *Flight Dynamics Principles*, 3rd ed.; Butterworth-Heinemann: Burlington, MA, USA, 2013.
32. Espinoza, E.S. Helicóptero Coaxial de Largo Alcance. Ph.D. Dissertation, Center for Research and Advanced Studies of the National Polytechnic Institute, Automatic Control, Mexico City, Mexico, 2013.
33. Czyba, R.; Lemanowicz, M.; Gorol, Z.; Kudala, T. Construction Prototyping, Flight Dynamics Modeling, and Aerodynamic Analysis of Hybrid VTOL Unmanned Aircraft. *J. Adv. Transp.* **2018**, *2018*, 7040531. [CrossRef]
34. Dai, X.; Ke, C.; Quan, Q.; Cai, K.Y. Unified Simulation and Test Platform for Control Systems of Unmanned Vehicles. *arXiv* **2019**, arXiv:1908.02704.
35. Plummer, A.R. Model-in-the-Loop Testing. *Proc. Inst. Mech. Eng. Part I J. Syst. Control Eng.* **2006**, *220*, 183–199. [CrossRef]
36. X-Plane. How X-Plane Works. 2020. Available online: https://www.x-plane.com/desktop/how-x-plane-works/ (accessed on 18 June 2023).
37. Thong, C.W.S. Modeling Aircraft Performance and Stability on X-Plane. *J. Undergrad. Eng. Res.* **2011**, 1–31.
38. Fayyaz, O. Flight Dynamics Simulation of Pilatus PC-9 Using X-Plane 10. *J. Undergrad. Eng. Res.* **2015**, 1–14.
39. X-Plane. Plane Maker Manual. 2022. Available online: https://developer.x-plane.com/manuals/planemaker/ (accessed on 10 May 2023).
40. PX4. Controller Diagrams. 2023. Available online: https://docs.px4.io/main/en/flight_stack/controller_diagrams.html (accessed on 10 May 2023).
41. X-Plane. Setting the Rendering Options for Best Performance. 2023. Available online: https://www.x-plane.com/kb/setting-the-rendering-options-for-best-performance/ (accessed on 28 June 2023).

Article

Design and Performance of a Novel Tapered Wing Tiltrotor UAV for Hover and Cruise Missions

Edgar Ulises Rojo-Rodriguez [1], Erik Gilberto Rojo-Rodriguez [1] and Sergio A. Araujo-Estrada [2] and Octavio Garcia-Salazar [1,*]

[1] Aerospace Engineering Research and Innovation Center, Faculty of Mechanical and Electrical Engineering, Autonomous University of Nuevo Leon, Apodaca 66616, Nuevo Leon, Mexico; edgar.rojordrg@uanl.edu.mx (E.U.R.-R.); erojor@uanl.edu.mx (E.G.R.-R.)

[2] Department of Aeronautical and Astronautical Engineering, University of Southampton, Southampton SO16 7PX, UK; s.araujo-estrada@soton.ac.uk

* Correspondence: octavio.garciasl@uanl.edu.mx

Abstract: This research focuses on a novel convertible unmanned aerial vehicle (CUAV) featuring four rotors with tilting capabilities combined with a tapered form. This paper studies the transition motion between multirotor and fixed-wing modes based on the mechanical and aerodynamics design as well as the control strategy. The proposed CUAV involves information about design, manufacturing, operation, modeling, control strategy, and real-time experiments. The CUAV design considers a fixed-wing with tiltrotors and provides the maneuverability to perform take-off, hover flight, cruise flight, and landing, having the characteristics of a helicopter in hover flight and an aircraft in horizontal flight. The manufacturing is based on additive manufacturing, which facilitates the creation of a lattice structure within the wing. The modeling is obtained using the Newton–Euler equations, and the control strategy is a PID controller based on a geometric approach on SE(3). Finally, the real-time experiments validate the proposed design for the complete regime of flight, and the research meticulously evaluates the feasibility of the prototype and its potential to significantly enhance the mission versatility.

Keywords: convertible UAV; tilting rotors; manufacturing; design; real-time experiments

Citation: Rojo-Rodriguez, E.U.; Rojo-Rodriguez, E.G.; Araujo-Estrada, S.A.; Garcia-Salazar, O. Design and Performance of a Novel Tapered Wing Tiltrotor UAV for Hover and Cruise Missions. *Machines* **2024**, *12*, 653. https://doi.org/10.3390/machines12090653

Academic Editors: Giuseppe Carbone and Luca Bruzzone

Received: 17 June 2024
Revised: 5 August 2024
Accepted: 9 September 2024
Published: 18 September 2024

1. Introduction

In recent years, unmanned aerial vehicles (UAVs) have become increasingly common in several civilian and military applications, including search and rescue, highway patrol, and inspecting infrastructure such as power lines, bridges, factories, buildings, exteriors, sewers, railroads, and wind turbines. There exist three main types of UAVs: multirotors, airplanes, and convertible UAVs (non-conventional configurations) [1].

The take-off and landing have historically presented difficulties for UAVs because of the limitations involved in each different configuration. In this sense, a fixed-wing UAV presents high aerodynamic performance and requires a runway in order to take off and land; however, a rotary-wing UAV suffers performance limitations in terms of endurance, range, and maximum forward speed. In order to combine these capabilities of the fixed-wing and rotary-wing UAVs, a solution is proposed in this paper: a tiltrotor configuration with a tapered wing.

Classic fixed-wing vehicles require dedicated runway infrastructure, limiting their operational reach [2]. Convertible aircrafts, on the other hand, can operate from confined spaces and eliminate the need for extensive runways. This translates to increased accessibility for remote locations, urban environments, and disaster zones where traditional landing strips might be unavailable or damaged [3]. While offering vertical agility similar to helicopters, convertible vehicles can transition to fixed-wing flight for extended range and higher cruise speeds compared to rotary-wing vehicles. This provides a significant advantage in terms of operational efficiency, particularly for applications such as long-distance cargo delivery, search and rescue missions over vast areas, or border patrol activities [4].

Research on convertible UAVs requires knowledge of rotary-wing and fixed-wing vehicles to combine hover and cruise flight properties; it means a CUAV that performs a complete flight: take-off, hover, cruise, and landing modes. Some convertible UAVs can be found in [5], computational fluid dynamics (CFD) simulation and aerodynamic characterization in [6], flight dynamics modeling and stabilization in [7], and additionally, the tilting mechanism was designed in [8]. The development of convertible vehicles, tiltrotor UAVs, tilt-wing UAVs, tail-body or tailsitter UAVs, has garnered significant interest in the scientific community [9–11]. The domain of convertible UAV indexing presents a multifaceted landscape. While a multitude of researchers have opted for bespoke designs with unique implementation strategies, this very diversity creates a significant challenge for comprehensive study. Nonetheless, several prominent approaches hold particular significance and warrant mention within the relevant literature. These approaches include a tiltrotor convertible UAV involving a quad-rotor design equipped with a tilting mechanism, as addressed in [12]. This mechanism provides the ability to dynamically change the direction of propulsion. This vehicle was stabilized with a nonlinear control for the complete regime of flight, and the autopilot was developed using a low-cost DSP embedded system to achieve real-time experiments. A prototype employing vectorized thrust was presented in [13], enabling the capability for motion without the need for corresponding body movement. This algorithm was validated in real-time experiments, showing the effectiveness of the proposed controller. In [14], the development of the transitioning vehicle called Cyclone was proposed, whose mission is to perform hover and horizontal flights considering a control with incremental nonlinear dynamic inversion. The real tests demonstrated the vertical take-off and landing capabilities of the vehicle. Furthermore, research on lightweight materials is crucial to optimize performance and range [15,16]. These advancements can have a ripple effect, benefiting the development of future generations of both conventional and unconventional aircraft.

In [17], a flight control system of a small tiltrotor UAV was proposed, and it is based on an improved mathematical model. The proposed controller is based on an eigenstructure assignment, and the proposed approach has been validated in a wind tunnel test and real-time flights. In [18], the authors presented a methodology to design a tiltrotor microair vehicle in order to perform hovering and cruise flight scenarios. Results showed the aerodynamic parameters of the proposed vehicle. The authors in [19] proposed a geometrically compatible integrated design to develop for the conformal rotor and nacelle of the distributed propulsion tilt-wing UAV. This methodology considered the complex geometric constraints and coordinated the aerodynamic efficiency of the rotor and nacelle, allowing a low drag. A tiltrotor UAV was presented in [20] whose configuration drives the attitude based on tilting rotors. The control strategy is based on a bounded smooth function, and it was implemented in real-time flights. In [21], a small trirotor test bed with tilting propellers was proposed to validate the flight control laws. The controller algorithm is based on a nonlinear dynamic inversion with two layers. The lower layer involves attitude stabilization, while the higher layer manages trajectory tracking. The authors in [22] worked on a robust adaptive mixing controller to achieve the trajectory tracking of a quad-tiltrotor convertible plane, and the mathematical model is obtained using Euler–Lagrange formalism. To validate the controller, the authors executed hardware-in-the-loop experiments. In [23], a model predictive controller was proposed for tiltrotor UAVs demonstrating the performance in real-time flights. The controller strategy considered a control allocation algorithm, and the model predictive control constitutes a unified nonlinear control for the convertible UAV that performs the complete flight. In [24], the authors presented a CFD (computational fluid dynamics) simulation for a tiltrotor UAVs in order to examine the flow fields on the fuselage and rotor under the transition mode of the vehicle. A transition strategy design based on optimization methods was proposed in [25];

the transition problem is solved using the optimal method with nonlinear programming. The optimization results showed that the transition strategy can manage the relationship between transition time, control input, and attitude stability. The work in [26] proposed a multidisciplinary optimization algorithm for preliminary convertible UAV design. This design is based on aerodynamic models, and it is validated using optimization techniques. Ref. [27] presented the design of a convertible UAV considering the parameters of the rotor, propeller, wing, and airfoil selection. The CUAV design was based on a flying wing and modified to add the tilting rotors. A basic PID controller was tested in real-time. In [28], the authors focus on a controller based on an MPC-based position for a tiltrotor tricopter VTOL UAV. The controller involves a conventional control in the outer loop, while the inner loop is an MPC controller. The simulation was executed for trajectory tracking under the realistic actuator limits. However, our proposed vehicle differs from those published in the literature, since it involves tiltrotor mechanisms combined with a tapered wing; in hover flight, the vehicle is controlled via the propulsion system providing vertical lifting acting against the gravity field. In horizontal mode, the convertible vehicle is airborne, so that the outer body surface (tapered wing) provides the lift force to maintain the horizontal flight.

The main contribution of this paper is a novel convertible tiltrotor UAV using a tapered wing in order to perform a hover flight as a helicopter and a horizontal flight as an airplane. The methodological process is outlined to achieve a comprehensive design to address the convertible configuration as well as the interaction between manufacturing and flight computer development. Real-time experiments are performed to validate system behaviors. The contribution of this paper is summarized as follows:

1. Development of a novel tapered wing tiltrotor UAV for hover and cruise missions.
2. A scheme of guidance, navigation, and control based on the special Euclidean group SE(3) for the convertible UAV.
3. The proposed convertible UAV is tested to obtain the performance in real-time flights.

The rest of this paper is organized as follows. Section 2 presents information about the UAV configuration, operational functions, performance, and manufacturing. Section 3 describes the equations of motion for the convertible UAV using the Newton–Euler formulation and proposes the geometric navigation based on the special Euclidean group SE(3) with a guidance frame and a saturated PID control. Section 4 presents the experimental platform and the autonomous navigation results of the convertible UAV. Finally, conclusions are given in Section 5.

2. Design Proposal

This section outlines a comprehensive design proposal for a convertible unmanned aerial vehicle (UAV) aimed at addressing the flight mission for both hover flight and cruise flight. Our proposal is motivated by the ever-increasing demand for versatile UAV systems capable of dynamically adapting to diverse and complex mission requirements. The subsequent sections address the configuration, actuation, manufacturing, and performance, elucidating the challenges encountered in the development and deployment and presenting significant potential for expanding the capabilities of UAVs. Figure 1 shows the schematic of the proposed convertible UAV and the operation efficiently under various flight conditions.

Figure 1. UAV mission proposal over different areas.

2.1. UAV Configuration

The proposed UAV configuration plays a pivotal role in its adaptability and flight capabilities. The main feature of this configuration is the use of frontal rotors with tilting capabilities and the rear rotor with a coaxial mechanism providing sufficient lift in the hover flight while the wing is a T-shaped design providing the lift force in the forward flight.

The frontal tiltrotors enable the UAV to transition seamlessly and continuously between hover and cruise modes, as shown in Figure 2. During take-off and hover, the rotors are oriented vertically, providing the necessary lift and control (right side of Figure 2). As the UAV transitions to forward flight, the rotors tilt horizontally, allowing the fixed wing to generate lift in relation to airspeed (left side of Figure 2). This dynamic configuration offers the best of both modes, combining the agility and versatility of a multirotor with the efficiency and endurance of a fixed-wing aircraft.

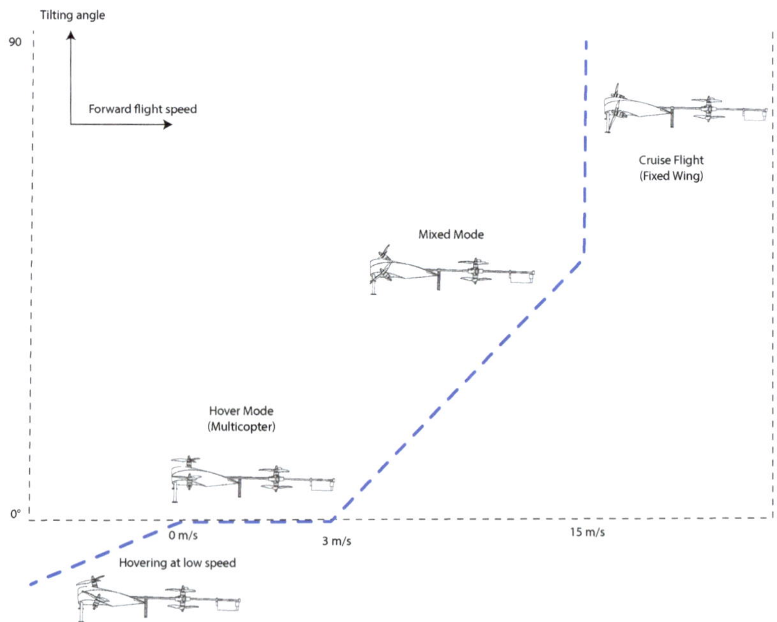

Figure 2. Flight phases for convertible UAV.

The T-shaped distribution of the frontal rotors ensures stability and control during the transition phase and fixed-wing flight. This configuration optimizes the UAV aerodynamic properties and reduces drag, improving performance and energy efficiency. Following this convention, the proposed configuration, as shown in Figure 3, involves the entire array of propulsion and structural fuselage.

Figure 3. Proposal of the convertible UAV.

2.2. Physical Parameters

This section presents the physical parameters of the platform, detailing its components in both configurations and highlighting the mechanisms that enable its convertibility. In the context of this convertible UAV design, a total weight estimation is required to reach the hover mode and to facilitate the analysis of the dual operational configuration. The first stage involves a multirotor configuration that supports the vehicle weight and enables flight control. However, the design also necessitates a fixed-wing configuration for forward flight. Consequently, appropriate airfoil selection and subsequent lift generation estimation are crucial considerations. It is important to note that the design is constrained by a maximum volume, with a boundary box of 0.065 m × 0.070 m × 0.020 m, to maintain a micro-UAV classification [29]. The proposed parameters are detailed in the Table 1.

Table 1. Aircraft physical parameters.

Parameter	Value
Span	0.04 m
Wing Root	0.024 m
Wing Tip	0.012 m
Wing Surface	0.075 m^2
Frontal Arms Length	0.025 m
C.G. to frontal	0.018 m
C.G. to bottom	0.016 m
Airfoil	FX-63
Incidence Angle	3°
Weight	420 g

These design parameters are chosen with the main objective of minimizing weight and enabling the complete regime of flight. Consequently, a total weight of less than 500 g and a compact volume are prioritized. These parameters significantly contribute to an aircraft maneuverability within a confined environment. Additionally, the wing surface selection process considers the airfoil type and its capacity to generate sufficient lift for sustained forward flight.

Note that equation $n = (L + F_I)/(W)$ presents the relationship between lift force, rotor force, and weight; a factor must be equal to or greater than 1 for the aircraft to fly. Considering load factors, this convertible UAV needs to be designed for two flight regimes. In multirotor mode, the aircraft is prepared for 4*G* to handle the fast maneuvers needed for stability and precision. However, for fixed-wing mode, the design needs to consider the different load factors experienced during cruise flight. This presents a design challenge for the convertible UAV. The structure needs to be robust enough to withstand ±4*G* accelerations in multirotor mode while also being lightweight for efficient fixed-wing flight. Additionally, the wing design, selected tapered form, of the convertible UAV is optimized for both high lift generation during multirotor operation and efficient aerodynamic performance during fixed-wing flight.

This design is developed based on additive manufacturing (AM) techniques to find lightweight parameters and integrate the entire mechanism into the internal structure, see Table 2. Different materials are used for various groups of parts, considering structural and impact reasons [16], as determined from structural optimization. The system is evaluated under static conditions, as this study focuses on representing maximum structural stress in dynamic environments. It uses maximum thrust for information on the internal structure, which informs optimization and deflection rejection.

Table 2. Materials specifications.

Part	Material
Wing Surface	ABS
Structural Frontal Arm (Left and Right)	Fiber Glass tube
Rotor Adapter	PETG
Tilt Mechanism	HIPS

Multiple studies are performed to develop and validate the effectiveness of geometrical optimization of this model, which is based on AM and also uses Solid Isotropic Material with Penalization (SIMP). SIMP is a powerful optimization algorithm determining the optimal material distribution within a design domain. SIMP generates lightweight structures with enhanced stiffness and strength by iteratively removing material from low-stress regions. When it is combined with 3D printing, this approach enables the fabrication of complex, lattice-like structures that would be infeasible using traditional manufacturing methods [30]. To effectively utilize SIMP optimization, it is essential to couple it with finite element analysis (FEA) and computational fluid dynamics (CFD) simulations. FEA provides accurate predictions of structural behavior under various loading conditions, while CFD enables aerodynamic performance evaluation. Figure 4 shows the mesh definition for the finite element method of the proposed CUAV. By iteratively refining the design based on simulation results, engineers can achieve optimal component performance.

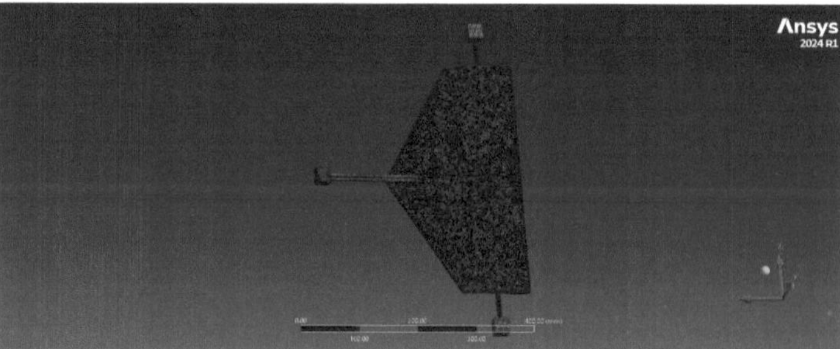

Figure 4. FEM mesh definition.

A maximum input force for thrust force is applied on each rotor base, which is also illustrated in Figure 5; it represents the maximum maneuver allowed by this type of aircraft. The center of gravity is selected as a fixed point for statics study, which is similar to a ground experiment on deformation effects shown in Figure 6, allowing for a geometrical profile of deformation. In Figure 7) it is shown the forces definition for cruise case. The structure deformation for cruise condition is depicted in Figure 8. A maximum deformation of 0.9258 mm on noncritical parts at the hover phase and 0.2921 mm for the cruise phase since these results are expected due to different forces distribution which is expected on lightweight structures. In this specific case, it is concentrated over the bottom part, which is only affected in hover mode and compensated by the control scheme.

The primary objective of this analysis is to evaluate the deformation characteristics and verify the structural integrity of the model under specified loading conditions. By utilizing ANSYS Mechanical, the objective is to identify potential weaknesses and ensure that the structure can withstand the applied forces without compromising its integrity.

The model created for this study is reduced to a mechanical representation of aircraft in order to simplify the model and develop a mesh of 2.5 mm of element size with a level of 7 at adaptive sizing resolution. These mesh elements were selected due to the minimum element size on the system with a 4.01 mm element, which is covered by the 1.25 mm change on elements selected on configurations.

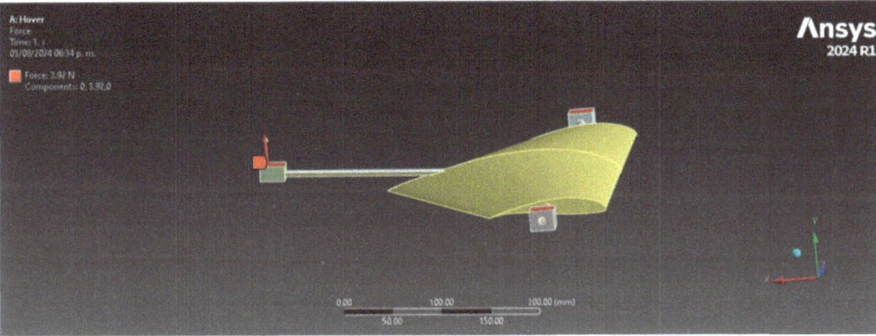

Figure 5. Forces definition for hover case.

For hover conditions, forces are placed at rotor points due to its multirotor nature, as seen in the red zones in Figure 5; for the maximum force developed by the rotor configuration, we selected 3.92 N for each rotor.

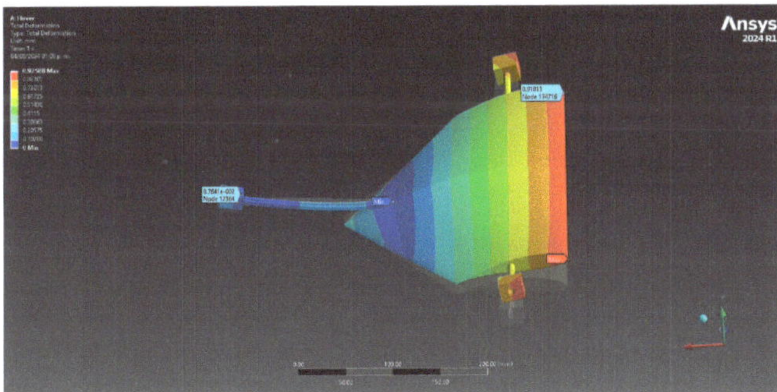

Figure 6. Structure deformation for hover condition.

Considering that the main structure is developed by ABS material and distributed to rotors by fiberglass tubes, it takes account of wing deformation, as shown in Figure 6. It has a probe illustrating the maximum deformation point at the inlet part. This behavior is ideal for this configuration due to the type of manufacture, which demonstrates the effectiveness of SIMP optimization, placing forces at required points. For this geometry, it is used to place internal structures on those points.

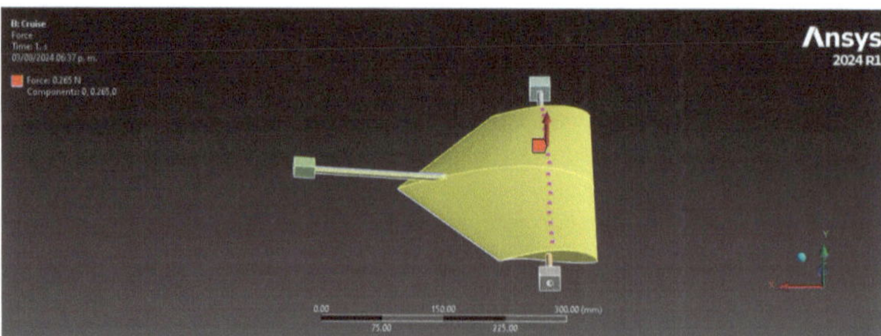

Figure 7. Forces definition for cruise case.

The cruise flight part of the study is used to verify wing effectiveness; in this case, it is developed as a force on a 1/4 part of the wing, which is later demonstrated in the CFD study. Addressing that point, a distribution is selected for force points at the wing surface. Figure 7 includes a distribution of the 14 m/s case, becoming a Gaussian distribution due to this wing geometry, with a total force of 4.24 N, which is the force required to lift the whole aircraft for cruise flight.

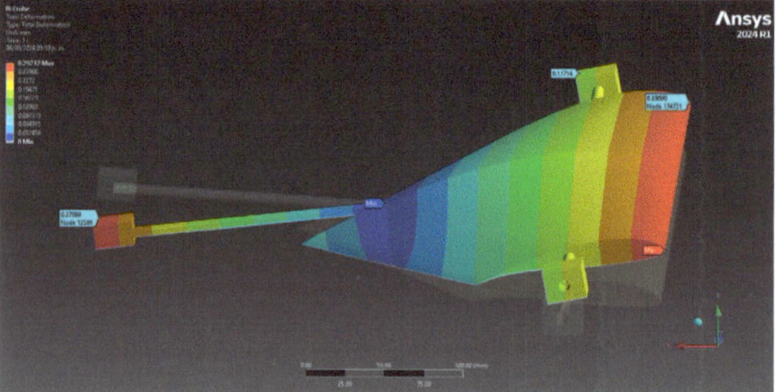

Figure 8. Structure deformation for cruise condition.

Actual results show the value 0.29098 mm as the maximum point of deformation, which is ideal for this micro-UAV case, making a structure that could handle the system and creating a force of 0.17714 mm on the rotor pads, which would not affect the final rotor force vector.

Considering the concentration of deformation shown in Figure 6, it is evident that the frontal rotors would experience the most deformation and equivalent stress, making it a critical point for examination. The frontal arms exhibit a maximum deformation of 0.92 mm under extreme conditions, although such conditions may not occur in real-world scenarios. However, given that these are made of a polymeric material of ABS nature, these can withstand this deformation while maintaining proper operation.

It is important to note that this structure is designed to withstand and exceed forces generated by aerodynamic conditions and rotor forces. This decision ensures the ability to accommodate dynamic behaviors without encountering issues. The accuracy and reliability of CFD simulations heavily depend on the mesh quality used to discretize the computational domain. This study highlights the significance of mesh refinement in capturing the intricate aerodynamic features of a wing. For this case, an adaptive mesh is selected. The outer air domain, which extends 150 mm from the wing, is crucial for capturing the far-field effects of the airflow, as shown in Figure 9. A coarser mesh is sufficient in this region to reduce computational cost while accurately predicting the overall flow behavior. Closer to the wing, within a 500 mm proximity, the mesh is refined to capture the near-field aerodynamic effects more accurately, using a body influence, which affects the 10 mm element size modification. The wing surface requires a highly refined mesh to capture the boundary layer effects and surface pressure distribution accurately. An element size of 1 mm is employed over the wing surface, which is essential for resolving the fine details of the flow around the wing, including the leading and trailing edges. To capture accurately the boundary layer development, 10 inflation layers are used on the wing surface, which are validated in Figure 10. These layers allow for a gradual transition from the wing surface to the free-stream, ensuring that the viscous effects are well resolved. The first layer of thickness is carefully chosen to capture the near-wall gradients accurately.

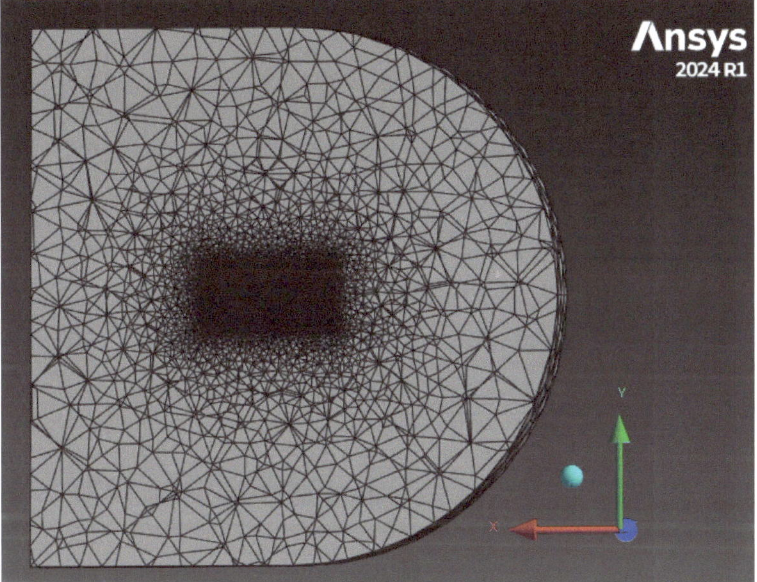

Figure 9. General CFD mesh.

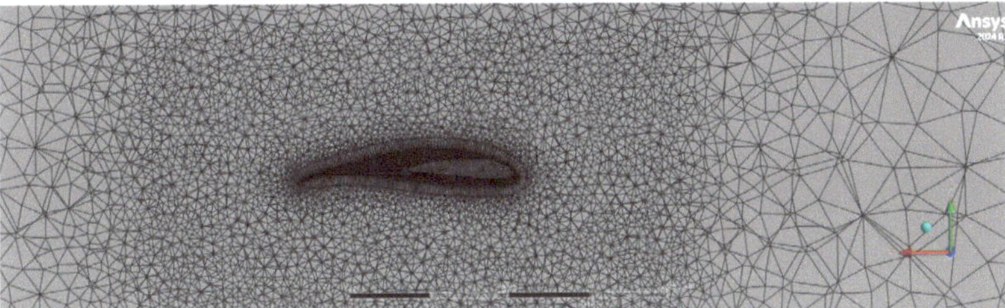

Figure 10. Detailed CFD mesh with the airfoil.

The system is analyzed under various scenarios, particularly at an airspeed of 14 m/s, as shown in Figure 11. It is observed that the system interacts with the air during cruise flights in a similar way to a flying wing thanks to the proposed design. This design minimizes external parts, exposing only the rotors and fixed wing, thus optimizing aerodynamic efficiency.

The CFD analysis employed a high-fidelity model encompassing both the wing and rotor geometries. The simulations are run with an inlet velocity of 14 m/s, representing typical cruise conditions. The selection of the FX63 airfoil for the wing is based on its well-documented performance characteristics.

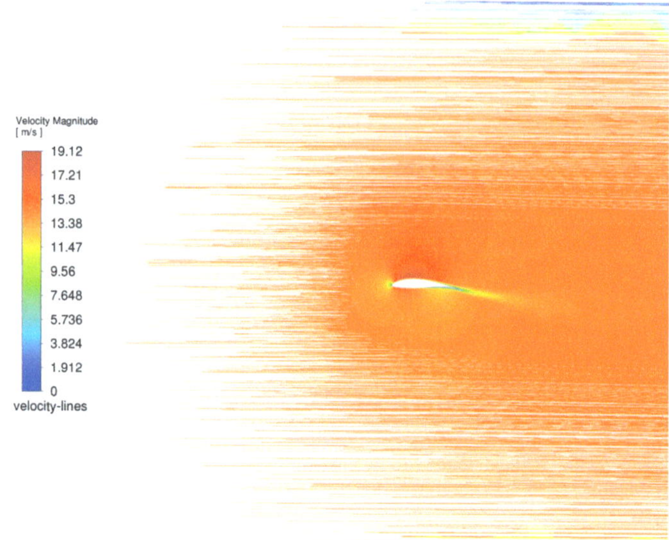

Figure 11. Airflow distribution.

Figure 11 presents contours colored by velocity magnitude, visualizing the flow behavior around the wing. The absence of significant deviations or swirling patterns in the contour suggests a predominantly laminar (non-turbulent) flow regime under the simulated conditions.

Furthermore, the analysis reveals that the airfoil design effectively accelerates the incoming flow. The velocity magnitude increases from the initial 14 m/s at the inlet to approximately 36 m/s over the wing surface. This acceleration is crucial for generating lift, which is a vital force for flight.

Understanding pressure gradients is vital for identifying regions of flow separation and potential stall. This information is crucial for designing airfoils that maintain smooth airflow, enhancing the performance and stability of the UAV; for that reason, it is analyzed in Figure 12. Pressure distribution data are essential for structural analysis. It helps determine the aerodynamic loads acting on the airfoil, which is necessary for ensuring the structural integrity and durability of the UAV. The CFD results reveal detailed pressure contours and distributions over the FX63-137 airfoil. High-pressure regions on the lower surface and low-pressure regions on the upper surface indicate the generation of lift. Areas of adverse pressure gradients highlight potential regions for flow separation, providing insights into the stall characteristics of the airfoil.

Multiple CFD studies were performed, and more prominent results are presented in Table 3, which mention different scenarios of this aircraft. The ideal scenario for this design is design point (DP) 9, where the aircraft flight is stable at the cruise case, considering that 3° is the incidence angle developed for this design.

Other cases are also analyzed as high angles of DP 10–13, showing that this aircraft could improve forces using more prominent cases, but with a drag consequence, which enables us to make future considerations for aggressive maneuvers.

Figure 12. Pressure distribution at cruise flight.

Table 3. CFD results.

Design Point	Angle of Attack	Airspeed (m/s)	Lift (N)	Drag (N)
DP 0	0	10	1.4831801	0.18688435
DP 1	0	5	0.33591906	0.049558448
DP 2	0	14	3.0199035	0.35773087
DP 3	0	20	6.3607885	0.71587954
DP 4	1	10	1.7161523	0.21013449
DP 5	2	10	1.9474965	0.23701694
DP 6	3	10	2.1824285	0.26784294
DP 7	10	10	3.8275471	0.59171158
DP 8	15	10	4.9289569	0.9282643
DP 9	3	14	4.3934829	0.52030514
DP 10	20	14	8.181392	2.3159387
DP 11	30	14	8.2654707	4.2134698
DP 12	45	14	7.5462137	7.0732765
DP 13	-10	14	-1.1898977	0.73946404

2.3. UAV Actuation

With rapid technological advancements, integrating complex actuation systems has revolutionized the efficacy and versatility of unmanned aerial vehicles (UAVs) across various mission profiles. One of the main advantages of UAV actuation lies in its ability to augment mission adaptability and responsiveness [31]. By incorporating dynamic actuation mechanisms, such as articulated wings, tilting rotors, or swiveling thrusters, UAVs can swiftly adapt to diverse environmental conditions and operational requirements. This agility enables UAVs to navigate challenging terrains, circumvent obstacles, and execute precision maneuvers with unparalleled efficiency.

The core of our convertible UAV design lies the innovative configuration of frontal rotors featuring tilting capabilities. These rotors are actuated by precision servomotors, enabling dynamic adjustments to their orientation. This pivotal feature facilitates the seamless transition between vertical take-off and landing (VTOL) operations and cruise flight, enhancing the versatility of the UAV.

For our prototype, a tilting mechanism based on the gear transmission of the servomotor force is developed. This mechanism directly controls the tilting angle, as depicted in Figure 13. Note that the tilting rotors are independent, allowing the system to be used as a differential one. V-22 aircraft were used as inspiration for rotor placement due to their performance and wing–rotor interaction, allowing better maneuverability.

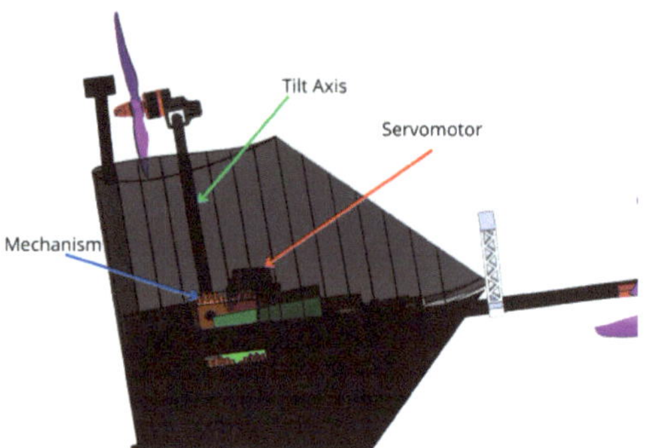

Figure 13. Tilting mechanism for rotor direction.

2.3.1. Hover Flight

Vertical Take-Off and Landing (VTOL) operations are crucial for unmanned aerial vehicles (UAVs), especially in scenarios where confined spaces or quick deployment are imperative. For our proposed vehicle in hover mode, the frontal rotors are strategically positioned vertically to generate the necessary lift and directional control, facilitating stable take-off, landing, and low-speed flight maneuvers. These rotors have precision movement capabilities, enabling orientation adjustments through tilting actions without the need for rotor speed variation. This innovative approach ensures efficiency and establishes a robust motion for hover mode control.

Furthermore, the differential control system governing the frontal rotors amplifies the CUAV agility and precision across both hover and cruise flight phases. By independently adjusting the tilt angles of each rotor, the CUAV gains precise control over pitch and yaw, facilitating seamless transitions between flight modes and empowering the vehicle to execute complex maneuvers with ease. This level of control versatility enhances operational fluidity and renders the CUAV adaptable to diverse mission requirements and environmental conditions. This condition stipulates that only saturated angles are applied to small motion, primarily utilizing the tilting mechanism for yaw motion while ensuring stability. The roll motion is obtained by the differential velocity of rotor 1 and rotor 2, and the pitch motion is obtained through the differential coupled rotors 1–2 and the coaxial rotors 3 and 4. The yaw motion is achieved by differential tilting rotors for 1 and 2; see Figure 14.

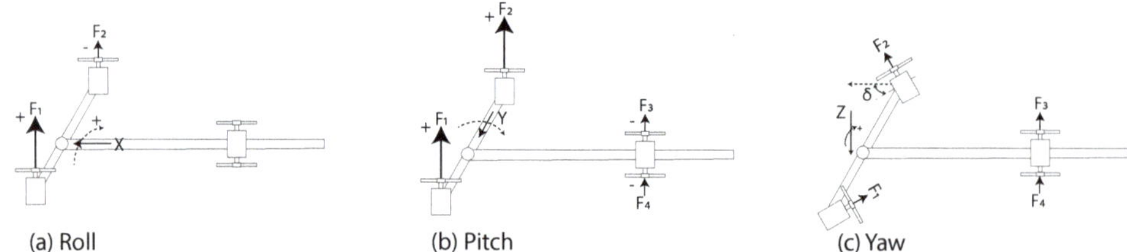

Figure 14. Diagram of tilting mechanism at low speed flight.

2.3.2. Cruise Flight

Once the UAV is airborne and ready to transition to cruise flight, the servomotors engage, facilitating the seamless transition of the rotors from vertical to horizontal orientation.

This pivotal moment marks the shift in operational dynamics as the UAV transitions from hover to fixed-wing mode. Unlike traditional aircraft configurations, where control surfaces such as ailerons, elevators, and rudders govern maneuverability, this CUAV employs a unique motion for the frontal tilting rotors.

This innovative approach highlights the control system and enhances maneuverability and responsiveness during flight. The rotor motions are defined in Figure 15, which enables responsiveness due to the strategy points used to perform an angle change of wing. By eschewing traditional control surfaces, the UAV achieves unprecedented agility and precision, enabling it to execute swift and intricate maneuvers with remarkable ease. The absence of control surfaces eliminates the associated mechanical complexities and aerodynamic constraints, allowing the UAV to push the boundaries of aerial maneuverability and operational performance.

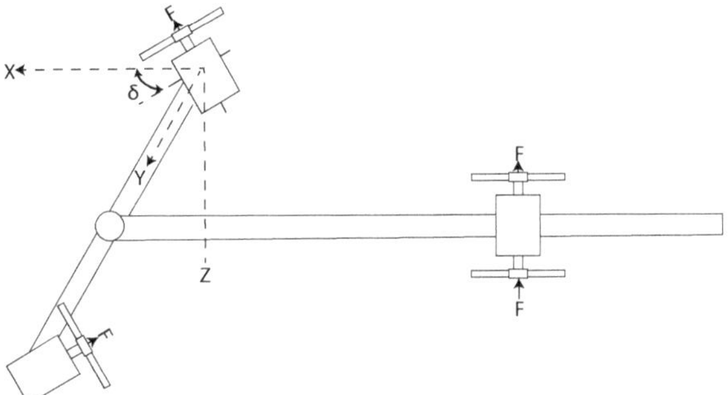

Figure 15. Tilting mechanism frame.

Furthermore, using a dual-mode capability, combining VTOL and fixed-wing flight, maximizes mission efficiency and versatility. During VTOL operations, the frontal rotors provide lift and control for take-offs, landings, and low-speed flight, ensuring operational flexibility in confined or challenging environments. Conversely, in cruise flight mode, the transition to horizontal rotor orientation optimizes aerodynamic efficiency, leveraging the fixed-wing configuration for sustained flight and extended mission endurance; see Figure 16.

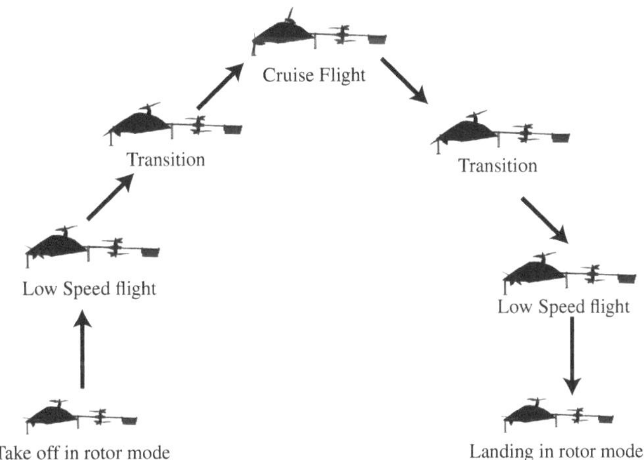

Figure 16. Hybrid mission.

2.4. Manufacturing

This innovative convertible UAV boasts a lightweight wing crafted using additive manufacturing, featuring a cutting-edge technique known as 3D printing. This method allows for the incorporation of multiple materials within the wing structure. By strategically integrating these materials, the engineers achieved an incredibly light weight without sacrificing strength. However, the benefits of additive manufacturing extend beyond the wing itself. This technology also facilitated the creation of a lattice structure within the wing. Lattice structures, resembling a complex web, offer exceptional strength-to-weight ratios, further contributing to the overall lightness of the UAV.

This lightweight design does not come at the expense of functionality. Thanks to additive manufacturing, the entire mechanism and avionics are seamlessly integrated within the interior of the wing. This ingenious approach frees up space and further streamlines the overall design of the convertible UAV.

The structural studies presented in Figures 6 and 8 aim to address the SIMP by strategically distributing bars along the wing to manage the forces. These studies were influenced by additive manufacturing techniques, specifically the interaction between walls and infill. However, in this case, these techniques served merely as inspiration. The structural points were determined based on the pressure distribution required at each point and a simplification of the SIMP algorithm, which only placed for this case a pressure point, filled by mechanical bars for this case. It was strategically distributed and filled with double the nozzle size to ensure a rigid structure.

As seen in Figure 17, this prototype utilized an improved and simplified structure that handles the forces as shown in Figure 6 and achieves maximum lightness.

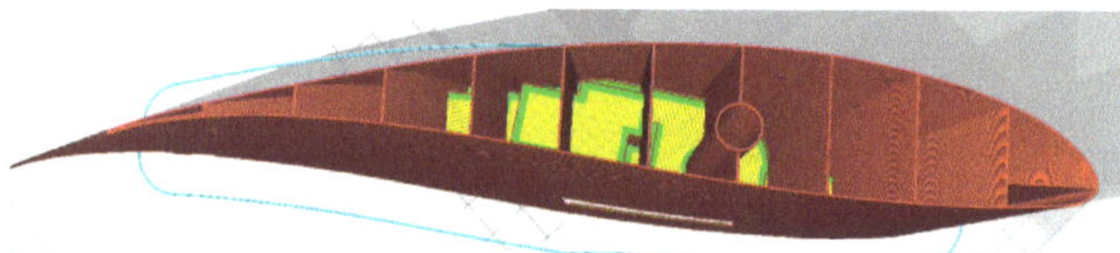

Figure 17. Internal 3D-printed wing structure.

Optimization and the use of multiple materials were key factors in the creation of the wing, as depicted in Figure 18. Several techniques were tested, but ultimately, a lattice polymeric structure, as seen in Figure 17, was chosen. Different patterns and materials exhibit different behaviors, but for this application, load distribution, as previously mentioned in Figures 6 and 8, guided the selection of the final approach.

In the realm of 3D printing, selecting the appropriate material is crucial for ensuring the quality and durability of the final product. After careful consideration, we chose ABS (acrylonitrile butadiene styrene) over PLA (polylactic acid), PETG (polyethylene terephthalate glycol), and other commonly used materials. This decision is based on the superior material stiffness and thermal deformation characteristics of ABS.

ABS is renowned for its excellent mechanical properties, particularly its high stiffness. This makes it an ideal choice for applications requiring durable and robust components. One of the significant advantages of ABS is its ability to withstand higher temperatures without deforming. ABS has a glass transition temperature of approximately 105 °C, which is significantly higher than PLA's 60 °C and PETG's 80 °C. This high thermal resistance ensures that ABS-printed parts maintain their shape and structural integrity under heat, making them suitable for a wider range of applications. ABS typically has a tensile strength of 40–50 MPa, while PLA ranges 37–50 MPa, and PETG ranges 48–55 MPa, giving us the best performance for this application

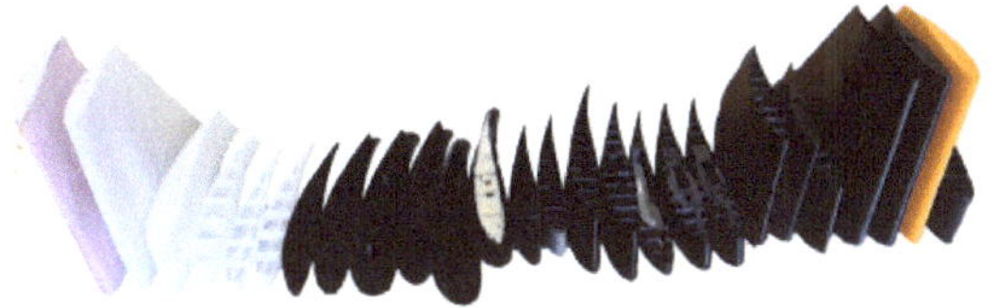

Figure 18. Multiple techniques used for wings.

2.5. Performance

Convertible aircraft, with their ability to transition between fixed-wing and rotary-wing modes, present unique challenges in performance evaluation. Unlike conventional aircraft, their performance is influenced by a complex interplay of factors, including airspeed, altitude, tilt angle, and payload. A comprehensive understanding of these interactions is crucial for optimizing aircraft design, operation, and mission planning.

The VN diagram, a fundamental tool in aerospace engineering, plays a crucial role in the design and operational planning of convertible UAVs. The VN diagram visually represents the relationship between an aircraft speed (V) and the load factor (N), providing engineers with critical insights into the aircraft's flight envelope. This diagram serves as a blueprint for understanding permissible operating limits across different flight modes for convertible UAVs.

This diagram provides a clear visualization of the operational limits of the airfoil under different load factors and velocities, as represented in Figure 19. The VN diagram analysis of our aircraft design demonstrates a well-defined range of operation, ensuring both safety and performance during various flight conditions. The evaluation indicates that the aircraft operates effectively within a speed range of 10 to 25 m per second (m/s), allowing for a versatile flight envelope.

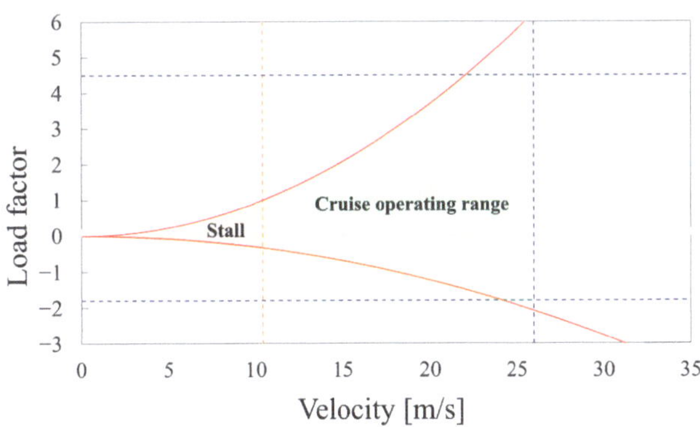

Figure 19. Flight envelope.

The stall speed of the aircraft (V_s) at 1G, or level flight, is calculated to be approximately 10 m/s. This speed marks the minimum velocity at which the aircraft can sustain level flight without stalling. On the other end, the maximum structural speed (V_{max}) is determined to be 25 m/s. Beyond this speed, there is a risk of structural damage, and the aircraft should not be operated at these velocities.

The positive load factor limit of the aircraft is evaluated to be +4.4G, while the negative load factor limit is −1.76G. These load factor limits indicate the maximum G-forces the aircraft can safely withstand during positive and negative maneuvers. The stall speed increases at higher load factors, such as during sharp turns or sudden climbs. For instance, at a 2G load factor, the stall speed increases to approximately 15.6 m/s, ensuring the aircraft remains stable and controllable even during aggressive maneuvers.

The VN diagram assessment confirms that the aircraft design provides a robust operational range from 10 to 25 m/s. This range not only supports stable and efficient cruise conditions but also accommodates various maneuvering needs, including steep turns and climbs, without compromising safety. For a convertible UAV with tilting rotors, the performance surface showcases how the available payload varies with different velocities and rotor tilt angles; see Figure 20.

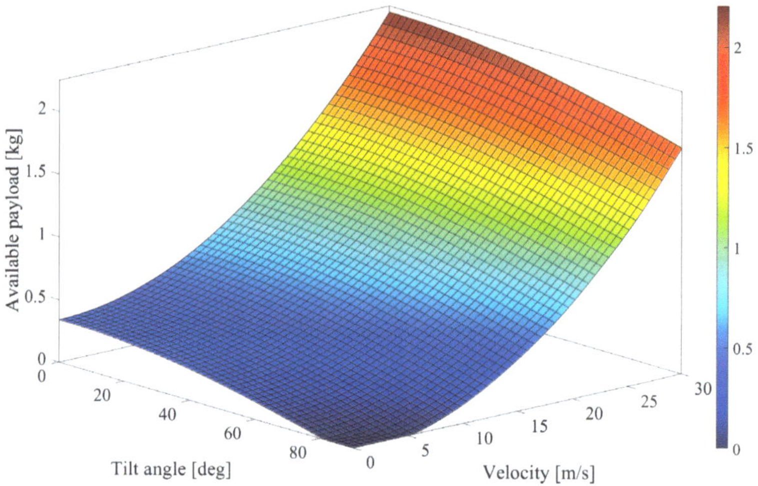

Figure 20. Performance surface for variation of the payload for each state.

The UAV is most efficient in forward flight, offering the highest payload capacity. This analysis optimizes the UAV operation for different missions, ensuring maximum payload capability while maintaining safe and efficient flight characteristics. The maximum payload is determined by the maximum hover take-off, limited by rotor forces, restricting the aircraft to 1.1 kg if the payload is at the center of mass for balanced force distribution. Figure 20 shows the system performance in each phase.

In hover mode with rotors fully tilted (0°), the UAV consumes more power, limiting the available payload to 0.4 kg. The UAV achieves better efficiency at a 45° rotor tilt and 20 m/s speed, allowing for a slightly higher available payload of 0.8 kg. The UAV reaches optimal efficiency in forward flight mode with rotors at 90° tilt, providing the highest available payload of 0.6 kg at 14 m/s.

3. Modeling and Stabilization

In this section, the equations of motion that govern the dynamic behaviors of CUAV are described. The mathematical framework, Newton–Euler equations, are used to model system dynamics, acknowledging the simplifications and assumptions inherent in these models.

The choice of reference frames is crucial in defining the 3D motion of the vehicle relative to its environment. The North–East–Down (NED) convention is considered, which is widely employed in aerospace applications, and the axes system involves the special Euclidean group SE(3). By examining the properties of SE(3) and its implications for reference frame transformations, the algorithm of guidance, navigation, and control is proposed. Note that those three reference frames are established to obtain the mathematical model of the aerial vehicles; see Figure 21.

For the vehicle orientation, Euler angles are defined as follows: ϕ is an angle defined between the x_B axis and a resultant plane from y_I and z_I; θ is an angle defined between the y_B axis and a resultant plane from x_I and z_I, and ψ is an angle defined between the z_B axis and a resultant plane from x_I and y_I. Those definitions allow aircraft to obtain attitude stabilization in a 3D space. In this sense, two angles that provide the information in the aerodynamic or wind frame are the angle of attack α and the sideslip angle β.

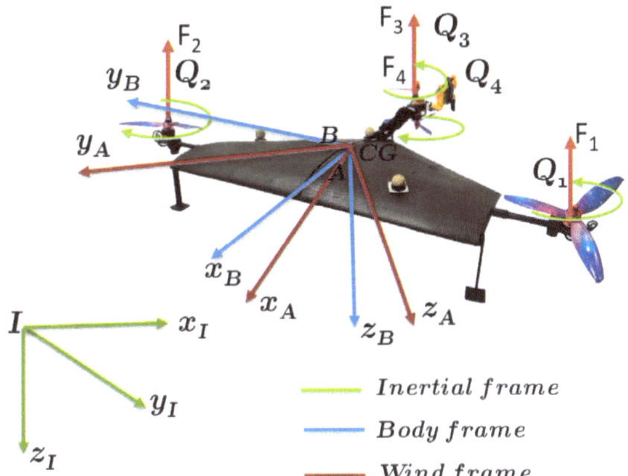

Figure 21. References frames of the CUAV.

For the tilting rotor system of proposed aircraft, an auxiliary tilting frame or rotating frame is defined about y_R, with x_{R_1}, z_{R_1}, x_{R_2}, and z_{R_2} as the principal axes, with tilting angles δ_{R_1} and δ_{R_2} as shown in Figure 22. The upwards position is $0°$, while the forward position is $-\frac{\pi}{2}°$ according to the NED (North–East–Down) convention and the right-hand rule.

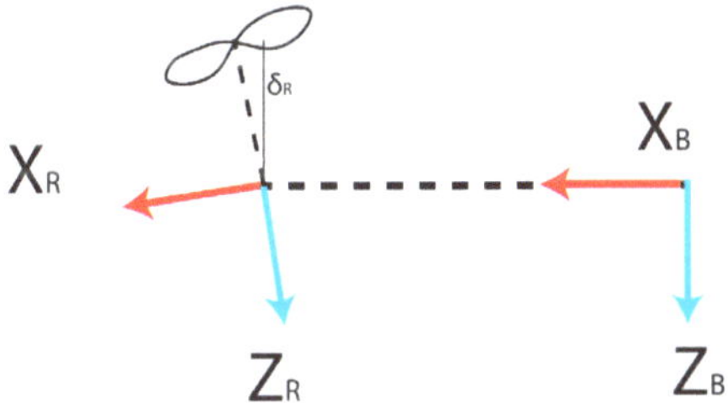

Figure 22. References frame for angular rotation.

3.1. Equations of Motion

The model of the vehicle considers an inertial fixed frame as $\mathcal{I} = \{x_{\mathcal{I}}, y_{\mathcal{I}}, z_{\mathcal{I}}\}$ and a body frame fixed attached to the center of gravity of the vehicle as $\mathcal{B} = \{x_B, y_B, z_B\}$. The wind frame $\mathcal{A} = \{x_A, y_A, z_A\}$ is considered during the forward flight [32] (see Figure 23). The configuration of the convertible UAV is defined by the location of the center of gravity and the attitude with respect to the inertial frame. Then, the configuration manifold is the special Euclidean group SE(3), which is the semidirect product of $\mathbb{R}^3$ and the special orthogonal group SO(3).

The Newton–Euler formulation, for a rigid body, is used in order to obtain the mathematical model as

$$\dot{\varsigma} = V \tag{1}$$

$$m\dot{V} = RF + mge_3 + D_{\varsigma}(t) \tag{2}$$

$$\dot{R} = R\hat{\Omega} \tag{3}$$

$$J\dot{\Omega} = -\Omega \times J\Omega + \tau_a + D_{\eta}(t) \tag{4}$$

where $\zeta = (x, y, z)^\top \in \mathbb{R}^3$ and $V = (v_x, v_y, v_z)^\top \in \mathbb{R}^3$ are the position coordinates and translational velocity relative to the inertial frame. $\eta = (\phi, \theta, \psi)^\top \in \mathbb{R}^3$ describes the rotation coordinates where ϕ, θ, and ψ represent the roll, pitch, and yaw or heading, respectively. e_1, e_2, and e_3 are the vectors of the canonical basis of $\mathbb{R}^3$ in $\mathcal{I}$. The rotation matrix, $R \in \mathrm{SO}(3) : \mathcal{B} \rightarrow \mathcal{I}$, satisfies the $\mathrm{SO}(3) = \left\{ R \mid R \in \mathbb{R}^{3 \times 3}, \det[R] = 1, RR^\top = R^\top R = I \right\}$ and is parameterized by the Euler angles ϕ, θ, and ψ. The rotation matrix is written as

$$R = \begin{pmatrix} c_\theta c_\psi & s_\phi s_\theta c_\psi - c_\phi s_\psi & c_\phi s_\theta c_\psi + s_\phi s_\psi \\ c_\theta s_\psi & s_\phi s_\theta s_\psi + c_\phi c_\psi & c_\phi s_\theta s_\psi - s_\phi c_\psi \\ -s_\theta & s_\phi c_\theta & c_\phi c_\theta \end{pmatrix}$$

where the shorthand notation of $s_a = \sin(a)$ and $c_a = \cos(a)$ is used. $\Omega = (p, q, r)^\top \in \mathbb{R}^3$ is the angular velocity in $\mathcal{B}$, where the heat map $\hat{\cdot} : \mathbb{R}^3 \rightarrow \mathfrak{so}(3)$ is defined by the condition that $\hat{a}b = a \times b$ for all $a, b \in \mathbb{R}^3$.

$$\hat{\Omega} = \begin{pmatrix} 0 & -r & q + \dot{\delta}_{R_1} \\ r & 0 & -p - \dot{\delta}_{R_2} \\ -q - \dot{\delta}_{R_1} & p + \dot{\delta}_{R_2} & 0 \end{pmatrix} \tag{5}$$

where δ_{R_1} and δ_{R_1} are the tilting angles.

The forces acting on the body frame are described as follows:

$$F = \begin{bmatrix} F_{x_B} \\ F_{y_B} \\ F_{z_B} \end{bmatrix} = \begin{bmatrix} 0 \\ 0 \\ -F_3 \end{bmatrix} + \begin{bmatrix} 0 \\ 0 \\ -F_4 \end{bmatrix} + \begin{bmatrix} -F_1 \sin(\delta_{R_1}) \\ 0 \\ -F_1 \cos(\delta_{R_1}) \end{bmatrix} + \begin{bmatrix} -F_2 \sin(\delta_{R_2}) \\ 0 \\ -F_2 \cos(\delta_{R_2}) \end{bmatrix} \tag{6}$$

where $F = (F_{x_B}, F_{y_B}, F_{z_B})^\top \in \mathbb{R}^3$ is the vector of the total forces in the x, y, and z axes, respectively. F_i is the lift force or thrust force of the propeller for $i = 1, 2, 3, 4$.

In hover mode ($\delta_{R_1} = 0$ and $\delta_{R_2} = 0$), (6) becomes

$$F = \begin{bmatrix} F_{x_B} \\ F_{y_B} \\ F_{z_B} \end{bmatrix} = \begin{pmatrix} 0 \\ 0 \\ -(F_1 + F_2 + F_3 + F_4) \end{pmatrix} \tag{7}$$

In cruise mode ($\delta_{R_1} \approx \frac{\pi}{2}$ and $\delta_{R_2} \approx \frac{\pi}{2}$), (6) becomes

$$F = \begin{bmatrix} F_{x_B} \\ F_{y_B} \\ F_{z_B} \end{bmatrix} = \begin{pmatrix} -(F_1 + F_2) \\ 0 \\ -(F_3 + F_4) \end{pmatrix} \tag{8}$$

For external forces, we include specially aerodynamics ones, defining them as

$$D_\zeta = \begin{pmatrix} d_{\zeta_1} \\ d_{\zeta_2} \\ d_{\zeta_3} \end{pmatrix} = RW^T \begin{pmatrix} D_a \\ Y_a \\ L_a \end{pmatrix} \tag{9}$$

where the rotation aerodynamic matrix $W : \mathcal{B} \rightarrow \mathcal{A}$ that transforms a force from the body frame to the aerodynamic frame is described as

$$W = \begin{pmatrix} c_\alpha c_\beta & s_\beta & s_\alpha c_\beta \\ -c_\alpha s_\beta & c_\beta & -s_\alpha s_\beta \\ -s_\alpha & 0 & c_\alpha \end{pmatrix}$$

where α is the angle of attack and β is the sideslip angle. L, Y, and D are the aerodynamic forces: lift, side force, and drag, respectively [32].

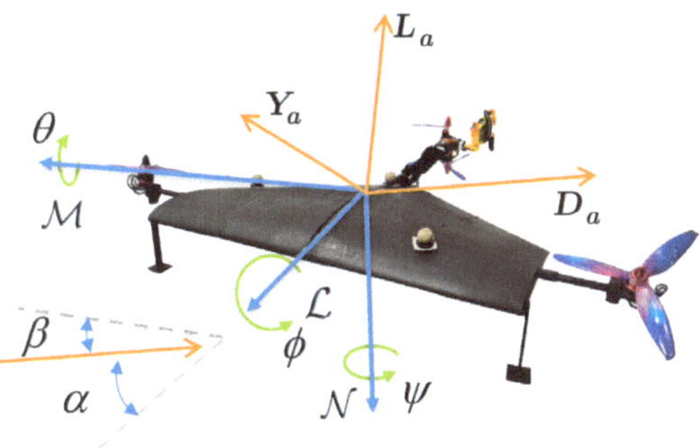

Figure 23. Aerodynamic forces of the CUAV.

In the context of torque analysis within the aircraft dynamics, the torque vector τ_a is defined at the center of gravity with a pivotal point corresponding to the body frame. This representation provides the rotational dynamics of the aircraft and is derived from the collective effects of the four-rotor forces. This torque is formulated as follows:

$$\tau_a = \begin{pmatrix} \tau_\phi \\ \tau_\theta \\ \tau_\psi \end{pmatrix} = \begin{pmatrix} d(F_1 - F_2) \\ l_2(F_3 + F_4) - l_1(F_1 + F_2) + \tau_{wing} \\ Q_1 + Q_3 - Q_2 - Q_4 + F_1\sin(\delta_{R_1}) - F_2\sin(\delta_{R_2}) \end{pmatrix} \tag{10}$$

where $\tau_{wing} = C_{m,wing}\frac{1}{2}\rho V^2 S_{wing} c_{wing}\Delta\alpha$ encompasses the contributions stemming from the frontal wing; here, $C_{m,wing}$ is the pitching moment coefficient of the wing, ρ is the air density, V is the airspeed of the aircraft, S_{wing} is the wing area, c_{wing} is the average chord length of the wing, and $\Delta\alpha$ is the change in angle of attack of the wing. $Q_i = \rho A_i r_i^3 c_{Q_i}\omega_i^2$, where A_i is the rotor disk area, r_i is the rotor radius, c_{Q_i} denotes the rotor shaft moment coefficient and ω_i denotes the angular velocity of the rotor i with $i = 1, 2, 3, 4$. d stands for arm length; l_1 and l_2 represent distances to the center of mass.

The moments acting on the aerial vehicle are described

$$D_\eta = \begin{pmatrix} d_{\eta 1} \\ d_{\eta 2} \\ d_{\eta 3} \end{pmatrix} = d_{\eta gyro} + d_{\eta aero} \tag{11}$$

The gyroscopic moment generated by the rotation of the airframe and the four rotors is described by

$$d_{\eta gyro} = \sum_{k=1}^{4}(-1)^{k+1}I_{r_k}[\Omega \times e_3\omega_k] \tag{12}$$

Finally, the aerodynamic moments presented on the airframe are described as

$$d_{\eta aero} = \begin{pmatrix} \mathcal{L} & \mathcal{M} & \mathcal{N} \end{pmatrix}^\top$$

where $\mathcal{L}$, $\mathcal{M}$ and $\mathcal{N}$ are the aerodynamic rolling, pitching and yawing moments, respectively [5,32].

Using Equations (1)–(4), a nonlinear set of equations can be described as

$$
\begin{aligned}
\dot{x} &= v_x \\
\dot{y} &= v_y \\
\dot{z} &= v_z \\
\dot{v}_x &= \frac{F_{x_B}}{m}\left(c_\theta c_\psi\right) + \frac{F_{y_B}}{m}\left(s_\phi s_\theta c_\psi - c_\phi s_\psi\right) + \frac{F_{z_B}}{m}\left(c_\phi s_\theta c_\psi + s_\phi s_\psi\right) + d_{\xi_1} \\
\dot{v}_y &= \frac{F_{x_B}}{m}\left(c_\theta s_\psi\right) + \frac{F_{y_B}}{m}\left(s_\phi s_\theta s_\psi + c_\phi c_\psi\right) + \frac{F_{z_B}}{m}\left(c_\phi s_\theta s_\psi - s_\phi c_\psi\right) + d_{\xi_2} \\
\dot{v}_z &= \frac{F_{x_B}}{m}\left(-s_\theta\right) + \frac{F_{y_B}}{m}\left(s_\phi c_\theta\right) + \frac{F_{z_B}}{m}\left(c_\phi c_\theta\right) + g + d_{\xi_3} \\
\dot{\phi} &= p + q\sin(\phi)\tan(\theta) + r\cos(\phi)\tan(\theta) \\
\dot{\theta} &= q\cos(\phi) - r\sin(\phi) \\
\dot{\psi} &= q\sec(\theta)\sin(\phi) + r\sec(\theta)\cos(\phi) \\
\dot{p} &= \left(\frac{J_{yy}-J_{zz}}{J_{xx}}\right)qr + \left(\frac{1}{J_{xx}}\right)\tau_\phi + d_{\eta 1} \\
\dot{q} &= \left(\frac{J_{zz}-J_{xx}}{J_{yy}}\right)pr + \left(\frac{1}{J_{yy}}\right)\tau_\theta + d_{\eta 2} \\
\dot{r} &= \left(\frac{J_{xx}-J_{yy}}{J_{zz}}\right)pq + \left(\frac{1}{J_{zz}}\right)\tau_\psi + d_{\eta 3}
\end{aligned}
\tag{13}
$$

Remark 1. *As the rotation of the four propellers on the convertible UAV is balanced, the gyroscopic moment will essentially be zero. The only cases in which gyroscopic moments will not be zero are if there is a significant difference in the RPM of the four motors and the presence of a strong sideways cross-wind.*

Remark 2. *The design of the convertible UAV is based on in a configuration that optimizes the aerodynamic properties and reduces drag forces, which provides steady flights. In addition, the wing involves a damping that reduces the transient or oscillatory motion, specifically unstable spiral roll.*

Based on the remarks, the disturbance terms D_η and D_ξ satisfy the linear growth bound as $\| D_\xi \| \leq c_\xi \ \forall t$ and $\| D_\eta \| \leq c_\eta \ \forall t$.

3.2. Guidance, Navigation and Control Algorithm

The guidance, navigation, and control of the convertible UAV is based on a geometric tracking control in SE(3) (special Euclidean group); see [33]. The control is a saturated proportional, integral, and derivative (PID) and provides smooth trajectory tracking based on SE(3) even in the presence of wind disturbances. For this purpose, the Equation (13) can be rewritten as

For this purpose, Equation (13) can be rewritten as

$$
\begin{aligned}
\dot{\zeta} &= V \tag{14} \\
\dot{V} &= u_n + d_\xi(t) \tag{15} \\
\dot{R} &= R\hat{\Omega} \tag{16} \\
\dot{\Omega} &= u_a + d_R(t) \tag{17}
\end{aligned}
$$

where $u_n \in \mathbb{R}^3$ and $u_a \in \mathbb{R}^3$ are virtual control inputs for the position and orientation dynamics. $d_\xi(t) = \frac{D_\xi(t)}{m}$ and $d_R(t) = J^{-1}[-\Omega \times J\Omega + D_\eta(t)]$.

For a smooth transition, a condition is defined in order to ensure that at each instant of time, at most one of the two control inputs is active. The geometric navigation considers a guidance frame that is designed to perform autonomous flights with a convergence to the contour of the task with small normal velocity.

$$
u_n = u_{n_1}g_1(t) + u_{n_2}f_1(t) \tag{18}
$$

$$
f_1(t) = \begin{cases} 0 & for \quad 0 \leq t < T_1 \\ 1 & for \quad T_1 \leq t \leq T_F \end{cases} \quad \text{with } g_1(t) = 1 - f_1(t) \tag{19}
$$

For hover flight, $0 \leq t < T_1$, the virtual control input u_{n_1} is defined as

$$u_{n_1} = ge_3 - \frac{RF}{m} \tag{20}$$

For cruise mode, $T_1 \leq t \leq T_F$, the virtual control input u_{n_2} is defined as

$$u_{n_2} = ge_3 - \frac{RF}{m} \tag{21}$$

For orientation dynamics, the virtual control input u_a is defined as

$$u_a = J^{-1}\tau_a \tag{22}$$

Remark 3. *The transition maneuver of the CUAV, from hover to cruise modes and vice versa, is smooth, and it starts when the vehicle reaches the hovering flight in the initial or actual waypoint, i.e., $F_{z_B} \approx mg$; after that, the transition starts, and the cruise mode is performed until the CUAV arrives to the final waypoint to return to the hovering flight.*

Definition 1. *A guidance frame $\mathcal{G} = \{f_g, b_g, n_g\}$ is a reference frame that consists of the control forward vector f_g, the control binormal vector b_g and the control normal vector n_g. This frame satisfies the NED (North–East–Down) system and considers the terminology from the names of the three unit vectors in the reference frame for a curve in $\mathbb{R}^3$.*

The three vectors are defined as follows (for more details, see [33]):

- The control normal vector n_g is defined as a function of the position and velocity errors.

$$n_g = \frac{ge_3 - u_n}{\|ge_3 - u_n\|} \tag{23}$$

- The control forward vector f_g is defined as a unit vector in the (n_g, t_d) plane and is orthogonal to n_g such as $n_g \cdot t_d > 0$ with $t_d = \frac{\dot{\xi}_d}{\|\dot{\xi}_d\|}$. Then

$$f_g = \frac{n_g \times e_1}{\|n_g \times e_1\|} \tag{24}$$

- The control binormal vector b_g is defined as

$$b_g = -(f_g \times n_g) \tag{25}$$

Definition 2. *A desired rotation matrix $R_d \in SO(3)$ is defined as $R_d = [f_g \ b_g \ n_g]$ corresponding to the reference frame or guidance frame where $f_g = R_d e_1$, $b_g = R_d e_2$ and $n_g = R_d e_3$.*

From [33], the next statements are well known.

1. $u_n \neq ge_3$;
2. n_g is a well-defined unit vector;
3. f_g is a well-defined unit vector;
4. $\{f_g \ b_g \ n_g\}$ is orthonormal and the matrix $R_d = [f_g \ b_g \ n_g]$.

Establishing a guidance frame enables the development of a control strategy that allows the introduction of u_n as an input while utilizing position references as feedback. In this case, a classical proportional, integral, and derivative (PID) control scheme is proposed. A navigation scheme using this type of control can be effective, providing accurate feedback and a sufficiently responsive system. For this purpose, the following PID saturated structure is utilized [34].

The position control for the CUAV is proposed as follows:

$$u_n = \mathrm{Sat}\left(k_{p_\zeta} e_\zeta + k_{d_\zeta} \dot{e}_\zeta + k_{i_\zeta} \int e_\zeta dt \right) \tag{26}$$

where $e_\zeta = \zeta_d - \zeta$, $\dot{e}_\zeta = \dot{\zeta}_d - \dot{\zeta}$ are the position and velocity errors. k_{p_ζ}, k_{d_ζ} and k_{i_ζ} stand for the diagonal and positive definite matrices. The stability analysis of this saturated control can be found in [35]

A similar procedure is used to propose the control to orientation dynamics, u_a, considering the rotation desired matrix $R = [f_g \ b_g \ n_g]$, which corresponds to the reference frame. Based on the group operation of SO(3), the attitude and the angular velocity errors are defined as $R_e = RR_d^\top$ and $e_\Omega = \Omega_d - \Omega$.

The orientation control is described as

$$u_n = \mathrm{Sat}\left(k_{p_\Omega} e_R + k_{d_\Omega} \dot{e}_\Omega + k_{i_\Omega} \int e_R dt \right) \tag{27}$$

where $e_R = \mathrm{Skew}(RR_d^\top)^\vee$, with $\mathrm{Skew}(A) = \frac{1}{2}(A - A^\top)$ and the operator $(\cdot)^\vee$ is the inverse of the "hat" operator. k_{p_Ω}, k_{d_Ω} and k_{i_Ω} stand for the diagonal and positive definite matrices [33].

A dual-layer controller architecture is utilized to enhance the navigation and attitude control of the convertible aircraft. This approach comprises two distinct layers. The first layer focuses on rate control, using velocity for translational motion and angular velocity for rotational motion. The second layer is dedicated to position control for translational motion and attitude control for rotational motion. By combining both layers, the system produces a final output that effectively guides the aircraft.

The dual-layer controllers follow the same structure previously mentioned but with a hierarchical arrangement. Instead of explicitly presenting the equations, the interaction between the two layers is emphasized for clarity. This arrangement ensures smooth integration and coordination between rate and position control, enabling precise and responsive navigation and attitude control for the convertible aircraft; see Figure 24.

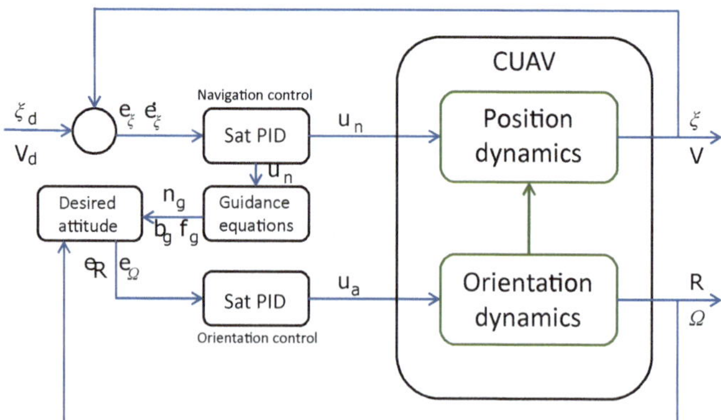

Figure 24. Guidance, navigation and control scheme for the CUAV.

4. Real-Time Validation

In order to validate the vehicle and the proposed GNC algorithm, a design of experiments (DoE) is executed in an indoor environment, performing the capabilities of the proposed system. The experiments focus on trajectory tracking tests, which facilitate the assessment of the system performance under consistent patterns and diverse movement combinations. However, specific missions are performed for convertible aircraft to evaluate the system across various scenarios.

The tests are carried out in the Navigation Laboratory at the Aerospace Engineering Research and Innovation Center of the Faculty of Mechanical and Electrical Engineering at the Autonomous University of Nuevo Leon. This laboratory features 16 VICON T-40 cameras to obtain the localization measurements; see Figure 25.

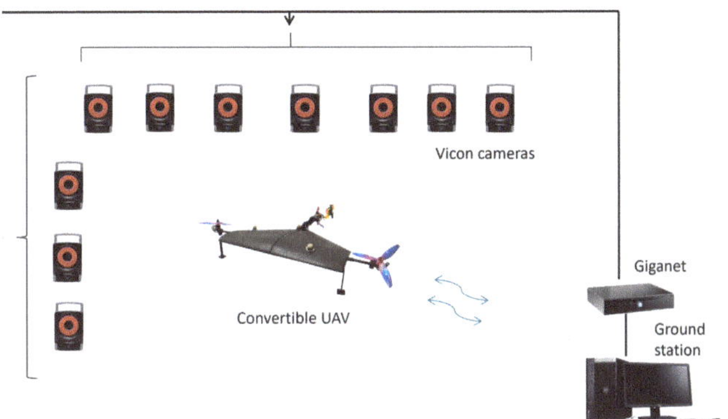

Figure 25. Tracking system.

The ground station receives and sends the information to the autopilot systems, and the interface is developed in order to graph the state variables of the system; see Figure 26.

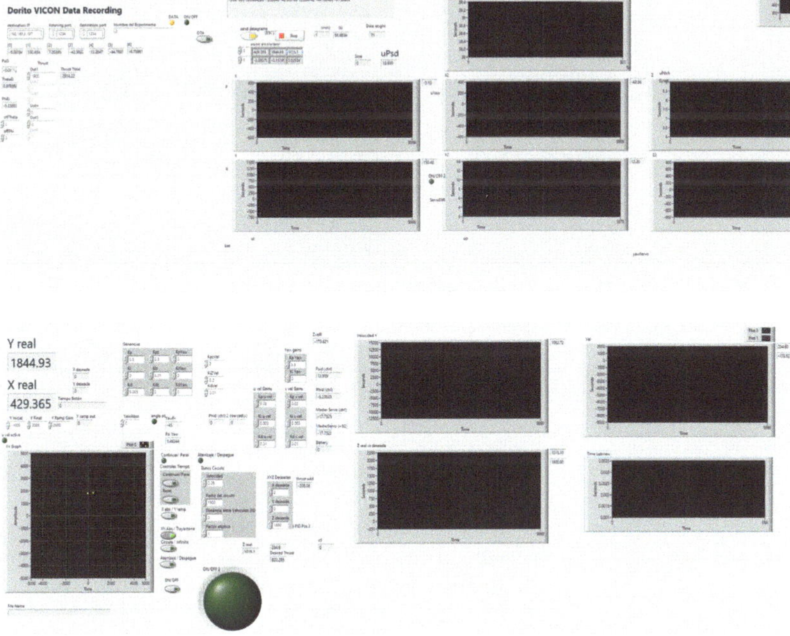

Figure 26. CUAV interface in the ground station.

The proposed CUAV is equipped with a low-cost avionics system developed by our laboratory, allowing us to access the whole state variables of the system. For the inner loop, the attitude is obtained via the estimation method, i.e., the complementary filter in SE(3), while for the outer loop, the position is obtained through estimation by the tracking system. The scheme of the aircraft is shown in Figure 27, providing a homemade autopilot in order to manipulate the complete systems for flights in real-time. For that reason, it is possible to debug the system on each of its steps, and Figure 28 illustrates the real-time experiments.

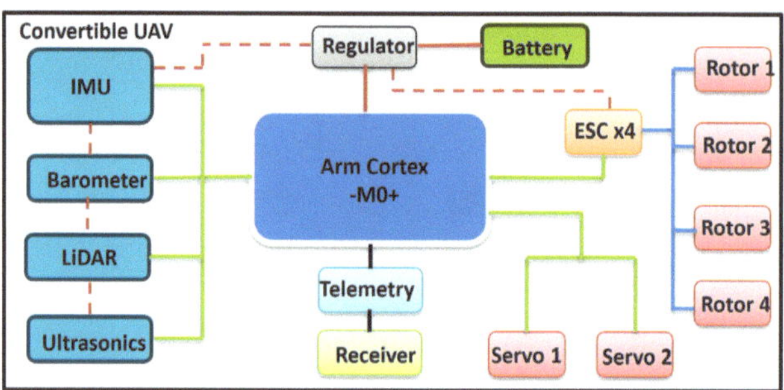

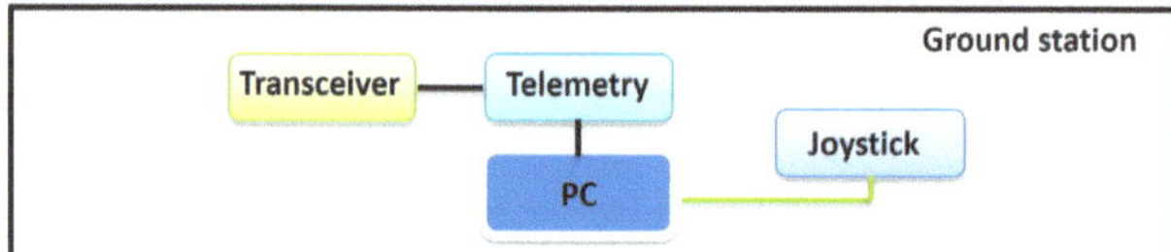

Figure 27. Flight computer scheme.

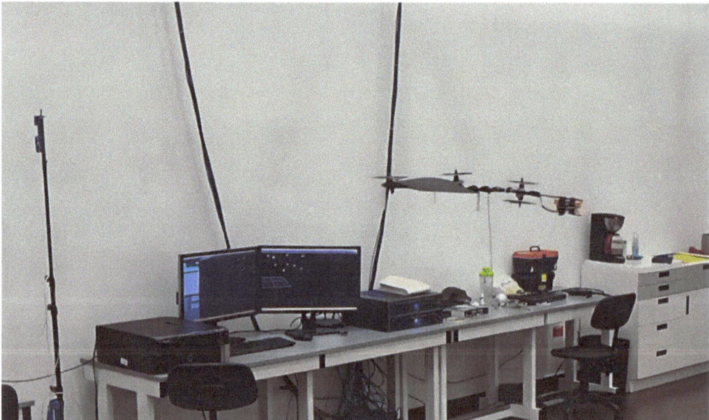

Figure 28. Real-time flight of the CUAV.

The experiments that are executed for this prototype enable the characterization of the system and the identification of unexpected behaviors. In this sense, specific paths and input ramp signals are selected for testing. Notably, these experiments are run in real-time, and the information about state variables is sent to the ground station. The camera system is essential for data tracking, requiring using a bounded environment for experimentation. Despite these constraints, the selected paths and input ramps provide valuable insights into the system performance and behavior under controlled conditions. The information is tracked during this experiment, including position data, sensor readings, and control inputs, which are recorded in a data file for analysis.

The experiments selected are those represented in Table 4.

Table 4. Design of experiments.

Experiment	Description
Circle	Circle pattern with tangent tracking, fixed Z, and multiple experiments development for dependence on velocity
Infinite	Complex pattern for whole system test; combined capabilities are tested
Tilting ramp	Tilting rotor test for control test, which required an input similar to a forward flight with a process of stopping at the end of the path
Fast line	Test for max linear velocity in controlled environment

4.1. Circle

For the circular trajectory, we selected a circle with a radius of 1750 mm, a height of 1800 mm, and a velocity of 0.03 rev/s. For this case, the tilting mechanism is assessed, which allows us to have that type of motion without using input on the rotors; those are only controlled by the mixer for stabilization purposes.

Geometrical errors, a natural consequence of the lower resolution on actuators, are a significant factor in our trajectory. These errors are visualized by a circle made of trajectories based online. Despite these challenges, the trajectory remains within correct values with some drift but an acceptable tracking error; see Figure 29.

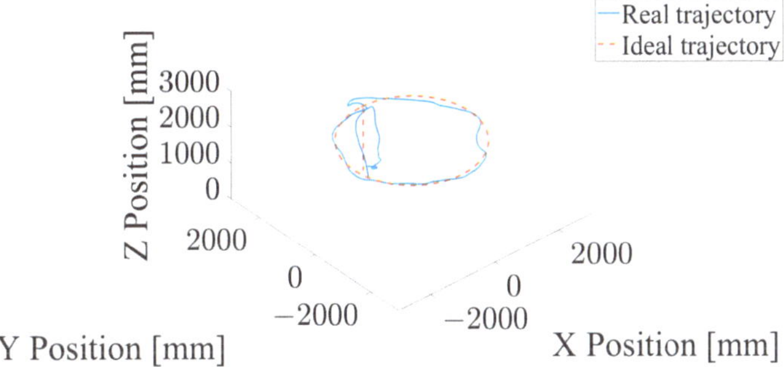

Figure 29. XYZ trajectory tracking, real vs. ideal.

One of the objectives of this design is to ensure that the pitch angle is as small as possible, especially so as not to interfere with the forces generated by the wing. As a result, it is shown that the system follows the system with 100 mm of maximum error and ±3 of angle error; see Figure 30.

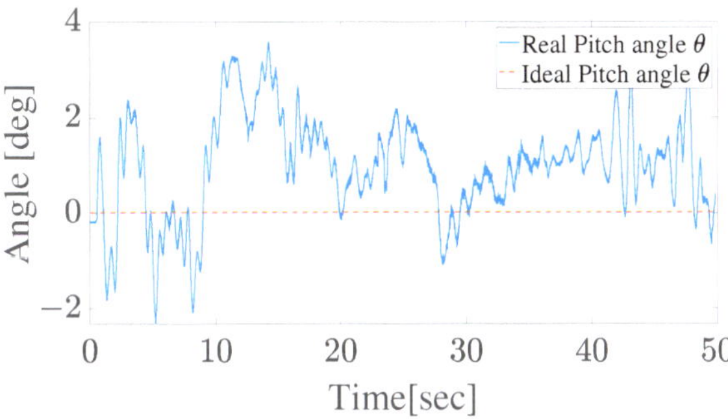

Figure 30. Pitch angle, real vs. ideal (low pitch).

4.2. Infinite

While the circular path experiment provides valuable insights into maintaining a constant radius, a separate experiment simulating an infinite straight path is equally important. This allows us to evaluate the system ability to handle long-distance navigation and drift correction, which is crucial for real-world applications like long-range surveillance. A complex trajectory is tested for an infinite trajectory, and a control signal response is illustrated; see Figure 31.

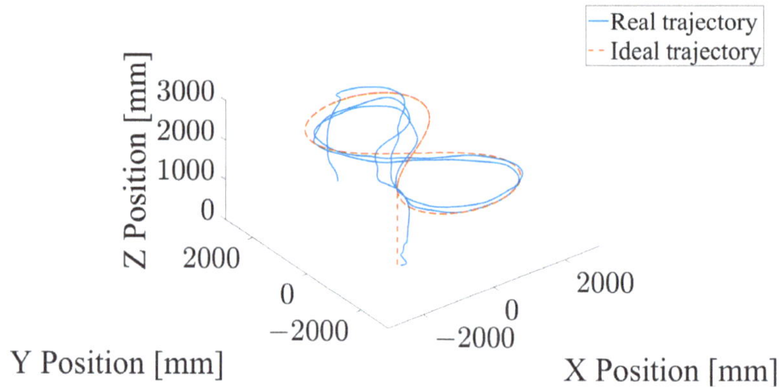

Figure 31. XYZ trajectory tracking, real vs. ideal.

As seen before, the CUAV accomplishes and follows this trajectory with a deviation of 200 mm; this behavior is expected because the system is performing a transition phase, which involves uncertainty to be solved for the proposed GNC algorithm. The form control field shows that the system is tracking signals; errors are expected from the external sensor data; see Figure 32.

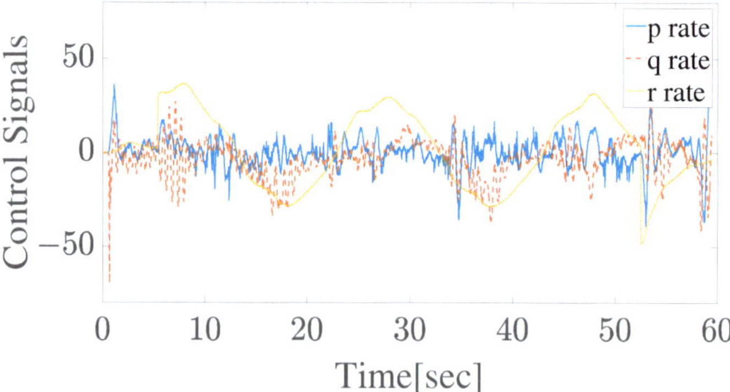

Figure 32. Control signals on infinite trajectories.

4.3. Tilting Ramp

This experiment shows the fast transition of the system on a tilting rotor, where the input is a fast change on tilting angles, as shown in Figure 33.

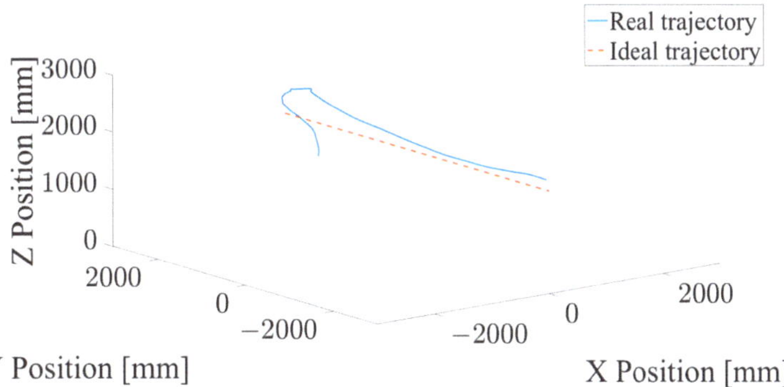

Figure 33. XYZ trajectory tracking, real vs. ideal.

For this case, a straight line is developed to change between hover and cruise flight. In this short-period experiment, the control scheme compensates for height loss. Also, it accelerates the CUAV, performing a maximum of 2 m/s, where it is realized that the system is still in hover mode, and rotors do not make the whole change. However, it is a short-period experiment showing the expected accomplishment of dynamics; see Figure 34.

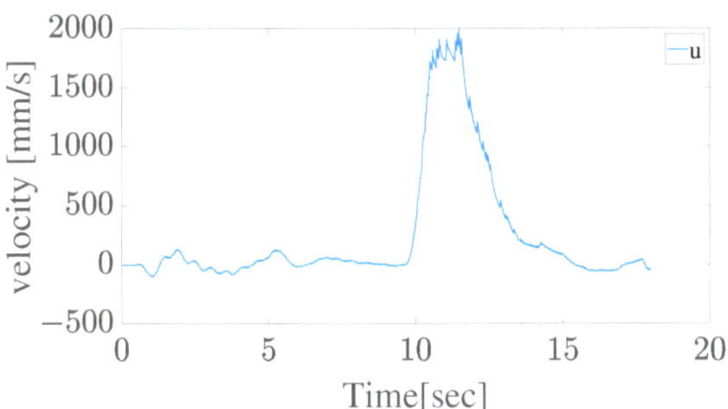

Figure 34. U velocity develop at ramp.

As seen in Figure 35, at 10 to 12 s, the system increases lift, which is directly seen as a decrease in the signal required for the system to maintain height.

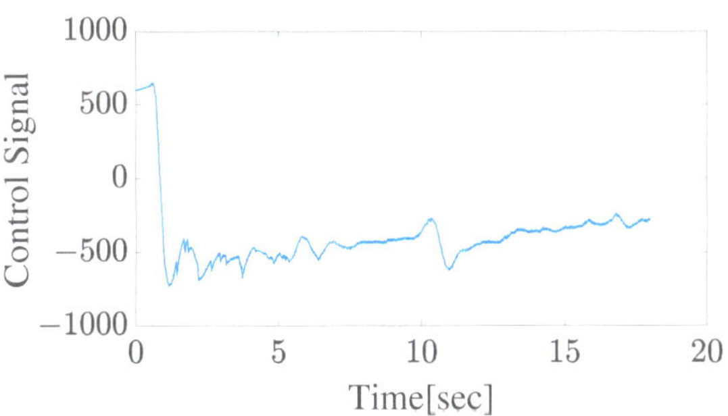

Figure 35. Lift behavior on global rotor forces.

4.4. Fast Line—Transition Mode

In this section, we present a development of an experiment where the aircraft is tested on maximum velocity conditions indoors. The line design is 5 m on the same axis, and the initial condition is used to execute a hover flight, followed by a fast line, and end with an instant decrease of velocity on the system; see Figure 36.

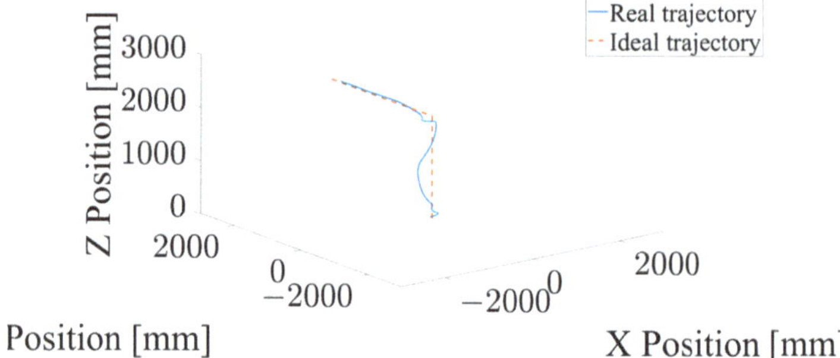

Figure 36. XYZ trajectory tracking, real vs. ideal.

The vehicle followed the trajectory even under demanding conditions. Some errors are expected, in this case, 300 mm on average; see Figure 37.

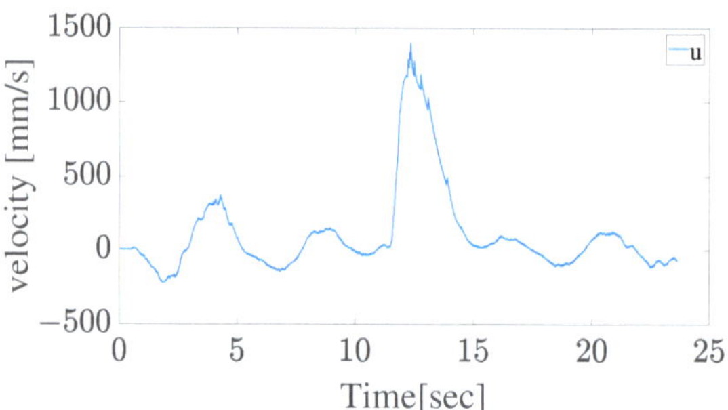

Figure 37. U velocity developed at the line.

Even the maximum condition on the system is required; just 3.7 m/s is reached due to space limitations. On a complete flight, the cruise phase is performed considering a change about 55°reached by the tilting rotors during the transition phase. In the transition phase, the vehicle reaches 90°on a complete transition, as can be seen in Figure 38.

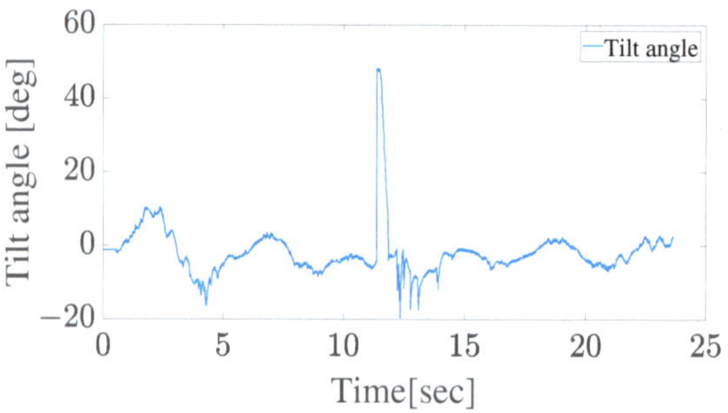

Figure 38. Servos transition phase.

In the following link, a test video is shown: https://youtu.be/h5RhDCh6QtQ?si=5VnO1xTrUZ-YuJhN accessed on 11 February 2024. It is important to mention that a circle experiment was developed in this video.

5. Conclusions

The design of a convertible UAV platform capable of executing hover and cruise flight missions was presented, having characteristics of a helicopter and an aircraft. The design was validated by taking into account the structural refinement and airfoil-based design methodology; each stage addressed critical aspects of design based on conceptual aerodynamics, mechanical properties, and material selection. Additive manufacturing was used to develop the proposed vehicle, considering the optimization techniques used to obtain a lightweight vehicle structure. The control strategy provided an effective performance for hover and cruise flights of the convertible UAV, and it was designed to ensure complete flight regimes. Notable achievements included reduced control authority reliance on rotors and effective lift generation by the main wing during cruise flights. Validation experiments, encompassing the convertible UAV approach, revealed promising results in trajectory tracking and efficient flight maneuvers.

Author Contributions: Conceptualization, E.U.R.-R.; methodology, E.U.R.-R.; software, E.U.R.-R. and E.G.R.-R.; validation, E.U.R.-R. and S.A.A.-E.; formal analysis, E.U.R.-R. and O.G.-S.; investigation, S.A.A.-E. and O.G.-S.; resources, O.G.-S.; data curation, E.G.R.-R. and S.A.A.-E.; writing—original draft preparation, E.U.R.-R.; writing—review and editing, E.U.R.-R. and S.A.A.-E.; visualization, E.G.R.-R. and S.A.A.-E.; supervision, O.G.-S.; project administration, O.G.-S.; funding acquisition, O.G.-S. All authors have read and agreed to the published version of the manuscript.

Funding: This research was supported by the Office of Naval Research Global through the grant number N62909-20-1-2030.

Institutional Review Board Statement: Not applicable.

Informed Consent Statement: Not applicable.

Data Availability Statement: The data presented in this study are available on request from the corresponding author.

Acknowledgments: The authors would like to thank the Aerospace Engineering Research and Innovation Center of the Faculty of Mechanical and Electrical Engineering at the Autonomous University of Nuevo Leon and the Department of Aeronautical and Astronautical Engineering at the University of Southampton for the facilities of this research.

Conflicts of Interest: The authors declare no conflicts of interest.

References

1. Mccormick, B.W. *Aerodynamics of V/STOL Flight*; Courier Corporation: North Chelmsford, MA, USA, 1999.
2. Semotiuk, L.; Józwik, J.; Kukiełka, K.; Dziedzic, K. Design and FEM Analysis of an Unmanned Aerial Vehicle Wing. In Proceedings of the 2021 IEEE 8th International Workshop on Metrology for AeroSpace (MetroAeroSpace), Naples, Italy, 23–25 June 2021; pp. 365–370.
3. Phung, D.K.; Morin, P. Modeling and Energy Evaluation of Small Convertible UAVs. *IFAC Proc. Vol.* **2013**, *46*, 212–219. [CrossRef]
4. Greenwood, W.W.; Lynch, J.P.; Zekkos, D. Applications of UAVs in Civil Infrastructure. *J. Infrastruct. Syst.* **2019**, *25*, 04019002. [CrossRef]
5. Lozano, R. *Unmanned Aerial Vehicles Embedded Control*, 1st ed.; John Wiley-ISTE Ltd.: Hoboken, NJ, USA, 2010.
6. Aláez, D.; Olaz, X.; Prieto, M.; Villadangos, J.; Astrain, J. VTOL UAV digital twin for take-off, hovering and landing in different wind conditions. *Simul. Model. Pract. Theory* **2023**, *123*, 102703. [CrossRef]
7. Lu, K.; Liu, C.; Li, C.; Chen, R. Flight Dynamics Modeling and Dynamic Stability Analysis of Tilt-Rotor Aircraft. *Int. J. Aerosp. Eng.* **2019**, *2019*, 5737212. [CrossRef]
8. Liu, N.; Wang, Y.; Zhao, J.; Cheng, X.; He, X.; Wu, K. A new inclining flight mode and oblique takeoff technique for a tiltrotor aircraft. *Aerosp. Sci. Technol.* **2023**, *139*, 108370. [CrossRef]
9. Bautista, J.A.; Osorio, A.; Lozano, R. Modeling and Analysis of a Tricopter/Flying-Wing Convertible UAV with Tilt-Rotors. In Proceedings of the International Conference on Unmanned Aircraft Systems, Miami, FL, USA, 13–16 June 2017.
10. Chamola, V.; Kotesh, P.; Agarwal, A.; Naren.; Gupta, N.; Guizani, M. A Comprehensive Review of Unmanned Aerial Vehicle Attacks and Neutralization Techniques. *Ad Hoc Netw.* **2021**, *111*, 102324. [CrossRef] [PubMed]
11. Ducard, G.J.; Allenspach, M. Review of designs and flight control techniques of hybrid and convertible VTOL UAVs. *Aerosp. Sci. Technol.* **2021**, *118*, 107035. [CrossRef]

12. Flores, G.; Escareno, J.; Lozano, R.; Salazar, S. Quad-Tilting Rotor Convertible MAV : Modeling and Real-time Hover Flight Control. *J. Intell. Robot. Syst.* **2012**, *65*, 457–471. [CrossRef]

13. de Oliveira, T.L.; Anglade, A.; Hamel, T.; Samson, C. Control of convertible UAV with vectorized thrust. *IFAC-PapersOnLine* **2023**, *56*, 9288–9293. [CrossRef]

14. Bronz, M.; Smeur, E.J.; de Marina, H.G.; Hattenberger, G. *Development of A Fixed-Wing mini UAV with Transitioning Flight Capability*; American Institute of Aeronautics and Astronautics: Reston, VA, USA, 2017.

15. Goh, G.; Agarwala, S.; Goh, G.; Dikshit, V.; Sing, S.; Yeong, W. Additive manufacturing in unmanned aerial vehicles (UAVs): Challenges and potential. *Aerosp. Sci. Technol.* **2017**, *63*, 140–151. [CrossRef]

16. Moon, S.K.; Tan, Y.E.; Hwang, J.; Yoon, Y.J. Application of 3D printing technology for designing light-weight unmanned aerial vehicle wing structures. *Int. J. Precis. Eng. Manuf.-Green Technol.* **2014**, *1*, 223–228. [CrossRef]

17. Yanguo, S.; Huanjin, W. Design of Flight Control System for a Small Unmanned Tilt Rotor Aircraft. *Chin. J. Aeronaut.* **2009**, *22*, 250–256. [CrossRef]

18. Hassanalian, M.; Salazar, R.; Abdelkefi, A. Conceptual design and optimization of a tilt-rotor micro air vehicle. *Chin. J. Aeronaut.* **2019**, *32*, 369–381. [CrossRef]

19. He, C.; Chen, G.; Sun, X.; Li, S.; Li, Y. Geometrically compatible integrated design method for conformal rotor and nacelle of distributed propulsion tilt-wing UAV. *Chin. J. Aeronaut.* **2023**, *36*, 229–245. [CrossRef]

20. Sanchez, A.; Escareño, J.; Garcia, O.; Lozano, R. Autonomous Hovering of a Noncyclic Tiltrotor UAV: Modeling, Control and Implementation. *IFAC Proc. Vol.* **2008**, *41*, 803–808. [CrossRef]

21. D'Amato, E.; Di Francesco, G.; Notaro, I.; Tartaglione, G.; Mattei, M. Nonlinear Dynamic Inversion and Neural Networks for a Tilt Tri-Rotor UAV. *IFAC-PapersOnLine* **2015**, *48*, 162–167. [CrossRef]

22. Campos, J.M.; Cardoso, D.N.; Raffo, G.V. Robust Adaptive Control with Reduced Conservatism for a Convertible UAV. *IFAC-PapersOnLine* **2023**, *56*, 4520–4526. [CrossRef]

23. Bauersfeld, L.; Spannagl, L.; Ducard, G.J.J.; Onder, C.H. MPC Flight Control for a Tilt-Rotor VTOL Aircraft. *IEEE Trans. Aerosp. Electron. Syst.* **2021**, *57*, 2395–2409. [CrossRef]

24. Wang, H.; Sun, W.; Zhao, C.; Zhang, S.; Han, J. Dynamic Modeling and Control for Tilt-Rotor UAV Based on 3D Flow Field Transient CFD. *Drones* **2022**, *6*, 338. [CrossRef]

25. Zhao, H.; Wang, B.; Shen, Y.; Zhang, Y.; Li, N.; Gao, Z. Development of Multimode Flight Transition Strategy for Tilt-Rotor VTOL UAVs. *Drones* **2023**, *7*, 580. [CrossRef]

26. Fayez, K.; Leng, Y.; Jardin, T.; Bronz, M.; Moschetta, J.M. *Conceptual Design for Long-Endurance Convertible Unmanned Aerial System*; American Institute of Aeronautics and Astronautics: Reston, VA, USA, 2021.

27. Carlson, S. A Hybrid Tricopter/Flying-Wing VTOL UAV. In Proceedings of the 52nd Aerospace Sciences Meeting, National Harbor, MD, USA, 13–17 January 2014.

28. Prach, A.; Kayacan, E. An MPC-based position controller for a tilt-rotor tricopter VTOL UAV. *Optim. Control. Appl. Methods* **2018**, *39*, 343–356. [CrossRef]

29. Valavanis, K.; Vachtsevanos, G. *Handbook of Unmanned Aerial Vehicles*, 1st ed.; Springer: Atlanta, GA, USA, 2015.

30. Rozvany, G. The SIMP method in topology optimization—Theoretical background, advantages and new applications. In Proceedings of the 8th Symposium on Multidisciplinary Analysis and Optimization, Long Beach, CA, USA, 6–8 September 2000.

31. Telli, K.; Kraa, O.; Himeur, Y.; Ouamane, A.; Boumehraz, M.; Atalla, S.; Mansoor, W. A Comprehensive Review of Recent Research Trends on Unmanned Aerial Vehicles (UAVs). *Systems* **2023**, *11*, 400. [CrossRef]

32. Stengel, R. *Flight Dynamics*, 1st ed.; Princeton University Press: Princeton, NJ, USA, 2004.

33. Garcia Salazar, O.; Rojo Rodriguez, E.G.; Sanchez-Orta, A.; Saucedo Gonzalez, D.; Muñoz-Vázquez, A. Robust Geometric Navigation of a Quadrotor UAV on SE(3). *Robotica* **2019**, *38*, 1019–1040. [CrossRef]

34. Ollervides-Vazquez, E.J.; Rojo-Rodriguez, E.G.; Rojo-Rodriguez, E.U.; Cabriales-Ramirez, L.E.; Garcia-Salazar, O. Two-layer saturated PID controller for the trajectory tracking of a quadrotor UAV. In Proceedings of the 2020 International Conference on Mechatronics, Electronics and Automotive Engineering (ICMEAE), Cuernavaca, Mexico, 16–21 November, 2020; pp. 85–91.

35. Garcia, O.; Sanchez, A.; Escareño, J.; Lozano, R. Tail-sitter UAV having one tilting rotor: Modeling, Control and Real-Time Experiments. *IFAC Proc. Vol.* **2008**, *41*, 809–814. [CrossRef]

Article

Variations in Finite-Time Multi-Surface Sliding Mode Control for Multirotor Unmanned Aerial Vehicle Payload Delivery with Pendulum Swinging Effects

Clevon Peris, Michael Norton and Sui Yang Khoo *

School of Engineering, Deakin University, Geelong, VIC 3216, Australia; ccperis@deakin.edu.au (C.P.); michael.norton@deakin.edu.au (M.N.)
* Correspondence: sui.khoo@deakin.edu.au

Abstract: Multi-surface sliding mode control addresses the limitations of traditional sliding mode control by employing multiple sliding surfaces to handle uncertainties, disturbances, and nonlinearities. The design process involves developing sliding surfaces, designing switching logic, and deriving control laws for each surface. In this paper, first, a robust finite-time multi-surface sliding mode controller will be presented and its performance analyzed by applying it to a multirotor subjected to a suspended payload, modeled in the form of a single pendulum, itself defined as a spatial (3D) dynamic model. Next, an adaptive finite-time multi-surface sliding mode controller will be derived—adding a variable adaptive parameter to the existing sliding surfaces of the robust finite-time control—and applied to the same system. It will be shown that the adaptive controller, with an adaptive parameter that adjusts itself based on the present value of the multi-surface sliding mode parameter, creates an improved fast finite-time convergence by obtaining an optimal settling time and minimizing undershoot of the multirotor state vector. Empirical verification of the effectiveness of the adaptive control will be carried out by presenting the control performances against a step response. It is also shown that the control may be utilized to approximate external disturbances—represented by the pendulum—and that with the application of control, the vehicle's motion may be stabilized and the payload swing suppressed. Lyapunov stability theory-based stability proofs for the controllers' designs are developed, showing the asymptotic stability of the output and uniform boundedness of the errors in the system dynamics. It is verified that the multi-surface sliding mode control can account for system uncertainties—both matched and mismatched—in addition to changes in internal dynamics and disturbances to the system, where the single pendulum payload is representative of the changes in dynamics that may occur to the system. Numerical simulations and characteristics are presented to validate the performance of the controllers.

Keywords: aerial robotics; unmanned aerial vehicles; multi-surface sliding mode control; adaptive control; control systems

Citation: Peris, C.; Norton, M.; Khoo, S.Y. Variations in Finite-Time Multi-Surface Sliding Mode Control for Multirotor Unmanned Aerial Vehicle Payload Delivery with Pendulum Swinging Effects. *Machines* **2023**, *11*, 899. https://doi.org/10.3390/machines11090899

Academic Editors: Octavio Garcia-Salazar, Anand Sanchez-Orta and Aldo Jonathan Muñoz-Vazquez

Received: 12 July 2023
Revised: 7 September 2023
Accepted: 8 September 2023
Published: 10 September 2023

1. Introduction

Multirotor systems in the present day have drawn major attention owing to their versatility in carrying out various applications [1]. Their ability to transport payloads from point to point opens their usage in many different fields, including agriculture, military, postage, and media. However, carriage of suspended payloads poses a challenge to multirotor operation as the system's dynamics are altered by the extra mass and swinging motion of the payload. The addition of a suspended payload adds an extra 2 Degrees of Freedom (DoF) to the existing 6 DoF multirotor UAV system. A multirotor-slung load system has different characteristics to that of a conventional multirotor, with the addition of a mass moment of inertia and the dynamics of the payload it carries. To this degree, the effects of payload mass, size, and distribution on the system's stability and control have previously been analyzed by researchers. Additional stability and control issues are introduced by the payload's swinging motion as they tend to cause disruptions and instability. Keeping the payload stable and maneuverable while maintaining the UAV's location and orientation poses a challenge in terms of control.

There have been several control strategies proposed to tackle this interesting problem. In Ref. [2], an Adaptive Linear Quadratic Regulator with input shaping is proposed, with an output weighing variation, where the work seeks to track desired trajectory of a quadcopter together with reducing the swing of the payload it carries. Hanafy et al. create a new genetic-based Fuzzy Logic Controller to reduce the oscillations caused by a quadrotor-suspended load in Ref. [3]. They use a genetic algorithm, attempting to optimize membership function distributions of the control inputs and outputs. In Ref. [4], a cable-suspended weight is tracked using the Lyapunov theory aided by backstepping methods to obtain a specific trajectory described using a time-parameterized position vector. The dynamics of both the quadrotor-slung load and cable direction, in addition to the coupling that they possess with the quadcopter, are taken as one single part of the nonlinear backstepping controller design. By developing a fractional order sliding mode control, El Ferik et al. [5] intend to propose a new method of controlling a quadrotor-slung load system, in addition to the swing angle of an imposed suspended load. A robust tracking control problem dealing with a quadrotor subject to system nonlinearities, coupling of inputs, aerodynamic irregularities, and external factors, such as wind, is analyzed through quantitative simulation studies in Ref. [6], focusing on position and attitude control. With tracking control and disturbance elimination of multirotor-slung load systems, Liu et al. [7] examine an adaptive hierarchical sliding mode controller, employing a sliding mode disturbance observer operating such that it converges in a fixed time interval. To address the issue of underactuated features, the projection operator is used to build the adaptive law, and the hierarchical sliding mode control concept is employed to establish the controller. In Ref. [8], a quadrotor subjected to disturbances is controlled by an integral non-singular hyperplane SMC scheme, which also assures precise tracking when these disturbances are present. With the consideration that wind gusts make the tracking operation of a drone more difficult, in order to complete tracking tasks when wind gusts emerge, a Proportional-Integral-Derivative (PID) control, which can tune itself to adjust to these changes, is derived in Ref. [9]. A supervisory control created with the Lyapunov method is introduced to display the stability of this approach when taking bounded disturbances into account. Cui et al. propose a Linear Active Disturbance Rejection Control, coupled with an input shaper, to create an attitude controller in Ref. [10]. In Ref. [11], the robust H control methodology known as linear matrix inequality (LMI), which is widely used, is chosen for the linear controller design. In addition, an observer is introduced for estimating load angles. Previous control strategies include Linear Quadratic Regulator (LQR) [12], fault tolerant flight control [13], geometric control [14], nonlinear model predictive control [15], and neural networks [16].

In recent years, multi-surface sliding mode control (MSSC) has become a promising control method. Aiming to improve robustness and tracking precision in dynamic systems, it tackles the constraints of standard sliding mode control (SMC) by introducing numerous sliding surfaces. In traditional sliding mode control, the system's states are guided down a desired trajectory by a single sliding surface. It can be subject to uncertainties, disturbances, and nonlinearities, which can result in performance degradation even though it can be beneficial in some situations. By separating the control space into numerous sliding surfaces, MSSC overcomes these difficulties. Various applications of MSSC, such as those in robotics, aerospace, automotive, and power systems, have been studied by researchers. For example, in robotics, where precise trajectory tracking is essential, MSSC has been used to operate manipulator arms and mobile robots. To manage uncertainties related to atmospheric conditions and outside disturbances, it has also been used in flight systems. MSSC offers improved robustness to uncertainty, where the control system can adjust to various operating situations by using several sliding surfaces, effectively reducing the effect of uncertainties on system performance. Additionally, it provides better tracking precision, as multiple sliding surfaces enable more accurate control and improved tracking of desired trajectories by allowing the control action to be tuned to areas of the control space. It has been used across different fields as an effective control method. A class of single-input, single-output (SISO) systems was the primary target for the development of multi-surface sliding control (MSSC). In this study, the MSSC is extended from SISO systems, such that it addresses a specific kind of Multi-Input Multi-Output (MIMO) system. In addition, the underactuated systems, which are analyzed in this work, may be altered into high-order MIMO nonlinear systems with mismatched uncertainties by changing the co-ordinate, as the MSSC is resilient against unknown disturbances. Then, this UAV control problem can be directly solved using the MSSC algorithm. Additionally, MSSC designs for UAV systems with suspended payloads are still in the early stages. Many control methods for such systems are created by decoupling the longitudinal and lateral dynamics separately due to the coupling effect of UAV dynamics [17]. Our approach provides a comprehensive control methodology by cascading the translational and rotational systems in addition to addressing the underactuated concerns. A reliable and desired control methodology for SISO systems has been demonstrated using MSSC [18] and high-order sliding mode control [19] for the regulation of underactuated nonlinear systems subjected to mismatched uncertainties, as in Refs. [20,21]. Numerous desirable characteristics of adaptive MSSC, including its resistance to both matched and mismatched uncertainties of several types, such as parametric variations and external disturbances, as well as the system's convergence to zero error, are useful for flight dynamic control. With this being considered, MSSC has been used in different applications as an effective control method. Ref. [22] addresses the issue of chained stabilization of a class of nonholonomic systems with known and unknown disturbances. The suggested design strategy is based on an adaptive MSSC applied to a discontinuous transformation of the perturbed nonholonomic system. In Ref. [23], a Takagi–Sugeno fuzzy model (TSFM)-based multi-objective resilient control scheme is proposed for a set of indeterminate nonlinear systems, which is then applied to a two-link robot system containing both matched and mismatched uncertainties. Hamood et al. track and control the height and angles of a 3-Degree-of-Freedom helicopter using a control strategy derived from the MSSC. In Ref. [24], Alqumsan et al. analyze the operation of continuum robots while taking mismatched uncertainties into account and propose a control strategy. Cosserat rod theory is used to construct the dynamic model of the system to develop the controller, and multi-surface sliding mode control is used to ensure robustness of the system against mismatched uncertainties. In Ref. [25], a multiple-surface sliding mode control based on disturbance observers is developed. It estimates the uncertainties as well as the derivatives of the virtual inputs.

Finite-time control is essential for control systems in terms of quick response to real-world situations, optimization of transient response, reduction in energy consumption, and efficacy of safety and performance. Previous finite-time control approaches have been proposed, such as a super-twisting algorithm [26], a nonlinear cascade structure [27], geometric control [28], and trajectory tracking control [29]. Taking these factors into consideration, this paper will propose both a robust and an adaptive finite-time multi-surface sliding mode control. The main contribution of this paper lies in the derivation and subsequent application of a finite-time multi-surface sliding mode control to a multirotor-slung load system. Unlike most existing studies, which consider planar (2D) dynamics for a pendulum, the slung load is modeled using the spatial (3D) dynamics of a single pendulum. While the dynamic model of the single pendulum slung load itself is defined, its effect upon the behavior of multirotor UAVs is unknown and hence, can be used to represent other such disturbances to the normal behavior of the UAV. These disturbances may arise due to external factors, such as wind and obstacles, or internal factors such as parametric changes occurring due to additions to the drone's structure, including the addition of a suspended payload itself.

The rest of this paper is outlined as follows. First, the dynamics of a multirotor will be derived, together with those of a single pendulum, representing a slung load to be transported by the multirotor. The dynamics of these two systems will cascade to form a single system (Sections 2.1 and 2.2). A multi-surface sliding mode control strategy will then be proposed and its stability verified in Section 2.3. An adaptive version of the same controller will then be derived in Section 2.4. In Section 3, both these controllers will individually be applied to the multirotor-slung load system for the stabilization of the multirotor on its travel path, along with the suppression of the slung load oscillations that take place. Their performance in different scenarios and with imposition of external disturbances will be presented. An analysis of the results obtained from these simulations will be analyzed in Section 4. Section 5 provides some conclusions.

2. Methodology

In this section, we define the dynamics of a multirotor and its suspended payload, which are then cascaded together to form a single system. This is the system upon which the proposed controllers will be applied and subsequently compared. The characteristics of the multirotor slung load system are as shown in Figure 1. Let $q_{trans} = [x \; y \; z \; \theta_x \; \theta_y]$ and $q_{rot} = [\varphi \; \theta \; \psi]$ denote the translational and rotational state vectors of the system, such that $q = [q_{trans} \; q_{rot}]$ represents the overall system state vector, with x, y, and z denoting the multirotor position along their respective co-ordinate axes, φ, θ, and ψ representing the roll, pitch, and yaw about each respective co-ordinate axis, and θ_x and θ_y being the payload tilt angle about the *x*-axis and *y*-axis, respectively. We take the mass of the multirotor to be M and the payload it is carrying to have a mass m.

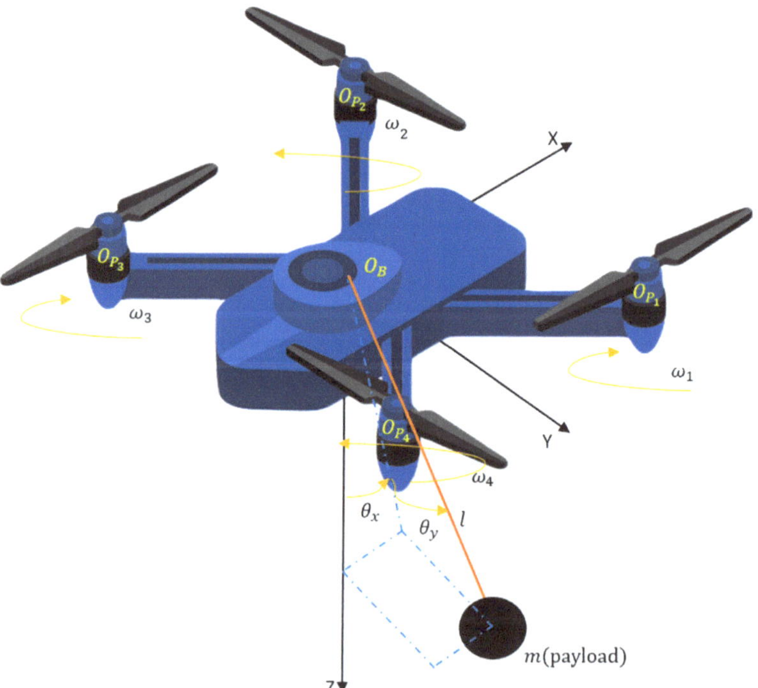

Figure 1. Characteristics of the Quadrotor with a suspended Single Pendulum.

2.1. Multirotor Dynamics

We consider a multirotor, specifically a quadrotor with 4 rotors, with the dynamics defined in our previous work [30]. R_1^2 represents the rotation of a frame 2 into a frame 1.

$$
\begin{aligned}
R_X(\theta) &= \begin{bmatrix} 1 & 0 & 0 \\ 0 & C(\theta) & -S(\theta) \\ 0 & S(\theta) & C(\theta) \end{bmatrix} \\
R_y(\theta) &= \begin{bmatrix} C(\theta) & 0 & S(\theta) \\ 0 & 1 & 0 \\ -S(\theta) & 0 & C(\theta) \end{bmatrix}. \\
R_z(\theta) &= \begin{bmatrix} C(\theta) & -St(\theta) & 0 \\ S(\theta) & C(\theta) & 0 \\ 0 & 0 & 1 \end{bmatrix}
\end{aligned}
\tag{1}
$$

These are the standard rotational matrices of the frame in terms of its roll, pitch, and yaw, where shorthand notation is used to complete the rotational matrix, such that $C(\theta) = \cos(\theta)$ and $S(\theta) = \sin(\theta)$.

$$
\begin{aligned}
R_{Pi}^b &= R_Z\left(\tfrac{(i-1)}{2}\pi\right) R_X(A_i) R_Y(B_i) \\
&= \begin{bmatrix} R_{P_{i11}}^b & R_{P_{i12}}^b & R_{P_{i13}}^b \\ R_{P_{i21}}^b & R_{P_{i22}}^b & R_{P_{i23}}^b \\ R_{P_{i31}}^b & R_{P_{i32}}^b & R_{P_{i33}}^b \end{bmatrix}
\end{aligned}
\tag{2}
$$

where

$$
\begin{aligned}
R_{P_{i11}}^b &= C_{\left(\frac{(i-1)}{2}\right)\pi} C_{B_i} - S_{\left(\frac{(i-1)}{2}\right)\pi} S_{A_i} S_{B_i}, \\
R_{P_{i12}}^B &= -S_{\left(\frac{(i-1)}{2}\right)\pi} C_{A_i}, \\
R_{P_{i13}}^b &= C_{\left(\frac{(i-1)}{2}\right)\pi} S_{B_i} + S_{\frac{(i-1)}{2}\pi} S_{A_i} C_{B_i}, \\
R_{P_{i21}}^b &= S_{\left(\frac{(i-1)}{2}\right)\pi} C_{B_i} + C_{\frac{(i-1)}{2}\pi} S_{A_i} S_{B_i}, \\
R_{P_{i22}}^B &= C_{\left(\frac{(i-1)}{2}\right)\pi} C_{A_i}, \\
R_{P_{i23}}^b &= S_{\left(\frac{(i-1)}{2}\right)\pi} S_{B_i} - C_{\frac{(i-1)}{2}\pi} S_{A_i} C_{B_i} \\
R_{P_{i31}}^b &= -C_{\left(\frac{(i-1)}{2}\right)\pi} S_{B_i}, \\
R_{P_{i32}}^B &= S_{A_i}, \\
R_{P_{i33}}^B &= C_{A_i} C_{B_i}, \\
O_{P_i}^b &= R_Z\left(\tfrac{i-1}{2}\right)\pi \begin{bmatrix} L \\ 0 \\ 0 \end{bmatrix} \\
&= \begin{bmatrix} C_{\left(\frac{(i-1)}{2}\pi\right)}L \\ S_{\left(\frac{(i-1)}{2}\pi\right)}L \\ 0 \end{bmatrix}
\end{aligned}
\tag{3}
$$

where $i = 1\dots, n$ (for this quadrotor, $n = 4$), and L is the multirotor arm's length, measured from the body frame center (O_B) to the center of a propeller ($O_{P_i}^b$). The torque acting on any propeller i, denoted as τ_{P_i}, is first obtained using Euler's angular momentum theory:

$$
\tau_{P_i} = I_{P_i}\dot{\omega}_{P_i} + \omega_{Pi} \times I_{Pi}\omega_{Pi} + \begin{bmatrix} 0 & 0 & k_c\omega_{P_{iZ}}\left|\omega_{P_{iZ}}\right| \end{bmatrix}^\top
\tag{4}
$$

where I_{Pi} represents each propeller's inertia matrix, ω_{Piz} denotes vector ω_{P_i}'s z-axis component, $k_c > 0$, the modulus of elasticity between ω_{Piz} and the counter-rotating torque about Z_{Pi} axis,

$$\omega_{Pi} = R_{P_i}^{B-1}\omega_B + \begin{bmatrix} \dot{\alpha_i} \\ \dot{\beta_i} \\ \hat{\omega_i} \end{bmatrix},\tag{5}$$

and $\hat{\omega_i}$ the angular velocity of the ith propeller. If τ_{Pi} is the ith propeller's produced thrust, we obtain

$$T_{Pi} = \begin{bmatrix} 0 & 0 & k_f\hat{\omega}|\hat{\omega}| \end{bmatrix}^T \tag{6}$$

where $k_f > 0$ is a fixed proportionality constant. Applying the fundamental theorem of mechanics for the body frame, together with Euler's angular momentum theory, we obtain

$$S_1 : m\begin{bmatrix} \ddot{X_b} \\ \ddot{Y_b} \\ \ddot{Z_b} \end{bmatrix} = R_W^B\begin{bmatrix} 0 \\ 0 \\ -Mg \end{bmatrix} + \sum_{i=1}^{4} R_{P_i}^B T_{Pi} + F_D \tag{7}$$

$$S_2 : I_b\dot{\omega}_b = \sum_{i=1}^{4}\left(O_{P_i}^B \times R_{P_i}^B T_{Pi} - R_{P_i}^B\tau_{Pi}\right) - \omega_b \times I_b\omega_b \tag{8}$$

where I_b is the inertial matrix governing the multirotor body, and F_D represents force due to drag.

Considering an n-propeller multi-copter, where $i = 1,\ldots,n$ and $n \geq 2$ and using the above equations, we acquire

$$S_3 : m\begin{bmatrix} \ddot{X_b} \\ \ddot{Y_b} \\ \ddot{Z_b} \end{bmatrix} = R_W^B\begin{bmatrix} 0 \\ 0 \\ -Mg \end{bmatrix} + \sum_{i=1}^{n} R_{P_i}^B T_{Pi} + F_D, \tag{9}$$

$$S_4 : I_b\dot{\omega}_b = \sum_{i=1}^{4}\left(O_{Pi}^b \times R_{P_i}^B T_{Pi} - R_{P_i}^B\tau_{Pi}\right) - \omega_b \times I_b\omega_b \tag{10}$$

2.2. Slung Load Dynamics

In this section, the slung load dynamic model is derived using Lagrange's modeling approach, in which we take the suspended payload to be a simple pendulum. When the payload is considered a point mass (m), the kinetic energy is only translated. The spatial angle of the cable can be used to determine the precise location of the payload by ignoring cable hoisting and keeping the cable length fixed at l. The pendulum displacements are characterized by θ_x and θ_y, which denote the rotation of the payload about the inertial x- and y-axes, respectively. A linear damping term is used to mimic the physical dampening of the pendulum swing. Using Kane's equation, the dynamical model is given by

$$M(q)\ddot{q} + C(q)\dot{q} + G(q) = JU \tag{11}$$

where $M(q)$ represents the mass–inertia matrix, $C(q)$ denotes the Coriolis force matrix, $G(q)$ represents a vector containing the forces of gravity acting upon the system, and $J \in R^{9\times4}$ is a Jacobian matrix for the system input vector $u \in R^4$. The payload orientation and position along the co-ordinate axes are first defined, aiding in deriving the Lagrangian model of the system. For this, we take the payload co-ordinates to be

$$X_1 = L_1\begin{bmatrix} \sin\theta_x\cos\theta_y \\ \sin\theta_x\sin\theta_y \\ \cos\theta_x \end{bmatrix} \tag{12}$$

Differentiating (12) with respect to time and accounting for angular velocity, we obtain

$$X_2 = L_1 \begin{bmatrix} \cos\theta_x\cos\theta_y\dot\theta_x - \sin\theta_x\sin\theta_y\dot\theta_y \\ \cos\theta_x\sin\theta_y\dot\theta_x - \sin\theta_x\cos\theta_y\dot\theta_y \\ -\sin\theta_x\dot\theta_x \end{bmatrix} \tag{13}$$

which gives us the velocity vectors of the hook.

Differentiating (13) and accounting for the angular acceleration, we have

$$X_3 = L_1 \begin{bmatrix} \sin\theta_x\cos\theta_y\ddot\theta_x^2 - \sin\theta_x^2\sin\theta_y\ddot\theta_y - 2\sin\theta_x\cos\theta_y\dot\theta_x^2 - 2\sin\theta_x\cos\theta_y\dot\theta_y^2 \\ \sin\theta_x\sin\theta_y\ddot\theta_x^2 - \sin\theta_x^2\cos\theta_y\ddot\theta_y - 2\sin\theta_x\sin\theta_y\dot\theta_x^2 - 2\sin\theta_x\sin\theta_y\dot\theta_y^2 \\ \cos\theta_x\ddot\theta_x - 2\cos\theta_x\dot\theta_x^2 \end{bmatrix} \tag{14}$$

Now, according to Lagrange's modeling equation,

$$\frac{d}{dt}\left(\frac{\partial L_p}{\partial \dot q_i}\right) - \left(\frac{\partial L_p}{\partial q_i}\right) = 0 \tag{15}$$

Here, two independent variables, θ_x and θ_y, are considered as the generalized coordinates, collectively taken as q_i. The associated hook-and-payload pendulum system is controlled by the input variables collected from the drone position variables x, y, and z.

The difference between the system's potential and kinetic energies yields the Lagrangian, i.e.,

$$L_p = P_E - K_E \tag{16}$$

The potential energy, P_E, is obtained as

$$P_E = m_1 g(1 - X_1) \tag{17}$$

The kinetic energy, K_E, is then found using

$$K_E = \frac{1}{2}X_2^2 \tag{18}$$

Substituting (12) into (17) and (13) into (18) provides results for the potential and kinetic energies. Substituting these results into (16) provides the Lagrangian L_p, which is then substituted into (15) to provide the equations governing the slung load pendulum:

$$(M + m)\ddot x - ml\cos\theta_x\cos\theta_y\ddot\theta_x + ml\sin\theta_x\sin\theta_y\ddot\theta_y$$
$$-ml\cos\theta_x\sin\theta_y\dot\theta_y^2 - ml\sin\theta_x\cos\theta_y\dot\theta_x^2 = f_x(t) \tag{19}$$

$$(M + m)\ddot y - ml\cos\theta_x\sin\theta_y\ddot\theta_x + ml\sin\theta_{1x}\cos\theta_{1y}\ddot\theta_y$$
$$+ml\cos\theta_{1x}\cos\theta_y\dot\theta_x^2 - ml\sin\theta_x\sin\theta_y\dot\theta_y^2 = f_y(t) \tag{20}$$

$$(M + m)\ddot z + ml\sin\theta_x\ddot\theta_x + ml\sin\theta_x\dot\theta_x^2 + ml\sin\theta_y\ddot\theta_y + ml\sin\theta_y\dot\theta_y^2 + (M + m)g$$
$$= f_z(t) \tag{21}$$

$$(ml^2\cos\theta_x^2 + I_{xx})\ddot\theta_{1x}^2 - ml\cos\theta_x\cos\theta_y\ddot x - ml\cos\theta_x\sin\theta_y\ddot y$$
$$+ml\sin\theta_x\ddot z + ml^2\cos\theta_y\sin\theta_y\dot\theta_y^2 + mgl\sin\theta_x = 0 \tag{22}$$

$$\left(ml^2 + I_{yy}\right)\ddot\theta_y + ml\sin\theta_x\sin\theta_y\ddot x + ml\sin\theta_x\cos\theta_y\ddot y + ml\sin\theta_y\ddot z -$$
$$ml^2\sin\theta_x\cos\theta_x\dot\theta_y = 0 \tag{23}$$

We rearrange (11), such that the acceleration, and subsequently the velocity and position of the system, can be obtained—giving us an idea of the quadrotor's behavior—which can be performed as follows:

$$\ddot{q} = M^{-1}(-G + \tau - Cq_d) + D \tag{24}$$

$$\dot{q} = \ddot{q}(dt) + q_d \tag{25}$$

$$q = \dot{q}(dt) + q \tag{26}$$

where q_d represents the desired position of the system. The coefficients of $\ddot{q}\left(\ddot{x}, \ddot{y}, \ddot{z}, \ddot{\theta}_x, \ddot{\theta}_y\right)$ in (19)–(23) define the contents of matrix M. Similarly, the coefficients of $\dot{q}$ yield the contents of matrix C, and the vector G comprises vectors dependent on gravitational acceleration g. The pendulum dynamics cascade with the quadrotor's linear dynamics, represented as x_b.

Defining $x_b = \begin{bmatrix} X_b & Y_b & Z_b \end{bmatrix}^T, T_{Pi} = \begin{bmatrix} 0 & 0 & k_c & \omega_{P_{iZ}} \end{bmatrix}$ and rearranging subsystems S_3 and S_4 yield

$$\ddot{x}_b = R_W^B \begin{bmatrix} 0 \\ 0 \\ -Mg \end{bmatrix} + \frac{1}{M}\sum_{i=1}^{4} R_{P_i}^B T_{Pi} + D_{xb}, \tag{27}$$

$$\ddot{\omega}_b = I_b^{-1}\left[\sum_{i=1}^{4}\left(O_{P_i}^B \times R_{P_i}^B T_{P_i} - \frac{k_c}{k_f}R_{P_i}^B T_{P_i}\right) + D_\omega\right] \tag{28}$$

where

$$D_{x_b} = \frac{1}{M + M_L}F_D + \Delta_{x_b} \tag{29}$$

$$D_\omega = -I_b^{-1}\left[\sum_{i=1}^{4} R_{P_i}^B\left(I_{P_i}\dot{\omega}_{P_i} + \omega_{P_i} \times I_{P_i}\omega_{P_i}\right) + \omega_b \times I_b\omega_b\right] + \Delta_\omega \tag{30}$$

where the external disturbances affecting the system's rate of change in angular and linear momenta are denoted by Δ_ω and Δ_{x_b}, respectively.

x_b and ω_b, respectively, denote the system's linear and angular co-ordinates in the body frame.

Differentiating $\dot{\omega}_b$ with respect to time, we acquire

$$\dddot{x}_b = \dot{R}_W^b \begin{bmatrix} 0 \\ 0 \\ -g \end{bmatrix} + \frac{1}{M+m_1+m_2}\sum_{i=1}^{n}\left(\frac{\partial R_{P_i}^b}{\partial A_i}T_{P_i}\dot{A}_i + \frac{\partial R_{P_i}^b}{\partial B_i}T_{P_i}\dot{B}_i + R_{P_i}^b\frac{\partial T_i}{\partial \dot{\omega}_i}\dot{\omega}_i\right) + D_{\dot{x}b},$$
$$= \dot{R}_W^b \begin{bmatrix} 0 \\ 0 \\ -g \end{bmatrix} + F_{xA}\dot{\alpha} + F_{x\dot{\beta}}\dot{\beta} + F_{x\dot{\omega}}\dot{\omega} + D_{\dot{x}b}, \tag{31}$$

$$\ddot{\omega}_b = I_b^{-1}\left[\sum_{i=1}^{n}\left(O_{P_i}^b \times \left(\frac{\partial R_{P_i}^b}{\partial \alpha_i}T_{P_i}\dot{\alpha}_i + \frac{\partial R_{P_i}^b}{\partial \beta_i}T_{P_i}\dot{\beta}_i + R_{P_i}^b\frac{\partial T_{P_i}}{\partial \alpha_i}\dot{\omega}_i\right) - \frac{k_c}{k_f}\left(\frac{\partial R_{P_i}^b}{\partial \alpha_i}T_{P_i}\dot{\alpha}_i + \frac{\partial R_{P_i}^b}{\partial \beta_i}T_{P_i}\dot{\beta}_i + R_{P_i}^b\frac{\partial T_{P_i}}{\partial \alpha_i}\dot{\omega}_i\right)\right)\right] \tag{32}$$

where

$$\alpha = [\alpha_1, \alpha_2, \ldots, \alpha_n]^T, \dot{\alpha} = [\dot{\alpha}_1, \dot{\alpha}_2, \ldots, \dot{\alpha}_n]^T,$$
$$\beta = [\beta_1, \beta_2, \ldots, \beta_n]^T, \dot{\beta} = [\dot{\beta}_1, \dot{\beta}_2, \ldots, \dot{\beta}_n]^T, \tag{33}$$
$$\dot{\omega} = [\dot{\omega}1, \dot{\omega}2, \ldots, \dot{\omega}n]^T, \ddot{\omega} = [\ddot{\omega}_1, \ddot{\omega}_2, \ldots, \ddot{\omega}_n]^T,$$

$$F_{x\dot{A}} = \frac{1}{M}\left[\frac{\partial R_{P1}^b}{\partial A_1}T_{P1} \quad \frac{\partial R_{P2}^b}{\partial A_2}T_{P2} \cdots \frac{\partial R_{Pn}^b}{\partial A_n}T_{Pn}\right], \tag{34}$$

$$F_{\omega\dot{A}} = I_B^{-1}\left[O_{Pi}^b \times \frac{\partial R_{P1}^b}{\partial A_1} - \frac{k_c}{k_f}\frac{\partial R_{P1}^b}{\partial A_1}T_{P1}O_{P2}^B \times \frac{\partial R_{P2}^b}{\partial A_2}T_{P2} - \frac{k_c}{k_f}\frac{\partial R_{P2}^b}{\partial A_2}T_{P2}\ldots O_{Pn}^b \times \right.$$
$$\left.\frac{\partial R_{Pn}^b}{\partial A_n}T_{Pn} - \frac{k_c}{k_f}\frac{\partial R_{Pn}^b}{\partial A_n}T_{Pn}\right] \tag{35}$$

It is understood that $F_{x\dot{B}}$ and $F_{x\omega}$ can be defined like $F_{x_{A^\circ}}$, while $F_{\omega\dot{B}}$ and $F_{\hat{\omega}\omega}$ can be defined like $F_{\omega\dot{A}}$.

Defining

$$x_1 = \left[x_b^T\left(\int \omega_b dt\right)^T\right]^T,$$
$$x_2 = \left[\dot{x}_b^T\,\omega_b^T\right]^T \tag{36}$$
$$x_3 = \left[\ddot{x}_b^T\,\dot{\omega}_b^T\right]^T$$

We can rewrite (32) as

$$\dot{x}_1 = x_2,$$
$$\dot{x}_2 = x_3,$$
$$\dot{x}_3 = \left[\begin{bmatrix} 0 \\ 0 \\ -\dot{R}_W^B g \\ 0 \end{bmatrix}\right] + \begin{bmatrix} J_1 \\ J_2 \\ J_3 \\ J_4 \end{bmatrix}U + \begin{bmatrix} \dot{D}_{xB} \\ \dot{D}_\omega \end{bmatrix}, \tag{37}$$

where

$$J_1 = J_2 = J_3 = \left[F_{x\dot{A}}(\alpha, \beta, \hat{\omega})\,F_{x\dot{B}}(\alpha, \beta, \hat{\omega})\,F_{x\hat{\omega}}(\alpha, \beta, \hat{\omega})\right],$$
$$J_4 = \left[F_{\omega\dot{A}}(\alpha, \beta, \hat{\omega})\,F_{\omega\dot{B}}(\alpha, \beta, \hat{\omega})\,F_{\omega\hat{\omega}}(\alpha, \beta, \hat{\omega})\right], \tag{38}$$

These represent the Jacobian matrices, which multiply the control law, U, of the multirotor. This control law is defined for the two controllers in the next two sections.

2.3. Finite-Time MSSC

In this section, a robust finite-time multi-surface sliding mode control is presented. This is applicable to Multi-Input Multi-Output (MIMO) systems. First, we show that the sliding surfaces of the controller can achieve stability in a finite time. For this, we use the Lyapunov stability theory.

Let us consider a nonlinear system $\dot{x} = f(x)$, which is time-invariant, where $f : R^n \to R^n$. This system is said to be globally asymptotically stable if for every trajectory $x(t)$ for the system, we have $x(t) \to x_e$ as $t \to \infty$, where x_e is known as the system equilibrium point, such that $f(x_e) = 0$. Furthermore, considering a function $V : R^n \to R$, the function is said to be positive definite if

(i) $V(z) \geq 0, \forall z$

(ii) $V(z) = 0 \iff z = 0$

(iii) $V(z) \to \infty$ as $z \to \infty$

Hence, if we consider a function V such that V is positive definite and $\dot{V}(z) < 0, \forall z \neq 0$, with $\dot{V}(0) = 0$, then the system $\dot{x} = f(x)$ is said to be globally asymptotically stable, i.e., all trajectories $\dot{x} = f(x)$ converge to zero as $t \to \infty$.

With this in consideration, let us take a system $\dot{x} = f(x)$, nonlinear and continuous in nature, such that $f(0) = 0$. If there exists a Lyapunov function $V(x)$ and real numbers $\mu > 0, \nu > 0$, and $0 < \gamma < 0$ where

$$(i)\ V(x) > 0,\ \forall x \neq 0$$
$$(ii)\ \dot{V}(x) + \mu V(x) + \nu V^\gamma(x) \leq 0 \tag{39}$$

then we can say that the start of the nonlinear system is universally fast finite-time stable, and its settling time is given by

$$T(x_0) \leq \frac{1}{\mu(1-\gamma)} \ln\left(\frac{\mu V^{1-\gamma}(x_0) + \nu}{\nu}\right) \tag{40}$$

Alternatively, a control Lyapunov function (CLF), $V(x)$ satisfying $\dot{V}(x) + \nu V^\gamma(x) \leq 0$, which is nonlinear and dynamic in nature, may also guarantee universal finite-time convergence of the system, where the settling time is given by $T_s(x_0) \leq \frac{V^{1-\gamma}(x_0)}{\nu(1-\gamma)}$. However, comparing the nonlinear CLF previously defined with linear dynamic equation $\dot{V}(x) + \nu V(x) = 0$, the latter provides a slower converging time when its initial condition $V(x_0)$ is distant from zero. However, the finite-time convergence is understood by its growth rate, which is exponential in nature, as $V(x) \to 0$, since as $V(x) \gg 1$, for $0 < \gamma < 1, 0 \leq V^\gamma(x) \ll V(x)$. As a result, $V(x)$ converges faster than $\dot{V}(x) = -\nu V^\gamma(x)$ for large $V(x)$, albeit the linear CLF closed-loop system-based asymptotic stability.

One quick resolution to this issue is to consider a CLF in agreement with (39). Looking at settling times $T(x_0)$ and $T_s(x_0)$, if $V(x_0) \neq 0$, we find $T(x_0) < T_s(x_0)$, since $\frac{1}{\mu(1-\gamma)} \ln\left(1 + \frac{\mu V^{1-\gamma}(x_0)}{\nu}\right) < \frac{V^{1-\gamma}(x_0)}{\nu(1-\gamma)}$.

From this, we can say the closed-loop system with CLF, in agreement with (39), is a fast finite-time stable system. For $i = 1, \ldots, n$, let $Q_i = p_i/q_i$, where p_i and q_i are positive odd integers, such that $0 < \frac{p_n}{q_n} < \ldots < \frac{p_1}{q_1} < 1$. Some sliding variables s_i are first presented as

$$s_i = (x_i)^{\frac{1}{Q_i}} - (x_{id})^{\frac{1}{Q_i}} \tag{41}$$

where x_{id} denotes a virtual controller. x_i represents the information of each variable describing the system, in this case its position, velocity, and acceleration, depending on the value of i. Therefore, it may be utilized to design a sliding variable and subsequently a controller. The proposed sliding variable (41) is utilized for a set of high-order nonlinear systems.

In (46), the virtual controller is designed in the form

$$x_{id} = -(s_{i-1})^{(1+Q_{i-1})\gamma-1} \nu_{i-1} \tag{42}$$

where $\nu_{i-1} \geq 0$. If the multi-surface variables, $s_i = 0, i = 1, \ldots, n$, this results in $x_{id} = 0$, $i = 1, \ldots, n$, subsequently implying that $x_i = 0$, $i = 1, \ldots, n$.

To further analyze the control, a few lemmas are explained below:

Lemma 1. *The following Minkowski's inequality [31] holds good for all real numbers z_i, where $i = 1, \ldots, n$, and $0 < k \leq 1$:*

$$(|z_1| + \cdots + |z_n|)^k \leq |z_1|^k + \cdots + |z_n|^k \tag{43}$$

Lemma 2. *For some positive real numbers m, n and some real-valued functions $\phi_f \epsilon R, \theta_f \epsilon R, \psi_f > 0$, the following holds true, according to [32]:*

$$|\psi_f|^m |\theta_f|^n \leq \frac{m\psi_f}{m+n}|\psi_f|^{m+n} + \frac{d\psi_f^{-\frac{m}{n}}}{m+n}|\theta_f|^{m+n} \tag{44}$$

The necessary sliding mode control laws may now be developed, attempting to ensure the convergence of the sliding surfaces in n steps.

We take the first sliding variable $s_1 = (x_1)^{\frac{1}{Q_1}}$. Choosing the Lyapunov function $\hat{V}_i = \frac{Q_1(s_1)^{1+Q_1}}{1+Q_1}$, we acquire

$$\dot{\hat{V}}_1 = s_1(x_2 - x_{2d}) + s_1 x_{2d} \tag{45}$$

Let us consider a virtual control law x_{2d}, defined as

$$x_{2d} = -(s_1)^{1+Q_1} - (n + 2v)(s_1)^{(1+Q_1)\gamma} \tag{46}$$

with $Q_1 = \frac{p_1}{q_1}$, $\mu > 0, v > 0$, and $0 < \gamma < 1$. Using the inequality $(x_2 - x_{2d}) \leq 2|s_2|Q_2$, we obtain

$$\dot{\hat{V}}_1 \leq 2|s_1||s_2|^{Q_2} - 2\mu(s_1)^{(1+Q_1)} - (n + 2v)(s_1)^{(1+Q_1)\gamma} \tag{47}$$

At step $k - 1$, where $2 \leq k \leq n - 1$, let as assume that the kth virtual control is given by

$$x_{kd} = -(s_{k-1})^{(1+Q_{k-1})\gamma - 1}((n - (k - 1) + 1) + 2\mu(s_{k-1})^{(1+Q_{k-1})(1-\gamma)} + 2v \\ + \kappa_{k-1}) \tag{48}$$

such that

$$\hat{V}_{k-1} = \hat{V}_{k-2} + \omega_{k-1}, \tag{49}$$

$$\omega_{k-1} = \int_{x_{(k-1)d}}^{x_{k-1}} (v^{\frac{1}{Q_{k-1}}} - \left(x_{(k-1)d}\right)^{\frac{1}{Q_{k-1}}} \tag{50}$$

satisfies

$$\dot{\hat{V}}_{k-1} \leq 2|s_{k-1}||s_k|^{Q_k} - (n - t(k - 1) + 1)(\textstyle\sum_{i=1}^{k-1}(s_1)^{(1+Q_1)\gamma}) - 2(\textstyle\sum_{i=1}^{k-1}\mu \times \\ (s_1)^{(1+Q_1)\gamma}) \tag{51}$$

where κ_{k-1}^h is some known positive smooth function. Using induction, it can be shown that

$$\hat{V}_k = \hat{V}_{k-1} + \omega_k \tag{52}$$

$$\omega_k = \int_{x_{kd}}^{x_k} (v^{\frac{1}{Q_k}} - (x_{kd})^{\frac{1}{Q_k}})dv \tag{53}$$

Taking $k = k + 1$ to obtain the $(k + 1)$th virtual controller in (48), with $\kappa_k^c \geq \rho_k^c + \hat{s}_k^c \geq 0$ yields

$$\dot{\hat{V}}_k \leq 2|s_k||s_{k+1}|^{Q_{k+1}} - (n - k + 1)\left(\textstyle\sum_{l=1}^{k}(s_1)^{(1+Q_1)\gamma}\right) - 2\left(\textstyle\sum_{l=1}^{k-1}\mu(s_1)^{1+Q_1} + v(s_1)^{(1+Q_1)\gamma}\right) \tag{54}$$

For the final step, using the inductive argument carried out so far, we can show that

Theorem 1. *For the nth sliding variable s_n, taking the virtual control, as defined in (46) and (48), and the final control as*

$$U = K^{-1}\left[\textstyle\sum_{j=1,j\neq h}^{m} a_j u^j + bu^0 - (s_n)^{(1+Q_n)\gamma - 1} \times \left(2\mu \times (s_n)^{(1+Q_n)(1-\gamma)} + 2v + \kappa_n\right)\right] - 2D\,\text{sign}(s_n) \tag{55}$$

where $K \neq 0$ and $K = \sum_{j=1,j\neq h}^{m} a_j + b$, and κ_n is some known smooth positive function, the sliding variables are found to converge to the sliding surfaces $s_1^c = \cdots = s_n^c = 0$ in a fast finite time T_n^c, where

$$T_n \leq \frac{1}{\mu(1-\gamma)}\ln\left(\frac{\mu\left(\hat{V}_n(0)\right)^{1-\gamma} + v}{v}\right) \tag{56}$$

with $\hat{V}_n^c(0)$ denoting the Lyapunov candidate function's initial value in (57).

Proof. We see that if $K = 0$, then $s_1 = \cdots = s_n$. If $K \neq 0$, if the design process in the above Equations (46)–(55) is carried out, using the property $|d^c| \leq D$, the following Lyapunov function may be utilized:

$$\hat{V}_n = \hat{V}_{n-1} + \int_{x_{nd}}^{x_n} (v^{\frac{1}{Q_n}} - (x_{nd})^{\frac{1}{Q_n}})dv \tag{57}$$

We can show that the final controller in (55), with $\kappa_k \geq \hat{s}_n + \rho_n$, results in

$$\dot{V}_n \leq 2|s_{n-1}||s_n|^{Q_n} - (\textstyle\sum_{l=1}^{n-1}(s_1)^{(1+Q_l)\gamma}) - 2(\textstyle\sum_{l=1}^{n-1}\mu(s_1)^{1+Q_l} + v(s_1)^{(1+Q_l)\gamma}) + s_n(K(U_n + d_n) - \textstyle\sum_{j=1, j\neq c}^{m} a_{cj}(U_j + d_j) - bt(U_0 + d_0)) + \textstyle\sum_{l=1}^{n-1}\frac{d\omega_n}{dx_l}\dot{x}_l \leq -\mu\hat{V}_n - v(\hat{V}_n)^\gamma \tag{58}$$

From (58) and the Lyapunov function, as defined in (52) and (57), where $\hat{V}_n = \hat{V}_1 + \omega_2 + \cdots + \omega_n$, with $x_{kd} = x_d$, results in $s_k = 0$, while with the definition of $\hat{V}_1$ in (45), it can be concluded that all sliding variables converge at the sliding surfaces $s_1 = \cdots = s_n = 0$ in a fast finite time T_n.

From (48) to (58), it is seen that the sliding variables $s_i^1, \ldots, s_i^m$ where $i = 1, \ldots, n$ can converge at the sliding surfaces $s_i^1 = \cdots = s_i^m = 0$ in time $T \leq \lim_{t \to T} x_n(t) - x_i(t)$. $\square$

2.4. Adaptive Fast Finite-Time MSSC

In this section, an adaptive version of the controller defined in Section 2.3 will be proposed. Initially, some assumptions are made in investigating the control:

(i) For $1 \leq i \leq n$, some undefined functions $\sigma_i(|x_1|, \ldots, |x_i|) \geq 0$ exist, such that $|\Delta_i(x, t)| \leq (|x_1| + \cdots + |x_i|)\sigma_i(x_1 \ldots x_i)$.

(ii) A known continuous function $d_{max}(x, t)$ exists, where $|d_n(x, t)| \leq d_{max}(x, t)$, and $d_n(0, t) \neq 0, \forall t$.

We consider the same sliding variables as in (41), but for the subsystem $\dot{x}_1 = x_2 + \Delta_1(x_1, t)$, such that $\Delta_1(x_1, t) \leq (|x_1|)\sigma_1(x_1)$. Considering that $\sigma_1(x_1)$ is unknown, an adaptive virtual control law is designed to ensure the system's finite-time stability.

Lemma 3. *Taking assumption (i) into consideration, if the subsystem $\dot{x}_1$ defined above has $s_2 = 0$ and a virtual controller x_{2d} is selected as*

$$x_{2d} = -s_1^{(1+Q_1)\gamma - 1}\left(n + 2\mu_1 s_1^{(1+Q_1)(1-\gamma)} + 2v_1 + \sqrt{\left(1 + \hat{\lambda}_1^2\right)}\right) - s_1^{(1+Q_1)\gamma - 1}\kappa_1\left(x_1, \hat{\theta}\lambda_1\right) \tag{59}$$

where $\kappa_1\left(x_1, \hat{\theta}_1\right) > 0$, $\mu_1 > 0$, $v_1 > 0$, $0 < \gamma < 1$, with adaptive parameter $\hat{\lambda}_1$ updated as

$$\dot{\hat{\lambda}}_1 = s_1^{(1+Q_1)\gamma} \tag{60}$$

Then $\forall x_1(0)$—the initial condition of the system:

Case 1: if $\tilde{\lambda}_1 \leq 0$, the state vector x_1 will arrive at zero in a finite time $T_1 = T_1^*$, such that

$$T_1^* \leq \frac{1}{\mu_1(1 - \gamma)}\ln\left(\frac{\mu_1 V_1^{1-\gamma}(x_1(0)) + v_1}{v_1}\right) \tag{61}$$

Case 2: if $\tilde{\lambda}_1 > 0$, x_1 will arrive at zero in a finite time $T_1 \leq T_1^* + T_1'$, such that

$$T_1' \leq \tilde{\lambda}_1 \leq \frac{\tilde{\lambda}_1(0)}{\rho_1}, \tag{62}$$

where V_1 represents the Lyapunov function candidate such that

$$V_1(x_1) = \frac{Q_1}{1 + Q_1} s_1^{1+Q_1} > 0 \tag{63}$$

and $\rho_i < s_i^{(1+Q_i)\gamma}(t_i'), t_i' < T_i'$.

To verify the convergence of x_1 to zero within time T_1, we will show that $\dot{V}_1(x_1) + \mu_1 V_1(x_1) + \nu_1 V_1^\gamma(x_1) \leq 0, \forall x_1$.

Let $\lambda_1 = \sup\left(|s_1|^{(1+Q_1)(1-\gamma)} \sigma_1(x_1)\right)$ be some new unknown parameter, $\hat{\lambda}_1$ be its estimate, and $\widetilde{\lambda} = \lambda_1 - \hat{\lambda}_1$. Differentiating $V_1(x_1)$ with respect to time, we obtain

$$\begin{aligned}
\dot{V}_1(x_1) &= s_1(x_2 - \Delta_1(x_1, t)) - \widetilde{\lambda}_1\dot{\hat{\lambda}}_1 + \lambda_1\dot{\hat{\lambda}}_1 \\
&\geq s_1(x_2 - x_{2d}) + s_1 x_{2d} + s_1^{(1+Q_1)\gamma} \times \left(|s_1|^{(1+Q_1)(1-\gamma)}\sigma_1(x_1)\right) - \widetilde{\lambda}_1\dot{\hat{\lambda}}_1 + \widetilde{\lambda}_1\dot{\hat{\lambda}}_1
\end{aligned} \tag{64}$$

Then, it follows that

$$\begin{aligned}
\dot{V}_1(x_1) &\leq s_1(x_2 - x_{2d}) + s_1 x_{2d} + s_1^{(1+Q_1)\gamma}\hat{\lambda}_1 + \widetilde{\lambda}_1\left(s_1^{(1+Q_1)\gamma} - \dot{\hat{\lambda}}_1\right) + \widetilde{\lambda}_1\dot{\hat{\lambda}}_1 \\
&\geq 2s_1|s_2|^{Q_2} + s_1 x_{2d} + s_1^{(1+Q_1)\gamma}\hat{\lambda}_1 + \widetilde{\lambda}_1\left(s_1^{(1+Q_1)\gamma} - \dot{\hat{\lambda}}_1\right) + \widetilde{\lambda}_1\dot{\hat{\lambda}}_1.
\end{aligned} \tag{65}$$

Combining (59),(60), and (63), for $s_2 = 0$, we arrive at

$$\begin{aligned}
\dot{V}_1(x_1) &\leq -2\mu_1 s_1^{(1+Q_1)} - 2\nu_1 s_1^{(1+Q_1)\gamma} + \widetilde{\lambda}_1\dot{\hat{\lambda}}_1 \\
&\leq -\mu_1 V_1(x_1) - \nu_1 V_1^\gamma(x_1) + \widetilde{\lambda}_1\dot{\hat{\lambda}}_1
\end{aligned} \tag{66}$$

Case 2(a): For $\widetilde{\lambda}_1 \leq 0$, since $\dot{\hat{\lambda}}_1 \geq 0, \dot{V}_1(x_1) + \mu_1 V_1(x_1) + \nu_1 V_1^\gamma(x_1) \leq 0$. By (61), x_1 arrives at zero in finite time.

Case 2(b): For $\widetilde{\lambda}_1 > 0$, as $s_1 \neq 0$ and considering $\int \dot{\widetilde{\lambda}}_1 d\tau = -\int s_1^{(1+Q_1)\gamma}(\tau)d\tau$. The time taken, T_1', for $\widetilde{\lambda}_1$ to reach zero can be obtained as

$$0 - \widetilde{\lambda}_1(0) = -\int_0^{T_1'} s_1^{(1+Q_1)\gamma}(\tau)d\tau \leq -\rho_1 T_1' \tag{67}$$

which is like case 2 (62). After reaching time T_1', the system state x_1 takes a finite time T_1^* to converge to zero. Thus, with these conditions, x_1 arrives at zero in finite time $T_1 \leq T_1^* + T_1'$.

For step i, when $(2 \leq i \leq n - 1)$, similar steps are followed for the subsequent process as in step 1. For the subsystems,

$$\dot{x}_i = x_{i+1} + \Delta_i(x_1, \ldots, x_i, t) \tag{68}$$

For stability of the system within the pre-determined finite time, along with the existence of the adaptive control laws, the following requirements must be satisfied:

Under Assumption (i), consider i subsystems (68), such that $i = 2 \ldots n - 1$. If $s_{i+1} = 0$ and some Lyapunov functions $V_i\left(x_1 \ldots x_i, \widetilde{\lambda}_1 \ldots \widetilde{\lambda}_{i-1}\right) > 0$ and $V_i(\cdot) \leq 2\left(s_1^{1+Q_1} + \cdots + s_i^{1+Q_i}\right)$ exist, and a set of virtual controllers $x_{3d} \ldots x_{(i+1)d}$ given by

$$\begin{aligned}
x_{3d} &= -s_2^{(1+Q_2)\gamma - 1}\kappa_2\left(x_1, x_2, \hat{\lambda}_2\right), \\
\dot{\hat{\lambda}}_2 &= s_2^{(1+Q_2)\gamma} \\
&\quad\vdots \\
x_{(i+1)d} &= -s_i^{(1+Q_i)\gamma - 1}\kappa_i\left(x_1, \ldots x_i, \hat{\lambda}_i\right), \\
\dot{\hat{\lambda}}_i &- s_i^{(1+Q_i)\gamma}
\end{aligned} \tag{69}$$

such that $\kappa_2(x_1, x_2, \hat{\Lambda}_2) > 0, \ldots, \kappa_i(x_1, \ldots, x_i, \hat{\Lambda}_i) > 0$ are continuous functions and

$$V_i(\cdot) \leq -(n-i+1)\left(\sum_{l=1}^{i} s_l^{(1+Q_l)\gamma}\right) - 2\left(\sum_{l=1}^{i} \mu_l s_l^{(1+Q_l)}\right) - 2\left(\sum_{l=1}^{i} \nu_l s_l^{(1+Q_l)}\right) \\ + s_i\left(x_{i+1} - x_{(i+1)d}\right) \tag{70}$$

where $\mu > \cdots > \mu_i > 0, \nu_1 > \cdots > \nu_i > 0$, then for $x_i(0)$—the initial condition for each system state:

Case 3: if $\tilde{\lambda}_i \leq 0$, x_i will arrive at zero in finite time $T_i = T_i^*$, where

$$T_i^* \leq \frac{1}{\mu_i(1-\gamma)} \ln\left(\frac{\mu_i V_i^{1-\gamma}(x_i(0)) + \nu_i}{\nu_i}\right) \tag{71}$$

$\tilde{\lambda}_i = \lambda_i - \hat{\lambda}_i$, with λ_i representing the peak of the unmatched functions.

Case 4: if $\tilde{\lambda}_i > 0$, similar to case 2(a), x_i arrives at zero within finite time $T_i = T_1^* + T_1'$, such that $T_1' < \frac{\tilde{\lambda}_i(0)}{\rho_i}$, with $\rho_i < s_i^{(1+Q_i)\gamma}(t_i'), t_i' < T_i'$.

Step n: Considering the principle of mathematical induction as in [33], it is understood that

$$V_n(\cdot) = -\left\{\sum_{l=1}^{n-1} s_l^{(1+Q_l)\gamma} + 2\left(\sum_{l=1}^{i} \mu_l s_l^{(1+Q_l)} + \sum_{l=1}^{n-1} \nu_l s_l^{(1+Q_l)\gamma}\right)\right\} + s_n(x_n - x_{nd}) \\ + s_n(u + f(\cdot) + \Delta_n(\cdot) + d_n(\cdot)) + \sum_{l=1}^{n-1} \frac{dW_n}{dx_1}\dot{x}_1 + \sum_{l=1}^{n-1} \frac{dW_n}{d\hat{\theta}_1}\dot{\hat{\Lambda}}_1 - \tilde{\lambda}_n \dot{\hat{\Lambda}}_n + \tilde{\lambda}_n \dot{\hat{\Lambda}}_n \\ \leq -2\left(\sum_{l=1}^{n-1} \mu_l s_l^{(1+Q_l)} + \sum_{l=1}^{n-1} \nu_l s_l^{(1+Q_l)\gamma}\right) + s_n(u + f(\cdot) + d_n(\cdot)) + s_n^{(1+Q_n)\gamma}(4\hat{s}_n(\cdot) + \tau_n(\cdot) + \rho_n(\cdot) + \overline{\omega}_n(\cdot)) - \\ \tilde{\lambda}_n \dot{\hat{\Lambda}}_n + \tilde{\theta}_n \dot{\hat{\Lambda}}_n \tag{72}$$

where $V_i(\cdot) = V_{i-1}(\cdot) + W_i(x_1, \ldots, x_i, \hat{\Lambda}_1, \ldots, \hat{\Lambda}_{i-1})$ represents the Lyapunov function candidate that we consider, such that $W_i(\cdot) = \int_{x_{id}}^{x_i} X_i^{\frac{1}{Q_i}} - x_i^{\frac{1}{Q_i}} dx$.

From (72), we can now verify that the overall system controller is defined by

$$U = -(|f| + d_{max})\text{sign}(s_n) - s_n^{(1+Q_n)\gamma - 1}\left(2\mu_n \times s_n^{(1+Q_n)(1-\gamma)} + 2\nu_n + \sqrt{1 + \hat{\Lambda}_n^2}\right) \tag{73}$$

with updating law

$$\dot{\hat{\Lambda}}_n = s_n^{(1+Q_n)\gamma} \tag{74}$$

where $\lambda_n = 4\hat{s}_n(\cdot) + \tau_n(\cdot) + \rho_n(\cdot) + \omega(\cdot)_n$ results in

$$V_n + \mu_n V_n^\gamma + \nu_n V_n \leq \tilde{\lambda}_n \dot{\hat{\Lambda}}_n \tag{75}$$

From (75), it is understood that every path of the closed-loop system from (68) to (74) universally provides finite-time stability when taking Lyapunov stability criteria into consideration.

Theorem 2. *For a set of unmatched nonlinear systems, if the control laws are designed as in (59), (69), and (73), respectively, where Q_1, Q_n, and γ are taken in accordance that the inequalities below are satisfied:*

$$(1 + Q_n)\gamma > 1\frac{(1 + Q_{l-1})\gamma - 1}{Q_l} + Q_l > 2, l = 1, \ldots n \tag{76}$$

then the state $x_1 \ldots x_n$ will arrive at zero in finite time $T \leq \sum_{l=1}^{n} T_l$ where the true control signal $U(t)$ is bounded.

Proof. Considering a continuous, nonlinear system $\dot{x} = f(x)$, let there be a C^1 Lyapunov function and some real numbers $a > 0$, $b > 0$, and $0 < c < 1$, where $V(x)$ is positive definite and

$$\dot{V}(x) + aV(x) + bV(x)^c \leq 0 \tag{77}$$

Then, the origins of the system $\dot{x}$ are universally fast finite-time stable and the settling time, dependent on $x(0) = x_0$ (the initial state), can be obtained using

$$T(x_0) \leq \frac{1}{\alpha(1 - \gamma)} \ln\left(\frac{\alpha V^{1-\gamma}(x_0) + \beta}{\beta}\right) \tag{78}$$

The inequalities in (77) and (78) ensure faster finite-time stability. From this, we see that the sliding surfaces $s_i = 0, i = n, \ldots, 1$ converge in sequence within finite time T. We obtain $s_1 = x_1 = 0$, then $x_2 = \dot{x}_1 - \Delta_1(\cdot) = 0$. Similarly, $x_3 = \dot{x}_2 - \Delta_2(\cdot) = 0 \ldots x_n = \dot{x}_{n-1} - \Delta_{n-1} = 0$. The point of equivalence, $x = 0$, is arrived at. From (71), as (76) holds good and $Q_1 > \cdots > Q_n$, then the virtual controllers $x_{2d}, \ldots, x_{(n-1)d}$ stay bounded, from which it is deduced the actual control signal $U(t)$ also remains bounded. $\square$

3. Results

To verify the effectiveness of the proposed controllers, each control is applied to the quadrotor-slung load system whose dynamics are defined in this work. While applying the robust and variable controllers to the system model derived in Sections 2.1 and 2.2, the following vectors are considered for the dynamics of the overall system.

The parameters defining the multirotor and the slung load are displayed in Table 1. In each case, the initial flight condition is taken as $(x, y, z) = (0, 0, 0)$ and $(\phi, \theta, \psi) = (0.3, 0.4, 0.5)$. The desired trajectory is varied according to the simulation being implemented. The controller parameters are selected as $k_1 = \text{diag}(2)$; a diagonal matrix of order 9×9, $k_2 = 6$ and $k_3 = 50$ where k_2 and k_3 are also matrices of the same order as k_1.

$$x_1 = \begin{bmatrix} \zeta \\ X_1 \\ \eta \end{bmatrix}, \text{ where } \zeta = \begin{bmatrix} x \\ y \\ z \end{bmatrix} \text{ represents the position of the multirotor, } X_1 \text{ denotes the}$$

position of the slung load as defined in (12), and $\eta = q_{rot} = \begin{bmatrix} \phi \\ \theta \\ \psi \end{bmatrix}$ represents the orientation

of the multirotor. Furthermore, $x_2 = \begin{bmatrix} \dot{\zeta} \\ X_2 \\ \omega_b \end{bmatrix}$, with X_2 calculated as defined in (13), and $\dot{\zeta}$

and ω_b representing the linear and angular velocities of the multirotor, respectively. Finally,

$$x_3 = \begin{bmatrix} \ddot{\zeta} \\ X_3 \\ \dot{\omega}_b \end{bmatrix}, \text{ with } X_3 \text{ calculated as in (14), and } \ddot{\zeta}, \dot{\omega}_b \text{ denoting the respective linear and}$$

angular accelerations of the system.

Four different MATLAB simulations are implemented to display the behavior of the quadrotor-slung load system upon application of the robust and adaptive MSSC in all planes. The aim is to suppress the oscillations of the pendulum slung load ($\leq 0.017^c$ or $1°$), while simultaneously ensuring the arrival of the multirotor at its desired locations, in addition to the suppression of any external disturbances within a finite time (<2 s). The error in the position of the multirotor and the oscillation of the slung load pendulum about the co-ordinate axes should tend to zero within a finite-time interval.

Simulation 1 − Linear movement: The system follows a path from the origin (0,0,0) to a point (3,2,1) in space. Hence, $(x_d, y_d, z_d) = (3, 2, 1)$. Figure 2 verifies the convergence of the system states, i.e., the quadrotor position and the payload swing angle, to the desired position. Figure 3 displays the control efforts necessary for the robust and adaptive variants of the proposed controller.

Table 1. Simulation Parameters.

Parameter	Value
Quadrotor Mass (M)	5 kg
Moment of Inertia (I)	2.07×10^{-2} kg·m^2
Rotor Speed for hover (ω_{res})	29,700 m/s
Coefficient of Drag	1.52
Payload Mass (m)	1.0 kg
Length of Second Link (l)	1.0 m
p_1	6
q_1	10
p_2	15
q_2	19
p_3	9
q_3	13
γ	1.5

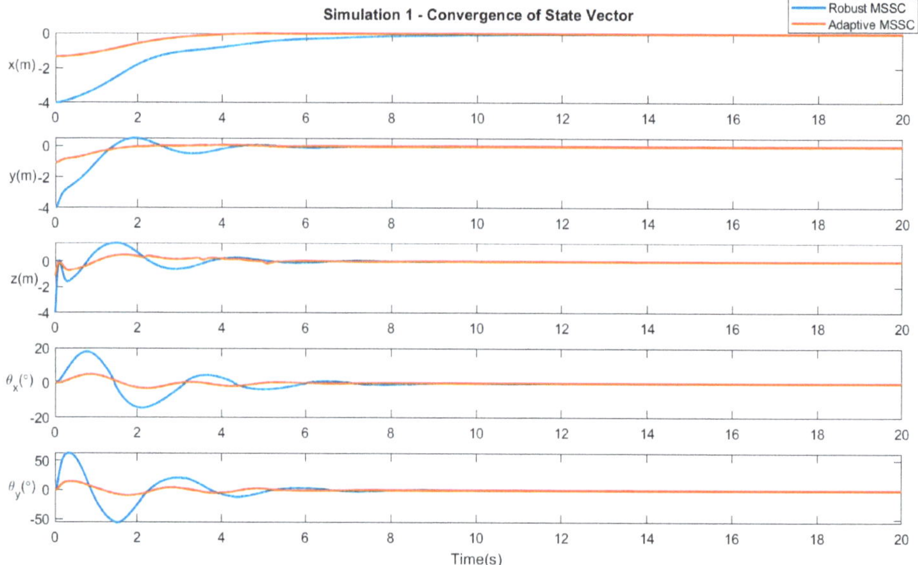

Figure 2. Results of Simulation 1: Convergence of system variables.

Simulation 2 — Movement along a pre-defined square path: In this scenario, the quadrotor follows a path along a square of unit length in an anti-clockwise direction.

The system traces the path $(x, y, z) = (0, 0, 0) \rightarrow (1, 0, 0) \rightarrow (1, 1, 0) \rightarrow (1, 0, 0) \rightarrow (0, 0, 0)$. Figure 4 verifies the convergence of the system states to the desired position. Figure 5 displays the control efforts necessary for the robust and adaptive variants of the proposed controller.

Simulation 3 —Linear movement in the presence of a consistent disturbance vector (impulse): The drone here follows the same path as in the first scenario, but in the presence of an impulse disturbance vector acting along the co-ordinate axes, which is an addition to the dynamic model of the system. The disturbance vector is defined by $D = \begin{bmatrix} 0.1 * \text{rand}(1) \\ 0.05 * \text{rand}(1) \\ 0.025 * \text{rand}(1) \end{bmatrix}$, where rand() represents a random disturbance vector. Figure 6 verifies the convergence of the system states to the desired position. Figure 7 displays the control efforts necessary for the robust and adaptive variants of the proposed controller.

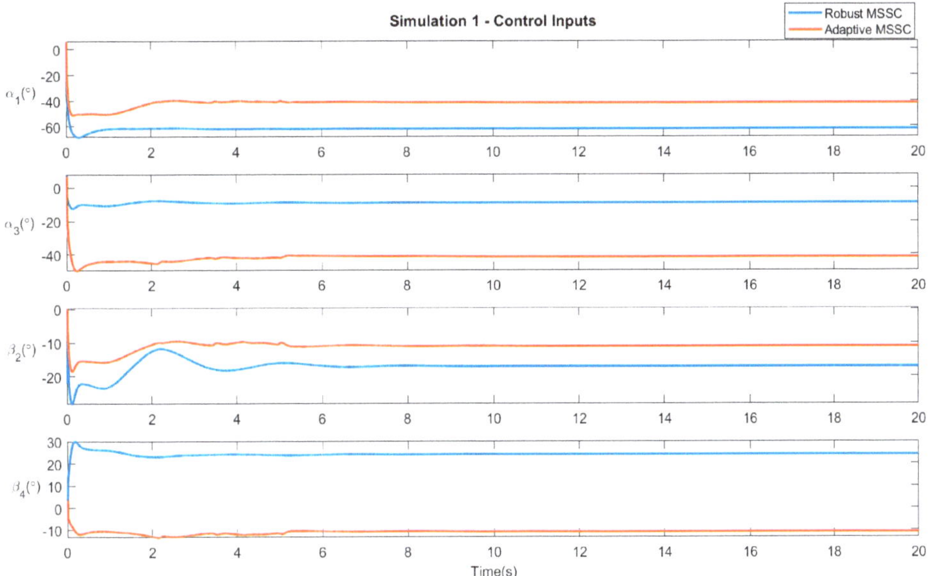

Figure 3. Results of Simulation 1: Control Inputs.

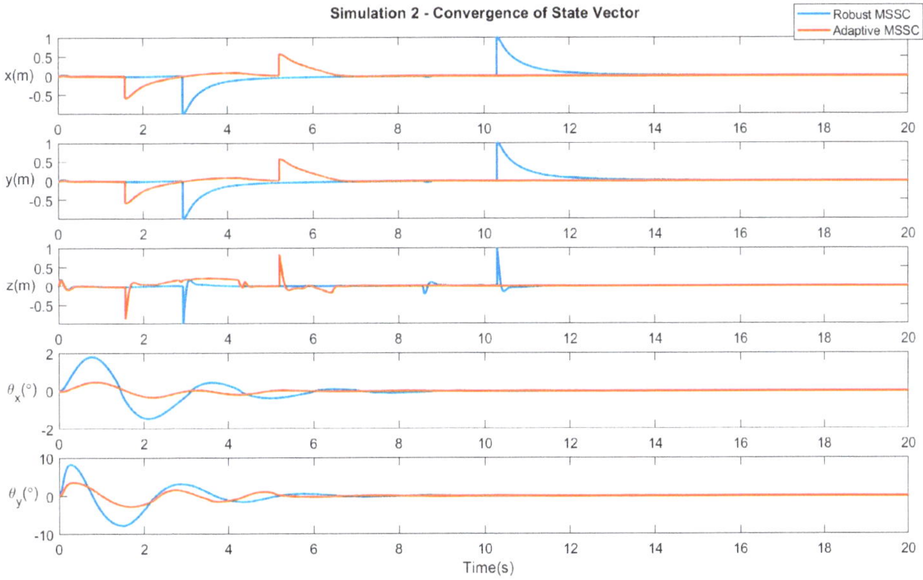

Figure 4. Results of Simulation 2: Convergence of system variables.

Simulation 4 —Random Disturbances at unspecified time intervals (intermittent disturbances): In the final simulation, the system is subjected to event-triggered disturbances, which are introduced at non-uniform time intervals. The magnitude of these disturbances is variant in nature, ranging from $-5 \times \text{rand}(1)$ to $5 \times \text{rand}(1)$ along each of the co-ordinate axes. Figure 8 verifies the convergence of the system states to the desired position. Figure 9 displays the control efforts necessary for the robust and adaptive variants of the proposed controller.

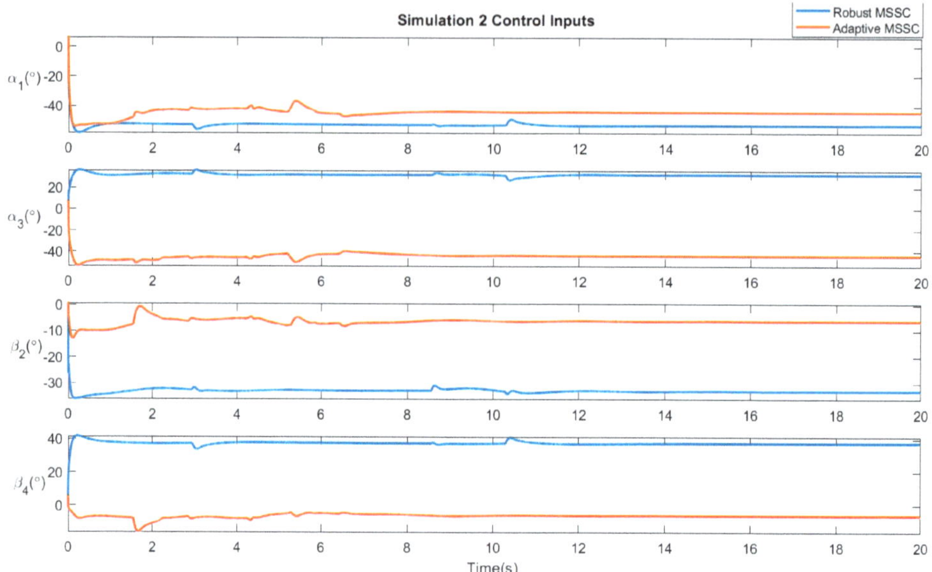

Figure 5. Results of Simulation 2: Control Inputs.

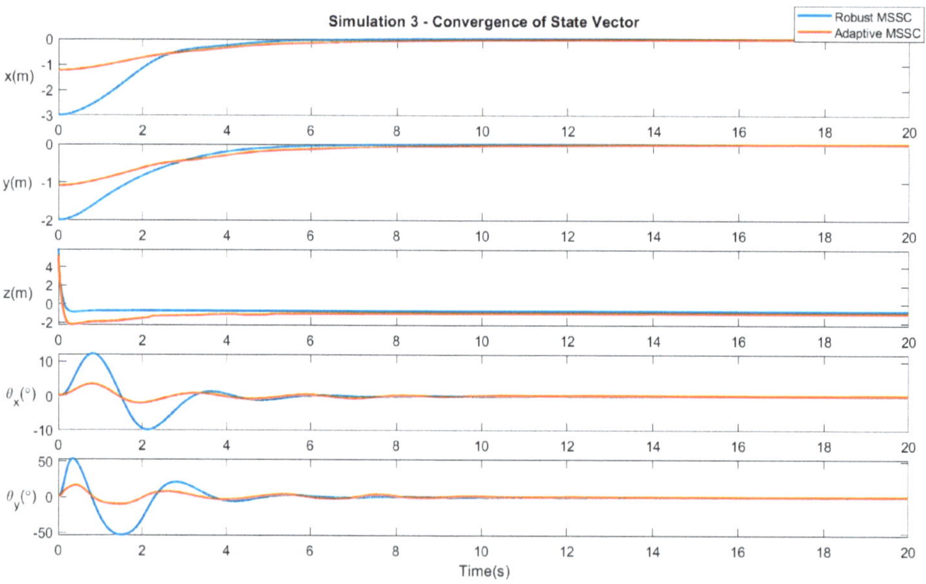

Figure 6. Results of Simulation 3: Convergence of system variables.

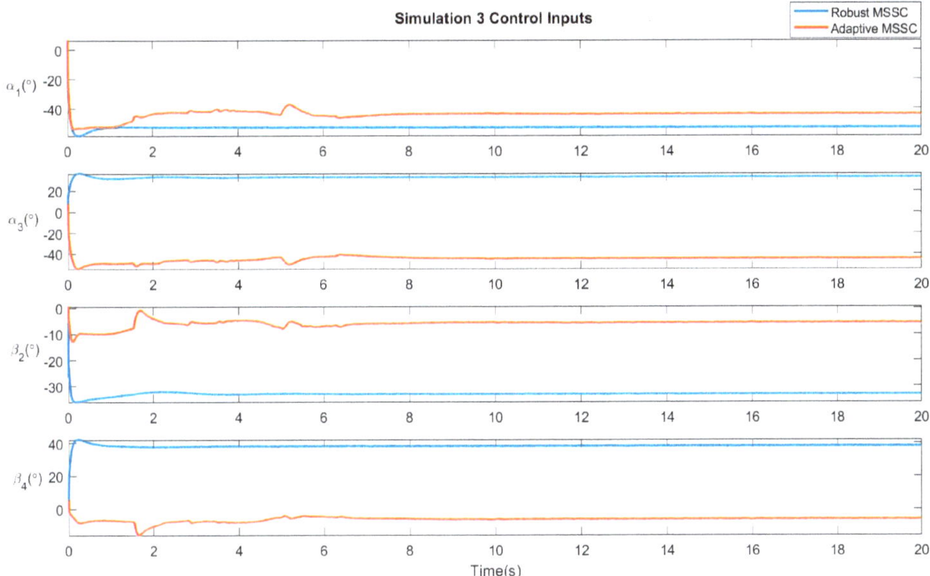

Figure 7. Results of Simulation 3: Control Inputs.

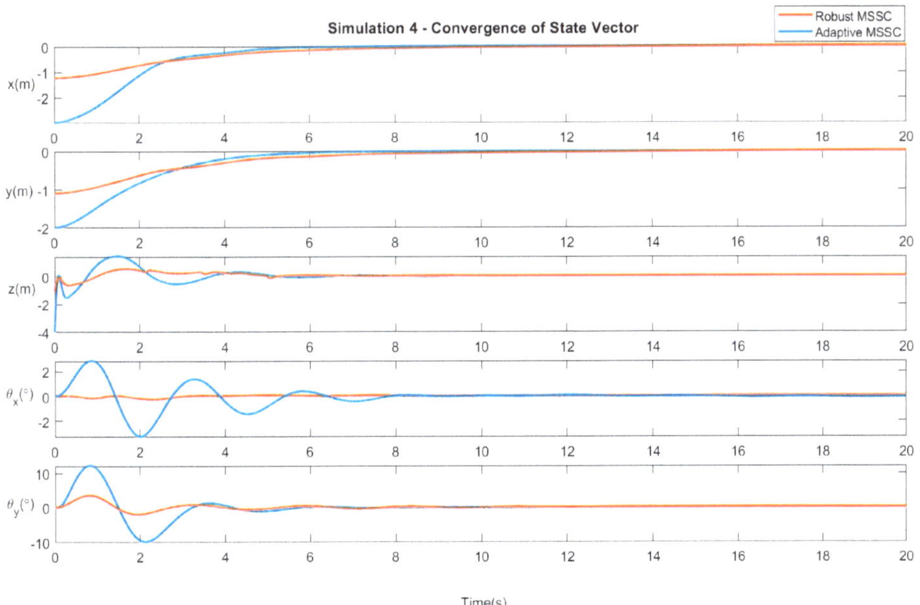

Figure 8. Results of Simulation 4: Convergence of system variables.

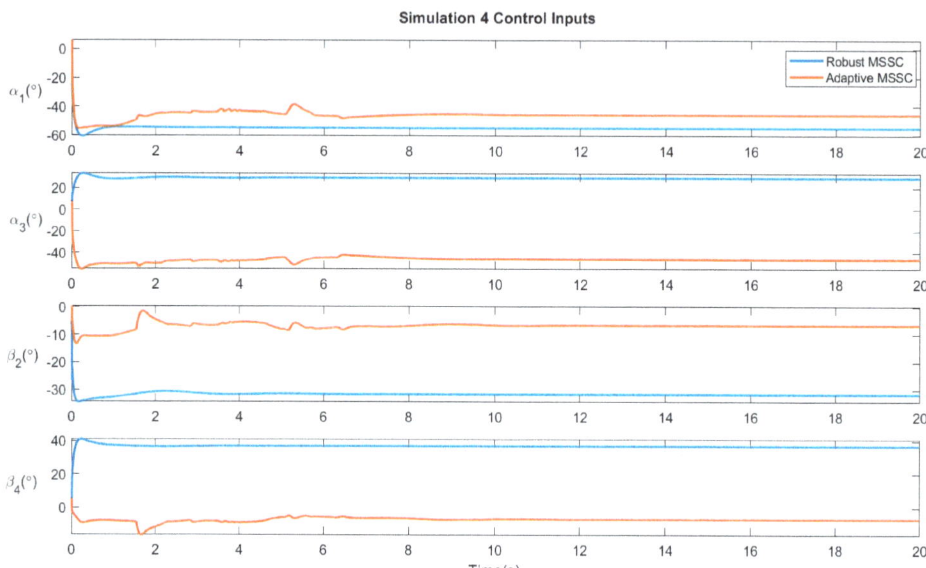

Figure 9. Results of Simulation 4: Control Inputs.

4. Discussion

The parameters defining the multirotor and the slung load are displayed in Table 1. Table 2 displays the step response characteristics of the system when subjected to the different controllers in their respective scenarios. The simulation results are displayed in Figures 2–9. Figures 2, 4, 6 and 8 verify the convergence of the system states, i.e., the quadrotor position and the payload swing angles, to the desired position. Figures 3, 5, 7 and 9 display the control efforts necessary for the robust and adaptive variants of the proposed controller. From Table 2, it can be seen that the adaptive control optimizes the overall control of the system by reducing the SettlingMin parameter by about 20%, in addition to reducing the undershoot by about 30%, hence increasing the overall stability of the system. This is due to the adaptive parameter $\tilde{\lambda}_1$, as defined in (60), which adjusts itself based on the difference between the desired and actual values of the system state, represented by sliding surface parameter s. When the system recedes or exceeds the desired position, the convergence rate of the system is adjusted by this parameter, as represented in (61) and (62). This results in an improvement of the settling time and, subsequently, an optimization of the system convergence. Hence, the adaptive control determines an optimal finite time of convergence for the system states. Looking at the control inputs supplied by the control systems in each case, the adaptive control makes several minute adjustments over the given time, while the robust controller quickly arrives at what it deems to be the necessary angle of the rotor. The optimization control helps in the elimination of chattering. Furthermore, the adaptive finite-time control also minimizes the oscillations of the swinging payload due to the quicker settling time, as seen in Table 3.

Table 2. Step response Characteristics of the System.

Parameter	Value							
	Simulation 1		Simulation 2		Simulation 3		Simulation 4	
	Robust	Adaptive	Robust	Adaptive	Robust	Adaptive	Robust	Adaptive
Rise Time	3.04×10^{-5}	1.84×10^{-5}	0.018	3.8×10^{-4}	0.0063	0.056	0.0033	0.052
Transient Time	8.23	9.13	10.5	11.72	8.43	9.34	8.31	9.7

Table 2. *Cont.*

Parameter	Value							
	Simulation 1		Simulation 2		Simulation 3		Simulation 4	
	Robust	Adaptive	Robust	Adaptive	Robust	Adaptive	Robust	Adaptive
Settling Time	19.95	20.00	19.98	20.00	19.89	19.95	18.87	19.86
Settling Min	−0.11	−0.06	−0.04	−0.03	−0.11	−0.07	−0.11	−0.07
Settling Max	0.03	0.03	0.04	0.02	0.02	0.01	0.03	0.02
Overshoot	9.85×10^3	1.46×10^3	4.08×10^3	7.13×10^2	4.76×10^3	3.38×10^3	9.21×10^3	3.97×10^3
Undershoot	1.25×10^4	0.7×10^3	4.04×10^3	1.02×10^2	5.87×10^3	3.35×10^3	4.15×10^3	3.72×10^3
Peak	0.14	0.079	0.04	0.04	0.138	0.07	0.14	0.076
Peak Time	0.78	0.84	0.75	0.81	0.77	0.83	0.78	0.83

Table 3. Quantitative Data of Simulations.

Parameter		Max		Mean		RMS	
		θ_x	θ_y	θ_x	θ_y	θ_x	θ_y
Simulation 1	Robust	18.07	11.22	0.092	2.2	2.6	2.9
	Adaptive	4.63	3.32	0.024	1.2	1.1	1.81
Simulation 2	Robust	10.24	11.17	0.049	3.5	2.1	4.8
	Adaptive	2.67	3.09	0.013	1.3	0.81	1.61
Simulation 3	Robust	12.07	52.74	1.1	6.35	9.1	7.17
	Adaptive	3.25	15.74	0.2	2.04	1.11	2.23
Simulation 4	Robust	2.83	12.07	0.19	6.25	9.1	7.05
	Adaptive	0.81	3.41	0.02	0.91	1.12	5.92

5. Conclusions

In this paper, the dynamic model of a multirotor UAV subjected to a slung load disturbance, modeled in the form of a single pendulum, is defined. Following this, robust and adaptive versions of a finite-time multi-surface sliding mode control are proposed and subsequently applied to the multirotor-slung load system. The proposed controllers are both effective in accounting for unknown system parameters and disturbances within a finite-time interval. The simulation results highlight the overall effectiveness of the adaptive finite-time MSSC over the robust finite-time MSSC, as the proposed adaptive control reduces the settling time by 20% and decreases the undershoot of the system dynamics by about 30%, in comparison to the robust control. It also provides an optimal finite-time convergence of the system state vector, hence increasing the overall stability of the control. With the disturbances in this case having only been simulated, there is scope for real-time uncertainties, such as wind, to be imposed upon an actual multirotor and for this control to be experimentally tested. Hence, future work will include the development and testing of an adaptive control to learn about and handle the matched and mismatched uncertainties that a multirotor UAV experiences in real time.

Author Contributions: Conceptualization, S.Y.K. and M.N.; methodology, C.P.; validation, C.P., M.N. and S.Y.K.; writing—original draft preparation, C.P.; writing—review and editing, M.N. and S.Y.K.; supervision, M.N. and S.Y.K. All authors have read and agreed to the published version of the manuscript.

Funding: This research received no external funding.

Data Availability Statement: No new data were created or analyzed in this study. Data sharing is not applicable to this article.

Conflicts of Interest: The authors declare no conflict of interest.

References

1. Kugler, L. Real-world applications for drones. *Commun. ACM* **2019**, *62*, 19–21. [CrossRef]
2. Tolba, M.; Shirinzadeh, B.; El-Bayoumi, G.; Mohamady, O. Adaptive optimal controller design for an unbalanced UAV with slung load. *Auton. Robot.* **2023**, *47*, 267–280. [CrossRef]
3. Omar, H.M.; Mukras, S.M. Developing geno-fuzzy controller for suppressing quadrotor slung-load oscillations. *Ain Shams Eng. J.* **2023**, *14*, 102051. [CrossRef]
4. Baraean, A.; Hamanah, W.M.; Bawazir, A.; Baraean, S.; Abido, M.A. Optimal Nonlinear backstepping controller design of a Quadrotor-Slung load system using particle Swarm Optimization. *Alex. Eng. J.* **2023**, *68*, 551–560. [CrossRef]
5. El Ferik, S.; Al-Qahtani, F.M.; Saif, A.-W.A.; Al-Dhaifallah, M. Robust FOSMC of quadrotor in the presence of slung load. *ISA Trans.* **2023**, *139*, 106–121. [CrossRef] [PubMed]
6. Roy, K.R.; Waghmare, L.M.; Patre, B.M. Dynamic modeling and displacement control for differential flatness of quadrotor UAV slung-load system. *Int. J. Dyn. Control* **2022**, *11*, 637–655. [CrossRef]
7. Liu, W.; Chen, M.; Shi, P. Fixed-Time Disturbance Observer-Based Control for Quadcopter Suspension Transportation System. *IEEE Trans. Circuits Syst. I Regul. Pap.* **2022**, *69*, 4632–4642. [CrossRef]
8. Labbadi, M.; Iqbal, J.; Djemai, M.; Boukal, Y.; Bouteraa, Y. Robust tracking control for a quadrotor subjected to disturbances using new hyperplane-based fast Terminal Sliding Mode. *PLoS ONE* **2023**, *18*, e0283195. [CrossRef] [PubMed]
9. Di Paola, V.; Goldsztejn, A.; Zoppi, M.; Caro, S. Design of a Sliding Mode-Adaptive Proportional-Integral-Derivative Control for Aerial Systems with a Suspended Load Exposed to Wind Gusts. *J. Comput. Nonlinear Dyn.* **2023**, *18*, 061008. [CrossRef]
10. Cui, C.; Shi, Y.; Wu, K.; Sheng, S. Research on Attitude Control of Unmanned Helicopter with Slung Load Combined Input Shaper and Linear Active Disturbance Rejection Control. In *Advances in Guidance, Navigation and Control*; Springer Nature: Singapore, 2003; pp. 5158–5168.
11. Yuan, X.; Ren, X.; Zhu, B.; Zheng, Z.; Zuo, Z. Robust H∞ Control for Hovering of a Quadrotor UAV with Slung Load. In Proceedings of the 2019 12th Asian Control Conference (ASCC), Kitakyushu, Japan, 9–12 June 2019; IEEE: Piscataway, NJ, USA, 2019; pp. 114–119.
12. Liu, L.; Chen, M.; Li, T. Disturbance Observer-based LQR Tracking Control for Unmanned Autonomous Helicopter Slung-load System. *Int. J. Control Autom. Syst.* **2022**, *20*, 1166–1178. [CrossRef]
13. Tan, L.; Shen, Z.; Yu, S. Adaptive fault-tolerant flight control for a quadrotor UAV with slung payload and varying COG. In Proceedings of the 2019 3rd International Symposium on Autonomous Systems (ISAS), Shanghai, China, 29–31 May 2019; IEEE: Piscataway, NJ, USA, 2019; pp. 227–231.
14. Zeng, J.; Sreenath, K. Geometric control of a quadrotor with a load suspended from an offset. In Proceedings of the 2019 American Control Conference (ACC), Philadelphia, PA, USA, 10–12 July 2019; IEEE: Piscataway, NJ, USA, 2019; pp. 3044–3050.
15. Gonzalez, F.; Heckmann, A.; Notter, S.; Zurn, M.; Trachte, J.; Mcfadyen, A. Non-linear model predictive control for UAVs with slung/swung load. In Proceedings of the IEEE International Conference on Robotics and Automation, Seattle, WA, USA, 26–30 May 2015; p. 1.
16. De La Torre, G.; Yucelen, T.; Johnson, E.N. Neuropredictive control and trajectory generation for slung load systems. In Proceedings of the AIAA Infotech@ Aerospace (I@ A) Conference, Boston, MA, USA, 19–22 August 2013; p. 5044.
17. Yang, K.; Kang, Y.; Sukkarieh, S. Adaptive nonlinear model predictive path-following control for a fixed-wing unmanned aerial vehicle. *Int. J. Control Autom. Syst.* **2013**, *11*, 65–74. [CrossRef]
18. Khoo, S.; Man, Z.; Zhao, S. Comments on "Adaptive multiple-surface sliding control for non-autonomous systems with mismatched uncertainties". *Automatica* **2008**, *44*, 2995–2998. [CrossRef]
19. Levant, A.; Livne, M. Exact Differentiation of Signals with Unbounded Higher Derivatives. *IEEE Trans. Autom. Control* **2011**, *57*, 1076–1080. [CrossRef]
20. Qian, D.; Yi, J.; Zhao, D. Multiple layers sliding mode control for a class of under-actuated systems. In Proceedings of the Multiconference on Computational Engineering in Systems Applications, Beijing, China, 4–6 October 2006; IEEE: Piscataway, NJ, USA, 2006; pp. 530–535.
21. Xu, R.; Özgüner, Ü. Sliding mode control of a class of underactuated systems. *Automatica* **2008**, *44*, 233–241. [CrossRef]
22. Ferrara, A.; Incremona, G.P.; Vecchio, C. Adaptive Multiple-Surface Sliding Mode Control of Nonholonomic Systems with Matched and Unmatched Uncertainties. *IEEE Trans. Autom. Control* **2023**, 1–8. [CrossRef]
23. Soltanian, F.; Shasadeghi, M.; Mobayen, S.; Fekih, A. Adaptive Optimal Multi-Surface Back-Stepping Sliding Mode Control Design for the Takagi-Sugeno Fuzzy Model of Uncertain Nonlinear System with External Disturbance. *IEEE Access* **2022**, *10*, 14680–14690. [CrossRef]
24. Abu Alqumsan, A.; Khoo, S.; Norton, M. Multi-surface sliding mode control of continuum robots with mismatched uncertainties. *Meccanica* **2019**, *54*, 2307–2316. [CrossRef]

25. Ginoya, D.; Shendge, P.; Phadke, S. Disturbance observer based sliding mode control of nonlinear mismatched uncertain systems. *Commun. Nonlinear Sci. Numer. Simul.* **2015**, *26*, 98–107. [CrossRef]
26. Yang, S.; Xian, B.; Cai, J.; Wang, G. Finite-time Convergence Control For A Quadrotor Unmanned Aerial Vehicle with A Slung load. *IEEE Trans. Ind. Inform.* **2023**, 1–9. [CrossRef]
27. Lv, Z.; Zhao, Q.; Li, S.; Wu, Y. Finite-time control design for a quadrotor transporting a slung load. *Control Eng. Pract.* **2022**, *122*, 105082. [CrossRef]
28. SGajbhiye, S.; Cabecinhas, D.; Silvestre, C.; Cunha, R. Geometric finite-time inner-outer loop trajectory tracking control strategy for quadrotor slung-load transportation. *Nonlinear Dyn.* **2021**, *107*, 2291–2308. [CrossRef]
29. Wang, M.; Chen, B.; Lin, C. Prescribed finite-time adaptive neural trajectory tracking control of quadrotor via output feedback. *Neurocomputing* **2021**, *458*, 364–375. [CrossRef]
30. Khoo, S.; Norton, M.; Kumar, J.J.; Yin, J.; Yu, X.; Macpherson, T.; Dowling, D.; Kouzani, A. Robust control of novel thrust vectored 3D printed multicopter. In Proceedings of the 2017 36th Chinese Control Conference (CCC), Dalian, China, 26–28 July 2017; IEEE: Piscataway, NJ, USA, 2017; pp. 1270–1275.
31. Hardy, G.H.; Littlewood, J.E.; Pólya, G. *Inequalities (Cambridge Mathematical Library)*; Cambridge University Press: Cambridge, UK, 1934.
32. Huang, X.; Lin, W.; Yang, B. Global finite-time stabilization of a class of uncertain nonlinear systems. *Automatica* **2005**, *41*, 881–888. [CrossRef]
33. Khoo, S.; Zhao, S.; Man, Z. Adaptive fast finite-time multiple-surface sliding control for a class of uncertain non-linear systems. *Int. J. Model. Identif. Control* **2012**, *16*, 392–400. [CrossRef]

Article

An Inverse Kinematics Approach for the Analysis and Active Control of a Four-UPR Motion-Compensated Platform for UAV–ASV Cooperation

Pedro Pereira [1,*] , Raul Campilho [2,3]   and Andry Pinto [1,4]

1   Faculdade de Engenharia da Universidade do Porto, 4200-465 Porto, Portugal
2   Instituto Superior de Engenharia do Porto, 4200-072 Porto, Portugal
3   INEGI, Pólo FEUP, 4200-465 Porto, Portugal
4   INESC TEC, Pólo FEUP, 4200-465 Porto, Portugal
*   Correspondence: pedro.arrojado.pereira@gmail.com

Abstract: In the present day, unmanned aerial vehicle (UAV) technology is being used for a multitude of inspection operations, including those in offshore structures such as wind-farms. Due to the distance of these structures to the coast, drones need to be carried to these structures via ship. To achieve a completely autonomous operation, the UAV can greatly benefit from an autonomous surface vehicle (ASV) to transport the UAV to the operation location and coordinate a successful landing between the two. This work presents the concept of a four-link parallel platform to perform wave-motion synchronization to facilitate UAV landings. The parallel platform consists of two base floaters connected with rigid rods, linked by linear actuators to a top mobile platform for the landing of a UAV. Using an inverse kinematics approach, a study of the position of the cylinders for greater range of motion and a workspace analysis is achieved. The platform makes use of a feedback controller to reduce the total motion of the landing platform. Using the robotic operating system (ROS) and Gazebo to emulate wave motions and represent the physical model and actuator system, the platform control system was successfully validated.

Keywords: parallel manipulator; unmanned aerial vehicle; inverse kinematics; screw theory; reciprocal screws; PID controller; robot operating system; wave-motion synchronization

Citation: Pereira, P.; Campilho, R.; Pinto, A. An Inverse Kinematics Approach for the Analysis and Active Control of a Four-UPR Motion-Compensated Platform for UAV–ASV Cooperation. *Machines* **2023**, *11*, 478. https://doi.org/10.3390/machines11040478

Academic Editors: Octavio Garcia-Salazar, Anand Sanchez-Orta and Aldo Jonathan Muñoz-Vazquez

Received: 7 March 2023
Revised: 10 April 2023
Accepted: 11 April 2023
Published: 14 April 2023

1. Introduction

Closed-loop mechanisms are parallel manipulators of a fixed and a moving platform connected by a set of kinematic chains [1]. The main advantages of a parallel manipulator are its high stiffness, compact size, capability for control with large bandwidth, robustness against external force and error accumulation, high dexterity, and suitability for accurate positioning systems [2]. The downsides of this type of mechanism are its limited workspace, and, because of the nonlinearity and complexity of the equations, a forwards dynamics solution is very challenging and complex to obtain. Concerning the design process of robotic mechanisms, serial kinematic chains are somewhat limited in their arrangements for a given motion pattern. This is not true for parallel mechanisms, which can have a very large variation of arrangements for a given motion pattern [3]. Some of the more famous examples of closed-chain mechanisms can be found in the Delta robot or the Stewart–Gough platform [4], illustrated in Figure 1. Evaluation of the degrees of freedom (DoF) of the end-effector constitutes one of the core issues in free-motion analysis of parallel mechanisms. In contrast to serial mechanisms, classical approaches, such as the traditional Kutzbach–Grübler criterion, are not valid for certain complex spatial parallel mechanisms [5]. According to the work of Merlet [6], the most widely used methods for structural synthesis of parallel mechanisms are divided into three main approaches, namely, graph theory, group theory, and screw theory. Kong and Gosselini [7] later introduced the concept of virtual chains for the type synthesis of parallel mechanisms. These authors also performed a mobility analysis using the screw-theory-based virtual-chain approach for numerous of different possible configurations of parallel robots. The theory of reciprocal screws was first introduced by Zhao et al. [8] and has since been used in many other works related to the field of parallel mechanisms and DoF estimation problems [9–11]. Regarding the development of parallel manipulators, there is a need to identify the workspace of these mechanisms, which frames the reachable positions of the centroid of the end-effector. The workspace of a parallel manipulator imposes limitations that depend on its practical application [12]. Determining the boundary of a workspace typically relies on the manipulator's pose parameter discretization. To establish the set of possible, reachable positions, the joint limits and a kinematic model of the mechanism must be established. This can be accomplished with either forward kinematics [13] or inverse kinematics [14] approaches. Due to the multiple limbs of parallel mechanisms, solutions for forward kinematics require numerical problem-solving, while inverse kinematics can be solved analytically [15]. Parallel manipulators have seen a wide variety of applications in different fields [16,17]. The advantages provided by parallel mechanisms potentiate the application of robotics for numerous applications, such as Jones and Dunlop's proposal of a closed-loop mechanism for satellite trackers [18]. Wapler et al. [19] used a Stewart platform to design a high-precision surgery mechanism with an accuracy of 20 µm.

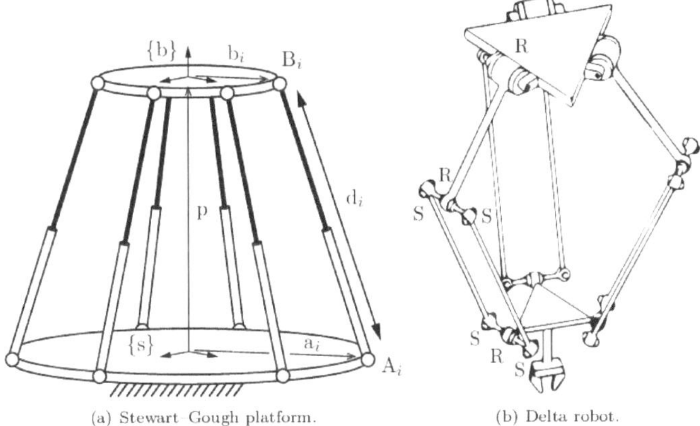

(a) Stewart Gough platform.

(b) Delta robot.

Figure 1. Popular parallel mechanisms [1].

Compared to traditional power generators on land, offshore wind energy technology presents the benefit of collecting more power due to stronger winds and the absence of landscape effects [20]. However, since offshore structures have difficult accessibility, implementation and maintenance can be time-consuming and costly [21]. Under these conditions, to reduce costs and avoid hazard to humans, underwater vehicles and inspection UAVs are an attractive option. Intervention autonomous underwater vehicles can be used together with imaging systems for inspection [22] and robotic arms for repairing [23]. UAVs can be equipped with imaging systems and perform pre-established paths for the inspection of critical parts of the structure [24]. Nonetheless, these technologies are yet in an incipient stage of evolution [25]. For offshore operations, UAVs often need a place to land closer to the structure, which ASVs could potentially provide. The motion caused by waves is mostly unpredictable, and, depending on sea conditions, it can produce large displacements of roll, pitch, yaw, sway, surge, and heave motions on a surface vehicle. Being so highly dependent on the weather and sea conditions, working in offshore environments often leads to a low working efficiency and economic loss due to unworkable conditions. The recent literature has explored the potential for cooperation between ASVs and UAVs in the automatic launch and recovery of the UAVs. These applications usually employ vision techniques, so the UAV is capable of detecting the landing pad located on an ASV. However, these techniques are still vulnerable to highly dynamic sea states [26]. To overcome the difficulties of landing the UAV, Li et al. [27] developed an ASV with an attitude-prediction controller designed by a bidirectional long-short term memory neural network and a proportional–integral–derivative (PID) pose controller for the UAV to guarantee real-time synchronization movement of the UAV–ASV. Their results showed that this type of cooperation effectively improves the landing accuracy. Furthermore, Liu et al. [28] explored the potential of parallel mechanisms for UAV landings in highly uneven and unstructured land terrain using a three-RRR-limb parallel mechanism that functions both as a manipulator and as adaptive landing gear. Outside the UAV-related research, some attempts at wave-motion compensation using parallel mechanisms are exemplified in the work of Guo et al. [29], which proposes a 4-CPS/RPS parallel mechanism with a classic PID controller to mitigate wave motion induced in a ship. Chen et al. [30] proposed a parallel manipulator driven by pitch and roll active motion compensation with a fuzzy sliding mode controller and a PID controller to diminish the risk of humans getting in and out of offshore wind turbine structures. Cai et al. [31] studied the concept of a ship-mounted Stewart platform for equipment such as cranes and drilling platforms to eliminate the effect of wave-induced ship motions, effectively increasing the workable time for offshore installations. A common issue many of these works face is that, due to the variability of sea conditions, a controller system may be adequate for a given sea state and inadequate for others.

The ATLANTIS project [32] aims to establish an infrastructure to allow the demonstration of key enabling robotic technologies for the inspection and maintenance of offshore wind turbines. One of the focuses of this project is the use of UAVs for blade inspection. Considering the distance of offshore structures to shore and the limited capabilities of UAV batteries it is necessary for UAVs to be recovered near the mission site after completing the inspection. Considering the set of advantages offered by a parallel mechanism, this work proposes a novel motion-compensated four-UPR parallel platform designed for the autonomous landing of UAVs in offshore environments based on the theory of reciprocal screws. There has been some work dedicated to a four-UPU parallel mechanism, which has Shöenflies type-motion (three translations plus one rotation) [33]. However, for the intended purposes, the platform presented in this work only needs two rotations around the x and y-axis to compensate for wave motions and one translation along the z-axis to position the landing platform. Thus, a variation of the four-UPU platform is introduced with the four UPR limbs. This variation is not described in the literature. Parallel mechanisms with fewer than six DoFs excel in terms of simpler structure, easier control, and lower cost. While this mechanism can be achieved with three limbs, a fourth one adds more rigidity and maintains the symmetry of the ASV. Having a permanently horizontal platform to land, the UAV is much less susceptible to land in an uneven position and topple, possibly damaging its sensors/structure or, in a worst case scenario, falling overboard. The first tests of this platform are to be conducted at the ATLANTIS test centre's buoy DURIUS. DURIUS is close to the coast, so the wave dynamics will not be as harsh as in offshore environments. To test the concept of the four-UPR mechanism for different wave states, this work implements a PID controller to establish a baseline performance, which provides a good solution for these conditions. This work addresses the difficult task of safely launching and recovering a UAV on a landing pad influenced by wave motions. The implementation of this platform offers more flexibility in UAV–ASV cooperative missions for offshore wind-turbine inspection.

This work is organized as follows: In Section 2, the problem and platform mechanism presented in this paper are described, as well as a DoF analysis based on the theory of reciprocal screws. Section 3 describes the systematic approach and solution for the inverse kinematics of a four-UPR parallel mechanism. From this solution, the limb lengths, position, and velocity of the passive joints are obtained for any given position and orientation of the mobile platform. Section 4 details the workspace and limits the analysis of the platform.

Section 5 presents the platform-simulation environment and the results of the implemented PID controller. Finally, Section 6 provides a discussion of the proposed work.

2. Platform Description

2.1. Problem Statement

One of the greatest limitations in UAV–ASV cooperative motions is in landing the UAV safely in highly dynamic sea states. Wave-motion compensation has been a topic of research for some time; however, there have been few attempts at tackling the issue of landing UAVs on a platform at sea. This is a difficult task, since highly dynamic waves and wind will induce significant angular displacements on the platform, and a bad landing can be devastating for the UAV. In this work, we attempt to tackle this challenge using a parallel mechanism that is capable of maintaining a horizontal position while the drone is trying to land. Consider the mission illustrated in Figure 2. Here, the main steps of the mission are as follows: the UAV takes off from the platform near the structure; it performs the inspection of the structure and registers relevant data using sensors, such as visual cameras and LiDARs (light detection and ranging); upon completing the inspection, the drone returns to the ASV using a perception system that combines visual, thermal, and three-dimensional LiDAR to detect an ArTuga equipped on the ASV [34]; once detected, the UAV attempts to land on the ASV's platform. This work addresses the first and last steps, which involve the launch and recovery of the UAV.

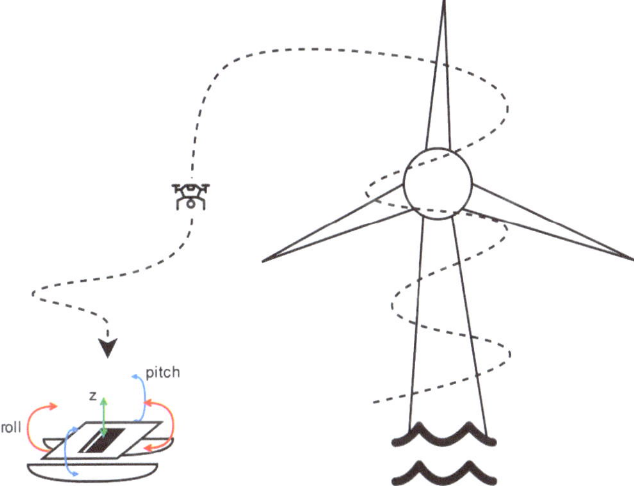

Figure 2. UAV–ASV cooperation for inspection and maintenance of offshore wind turbines. Mission schematic.

The proposed ASV has a 4-UPR mechanism depicted in Figure 3 with a top mobile platform for the landing of UAVs linked to the floating base platform with four limbs. Each limb consists of a revolute joint (R-joint), followed by a prismatic joint, and ending in a universal-joint (U-joint). The limb length is characterized by l_i. This mechanism is capable of translational movement along the z axis and rotational movements along the x- and y-axis. The orientation and position of the platform is provided by an inertial measurement unit (IMU) installed in the base platform. The centre point of the global coordinate system is located at origin O (centre of the base platform).

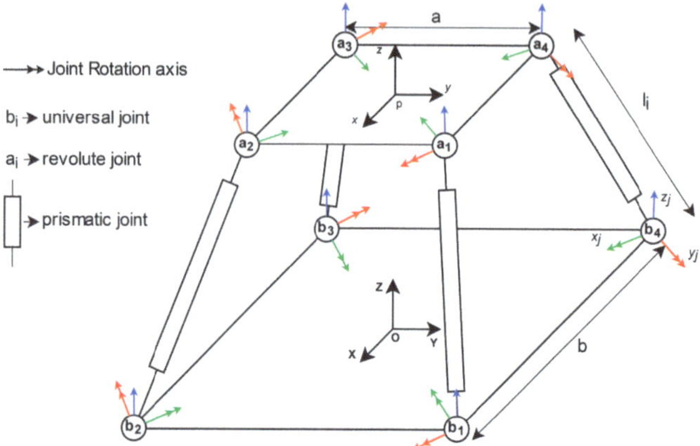

Figure 3. The 4-UPR platform mechanism schematics.

The top and bottom joints of the platforms form a square with side a and b, respectively. The z-axis is perpendicular to the platform, and the x-axis and y-axis are oriented in such a way that the positions of the b_i joints are given by:

$$
\begin{aligned}
b_1 &= \begin{bmatrix} b/2 & b/2 & 0 \end{bmatrix}^T \\
b_2 &= \begin{bmatrix} b/2 & -b/2 & 0 \end{bmatrix}^T \\
b_3 &= \begin{bmatrix} -b/2 & -b/2 & 0 \end{bmatrix}^T \\
b_4 &= \begin{bmatrix} -b/2 & b/2 & 0 \end{bmatrix}^T
\end{aligned}
\tag{1}
$$

It is assumed that a_i is the position of the upper platform joints relative to the origin frame. The mobile frame of reference, p, located at the centre of the upper platform, is oriented with the Z-axis perpendicular to the mobile platform, and the X-axis and Y-axis are oriented in such a way that the positions of the a_i joints relative to p are:

$$
\begin{aligned}
pa_1 &= \begin{bmatrix} a/2 & a/2 & 0 \end{bmatrix}^T \\
pa_2 &= \begin{bmatrix} a/2 & -a/2 & 0 \end{bmatrix}^T \\
pa_3 &= \begin{bmatrix} -a/2 & -a/2 & 0 \end{bmatrix}^T \\
pa_4 &= \begin{bmatrix} -a/2 & a/2 & 0 \end{bmatrix}^T
\end{aligned}
\tag{2}
$$

For a proper workspace definition, the limits of each joint need to be established. The prismatic joint is representative of an electric actuator with a retracted cylinder size of 0.7225 m, with a stroke of 0.4 m, defining l_i as $0.7225 \geq l_i \geq 1.1225$ m. The angle between the two parts that constitute a U-joint can be represented by $\beta_i = \begin{bmatrix} roll & pitch & 0 \end{bmatrix}$, where roll and pitch represent the angle around the local joint frame's x_i and y_i axis, respectively. The U-joint roll and pitch angles can be defined by the dot product of the limb frame x- and y-axis with the joint frame x- and y-axis. The limb frame is a rotation of the joint frame, where z is axial to the limb. The joint limit established for the U-joints is 30 degrees, as it is a common limit in standard models, defining β as $-45 \geq \beta \geq 45$ degrees.

As shown in Figure 4, the base platform consists of two floaters connected by rigid rods. The U-joint's position at the base, b_i, and the R-joint's position at the upper platform are adjustable, so the length of a and b can be optimized for greater mobility and angular velocity. A key point for establishing the configuration of the centroid p is the position in the global frame Z-axis. A lower centre of gravity will allow the platform not to topple with more aggressive wave amplitudes. Another important aspect to consider is the distance between the platforms, in order to avoid a collision between the two bodies when rotating the top part. The top platform is a square with size 2×2 m². To avoid a collision, a relationship between the platform orientation and the z position of the platform needs to be established. Consider α the angle between the z'-axis and z-axis of the frame p, where the z-axis is the z-vector of the platform initial orientation, given by $\begin{bmatrix} 0 & 0 & 1 \end{bmatrix}$, and the z' vector is the z-axis after a rotation. Thus, α can be obtained by the dot product of z and z' as:

$$\alpha = \arccos \frac{z \cdot z'}{\|z\| \times \|z'\|} \qquad \|z\| > \sin(\alpha) \times \sqrt{2} \tag{3}$$

Figure 4. 3D model of the 4-UPR platform mechanism.

2.2. Degree of Freedom Estimation

To determine the quantity, type, and direction axes of the end-effector DoFs for the 4-UPR platform presented in this work, the applied method relies on screw theory and reciprocal screw systems. In this method, the constraint space of the end-effector is studied according to the terminal constraint space of its kinematic chains. As previously mentioned, this method was first developed by Zhao et al. [8]. A more mature demonstration is presented in the book by the same author [5]. Let us consider the kinematic chain presented in Figure 5, which represents the ith limb of the platform.

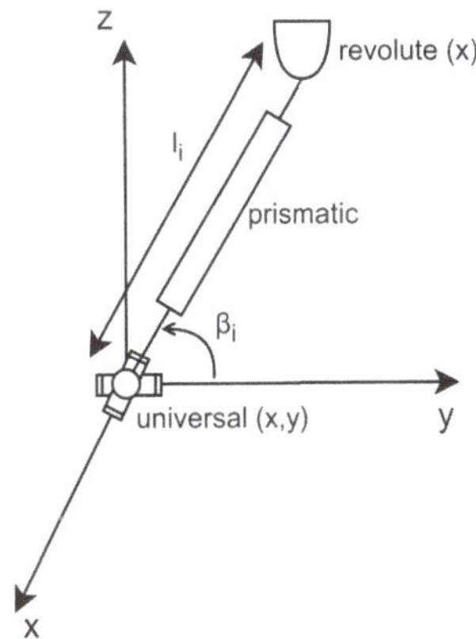

Figure 5. Kinematic chain of the ith limb of the platform.

The Plücker coordinates of the kinematic pair of the joints belonging to the limb can be written as:

$$\$_1 = \begin{bmatrix} 0 & 1 & 0; & 0 & 0 & 0 \end{bmatrix}^T \tag{4}$$

$$\$_2 = \begin{bmatrix} 1 & 0 & 0; & 0 & 0 & 0 \end{bmatrix}^T \tag{5}$$

$$\$_3 = \begin{bmatrix} 0 & 0 & 0; & 0 & \cos\beta_i & \sin\beta_i \end{bmatrix}^T \tag{6}$$

$$\$_4 = \begin{bmatrix} 1 & 0 & 0; & 0 & -l_i \times \sin\beta_i & l_i \times \cos\beta_i \end{bmatrix}^T \tag{7}$$

Now, the Plücker coordinates of the limb, $\$_{l_i}$, are defined as:

$$\$_{l_i} = \begin{bmatrix} \$_1 & \$_2 & \$_3 & \$_4 \end{bmatrix}^T \tag{8}$$

According to reciprocal screw theory, two screws that satisfy Equation (9) are a pair of reciprocal screws. According to the physical meaning of reciprocal screws, the reciprocal of $\$_{l_i}$ denotes the constraints forces applied to the kinematic chain.

$$\$^T \Delta \$^r = 0 \tag{9}$$

$$\Delta = \begin{bmatrix} 0 & I \\ I & 0 \end{bmatrix} \tag{10}$$

where Δ is a 6×6 matrix. The solution for $\$_{l_i}^r$ can be obtained with linear algebra methods.

$$\$_{l_i}^r = \begin{bmatrix} 1 & 0 & 0; & 0 & 0 & 0 \\ 0 & 0 & 0; & 0 & 0 & 1 \end{bmatrix}^T \tag{11}$$

which denotes a pure couple that is parallel to the z-axis and a force parallel to the x-axis. The movement constrained by $\$_{l_i}^r$ is given by $\$_{l_i}^m = \begin{bmatrix} 0 & 0 & 1; & 1 & 0 & 0 \end{bmatrix}^T$, denoting a rotation parallel to the z-axis and a translation along the x-axis. In the global frame of reference, $O - XYZ$ (see Figure 3), the constrained motion screw can be represented as:

$$\$_{l_i}^m = \begin{bmatrix} 0 & 0 & 1; & \cos\alpha_i & \sin\alpha_i & 0 \end{bmatrix}^T \tag{12}$$

where α_i is the angle of the local x-axis of the ith limb to the global frame of reference. The constraints of the end-effector are directly related to the terminal constraints of the kinematic chains. It is possible to find the constraints of the end-effector by analysing the span ($\$_P^C$) of its constrained motions:

$$\$_P^C = span\left\{ \begin{bmatrix} \$_{l_1}^m & \$_{l_2}^m & \$_{l_3}^m & \$_{l_4}^m \end{bmatrix}^T \right\} \tag{13}$$

$$\$_P^C = \begin{bmatrix} 0 & 0 & 1; & 0 & 0 & 0 \\ 0 & 0 & 0; & 1 & 0 & 0 \\ 0 & 0 & 0; & 0 & 1 & 0 \end{bmatrix} \tag{14}$$

The DoF of a mechanism are given by $F = 6 - d$, where d is the number of dimensions that all the reciprocal screws can be spanned in the normal linear spaces. It is clear that this parallel mechanism has 3 DoF and can rotate about the X and Y-axis and translate in the Z-axis.

3. Inverse Kinematics

Inverse kinematics refers to the construction of the kinematic equations, so that, for a given end-effector position, the joint variables can be established. The orientation of the upper platform is defined by a rotation matrix around the roll (θ_x) and pitch (θ_y) angles, provided by an IMU on the bottom platform. Since it is not possible, nor advantageous, to control the yaw of the platform, it is assumed that θ_z is always zero. The rotation matrix is given by:

$$R = \begin{bmatrix} R_y R_x \end{bmatrix} = \begin{bmatrix} c\theta_y & s\theta_x \times s\theta_y & s\theta_y \times c\theta_x \\ 0 & c\theta_x & -s\theta_x \\ -s\theta_y & c\theta_y \times s\theta_x & c\theta_y \times c\theta_x \end{bmatrix} \tag{15}$$

where c and s are the cosine and sine trigonometric functions. Considering the DoFs of the platform, at any time instance, the state of the upper platform can be represented with the orientation θ and the centroid position P, in global coordinates, given by:

$$P = \begin{bmatrix} 0 & 0 & P_z \end{bmatrix}$$
$$\theta = \begin{bmatrix} \theta_x & \theta_y & 0 \end{bmatrix} \tag{16}$$

The position of the upper U-joints, a_i, can be solved as:

$$a_i = P + R \times pa_i \tag{17}$$

Substituting Equations (2), (15), and (16) in Equation (17) gives the position of all top platform joints. Since the lower U-joint's position is a known constant, the length of each limb (l_i) can be determined as:

$$l_i = \sqrt{[a_i - b_i][a_i - b_i]^T} \tag{18}$$

This parameter defines the necessary length of the prismatic cylinders, and therefore the control input, for the desired platform position and orientation. The vector that represents the ith limb is given by:

$$S_i = a_i - b_i \tag{19}$$

The unit vector that defines the ith limb is:

$$s_i = S_i / l_i \tag{20}$$

Considering $\dot{\theta} = [\dot{\theta}_x, \dot{\theta}_y, 0]^T$ as the angular velocity and $\dot{P} = [0, 0, \dot{P}_z]^T$ the linear velocity of the centroid p, the velocity of each limb can be determined as:

$$\dot{S}_i = \dot{\theta} \times a_i + \dot{P} \tag{21}$$

The sliding velocity, given by the velocity component along the limb, can be determined as:

$$\dot{l}_i = s_i \cdot \dot{S}_i \tag{22}$$

The inverse kinematics analysis enables the study of how the different design parameters of the platform influence the constrictions of movement. The different design parameters are considered as:

- Side a of the top platform, which denotes the position of the top joints. Smaller values of a mean more angular movement is achieved for shorter strokes, which can be explored to compensate for slower linear actuators. On the other hand, small values of a can lead to some instability regarding the stiffness of the mobile platform.
- Side b of the base platform, which denotes the position of the bottom joints. By distancing the bottom joints from one another, it is expected that the mechanism is more stable for harsher dynamic conditions. However, the steepness of the linear actuators also influences the angular speed of the mobile platform.
- Position along the Z-axis. This design parameter is adaptable even after construction, as it depends entirely on the linear actuator's stroke position. The major influence of this parameter is on the joints and collision limits of the mobile platform, and the position of the centre of mass of the mechanism.

4. Workspace Analysis

The workspace of a robot is defined by the volume of space reachable by its end-effector. For this parallel mechanism, let us consider that the end-effector is defined by the upper platform's centroid, p, and the position of the top joints, a_i. Due to the nature and application of this mechanism, it is not necessary to control the z-position while compensating for the wave motions. However, due to the limits of the platform, be it either through collision within itself or the joint limits established in Section 2, the position of the platform in the Z-axis can be studied to conciliate the platform stability with the range of motion. Using the inverse kinematics analysis performed in Section 3, the joints and platform positions, as well as the limbs length and orientations, can be calculated. The bottom U-joint angles, β, can be obtained through the s_i unit vectors:

$$s_i = \begin{bmatrix} s_{xi} & s_{yi} & s_{zi} \end{bmatrix} \tag{23}$$

$$\beta_{xi} = \arccos(s_{xyi} \cdot s_i) \tag{24}$$

$$s_{xyi} = \frac{\begin{bmatrix} s_{xi} & s_{yi} & 0 \end{bmatrix}}{\left\| \begin{bmatrix} s_{xi} & s_{yi} & 0 \end{bmatrix} \right\|} \tag{25}$$

$$\beta_{yi} = \arccos\left(s_{xyi} \cdot \begin{bmatrix} 0 & 1 & 0 \end{bmatrix}\right) - \frac{\pi}{4} \tag{26}$$

In Equation (26) we subtract $\pi/4$, since the local y-axis of the joint is at 45 degrees to the global Y-axis. Due to the disposition of the platform and the symmetrical nature of the joint's disposition the top platform R-joint angles, γ_i, can be obtained as:

$$\gamma_i = \arccos\left(\frac{p - a_i}{\|p - a_i\|} \cdot s_i\right) \tag{27}$$

The superscript "$(\cdot)^0$" refers to the initial time instance. For the workspace analysis, the required numerical elements are the size of a and b and the position of z^0, as well as the limits of each joint. The workspace of the platform at position z^0 can be defined through a sweep, where the limits are computed through each iteration of possible roll and pitch combinations. To ease the computation load, the limits for non-composite movements (only roll or only pitch) were first determined for $z^0 = 0.5$ m, $a = 0.2$ m, and $b = 1.25$ m. The choice of z^0 is determined by the maximum angle required to stabilize the upper platform for a specific wave height. To determine the optimal value of z^0, an iterative process was performed using the simulator presented in Section 5. Considering the roughest wave parameters used in the simulations, in a sea state of 3 on the Douglas sea scale, this angle was found to be 20 degrees, which justifies z^0 to be 0.5 m. The results of the sweep are shown in Table 1. For this mechanism, the limits of the roll and pitch of the end-effector are identical. In addition, the lower and upper limits are symmetrical.

Table 1. Range of motion for the roll and pitch of the end-effector at $z = 0.5$ m.

Motion Variable	Range of Sweep	Verified Limit	Units
Pitch	$-\pi/2$ to $\pi/2$	$-\pi/9$ to $\pi/9$	rad
Roll	$-\pi/2$ to $\pi/2$	$-\pi/9$ to $\pi/9$	rad

For the set values of z^0, a, and b described above, the workspace of the upper platform a_i U-joints can be seen in Figure 6. The edges of the upper platform (2×2 m^2) are represented by a green outline in Figure 6a. For better clarification, the same plot can be observed in Figure 6b, with more resolution and without the edges. The projected 2D range of motion of the U-Joint a_1 can be seen in Figure 7.

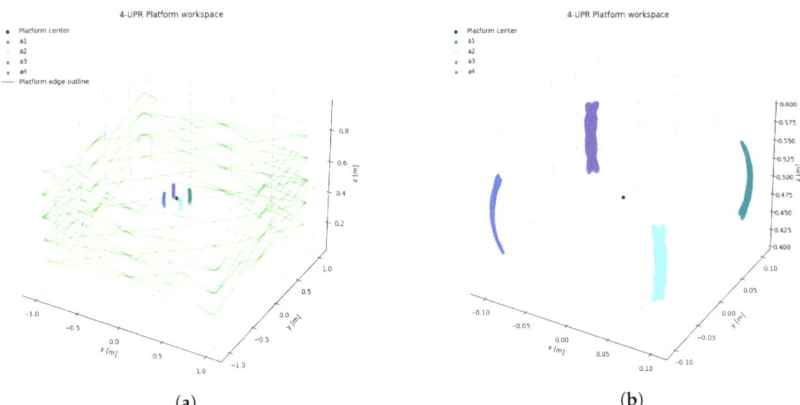

(a)   **(b)**

Figure 6. Workspace analysis of a 4-UPR platform for $z = 0.5$ m: (**a**) Workspace with platform outline; (**b**) Workspace without platform outline.

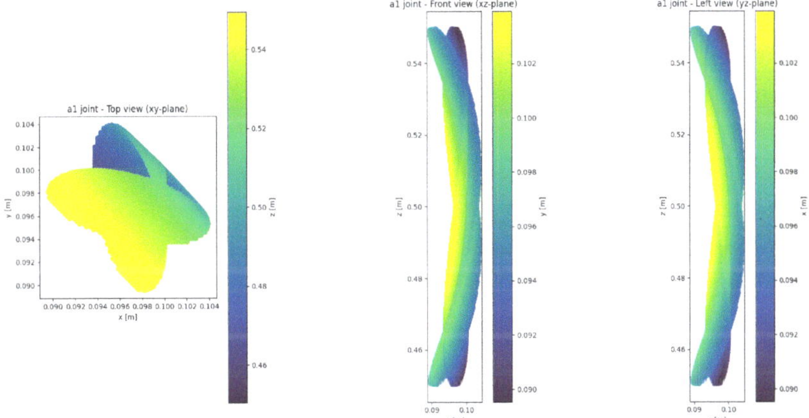

Figure 7. Workspace analysis for optimum configurations projected in 2D space for $z = 0.5$ m.

The workspace of the platform shows that the limits of the roll and pitch of the platform change with one another. For example, in an orientation where the platform has a pitch of $\pi/9$ degrees, the roll limit is at 0 degrees. This relationship can be analysed with a sweep, where the resulting plot can be seen in Figure 8b. The filled area shows the possible combinations for orienting the platform. That area can also be represented by Equation (28).

$$\sqrt{\theta_x^2 + \theta_y^2} \leq \frac{\pi}{9} \tag{28}$$

Due to the property of the upper and lower limits of this specific case, one can quickly identify the radius of the circle as either the pitch or roll upper limit in the case of $z^0 = 0.5$ m. The only constraint in this configuration of the mechanism is the collision between the top and bottom platforms (Equation (3)). It is correct to assume that a higher z^0 will mean a higher platform mobility, at least until the joints start reaching their limits and at the cost of a higher centre of mass, meaning less stability. This relationship is demonstrated in Figure 8a.

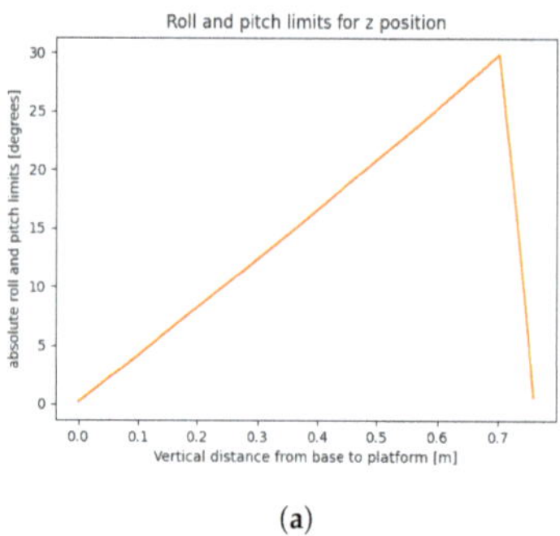
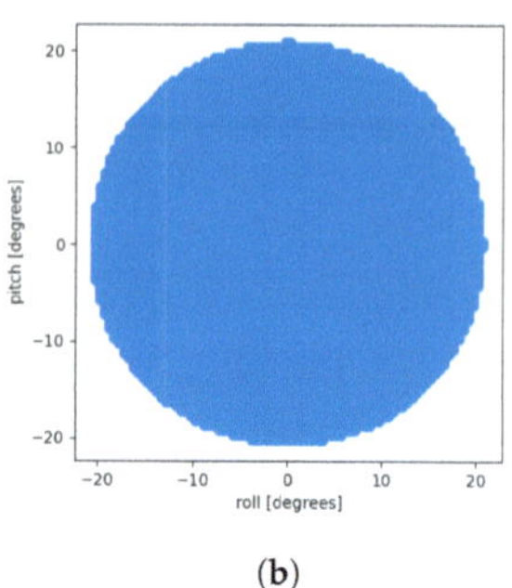

(**a**) (**b**)

Figure 8. Roll and pitch limit analysis: (**a**) Roll and pitch limit variation with distance between platforms; (**b**) Allowed platform orientation in degrees for $z = 0.5$ m.

5. Simulation and Results

To validate the design and evaluate the concept of the four-UPR platform, a demonstration is presented in a simulated environment, recurring to the ROS framework and the Gazebo simulator to model the platform joints and links and simulate the control of the mechanism and the wave motions. A screenshot of the simulator can be seen in Figure 9.

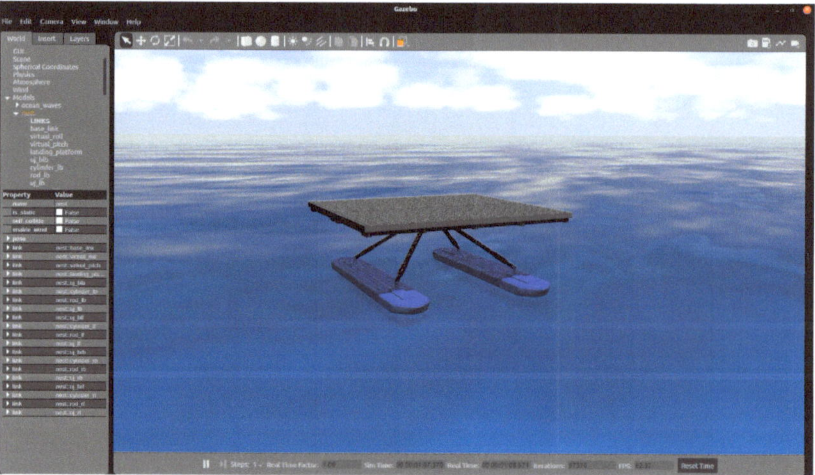

Figure 9. Screenshot of the simulator in Gazebo 11.

The robot is modelled in a Unified Robot Description Format (URDF), an XML format to represent robots. However, due to the tree-like structure of the URDF format, it is impossible to model a parallel mechanism without some workarounds. To solve this issue, a virtual chain was created to link the top platform to the bottom one, and the four kinematic chains were then connected to the top platform via a script that continuously calculates the joint angles over time using the equations described in Sections 2 and 3, recurring to static links that represent s_i^0, to have a static reference frame within the robot (see Figure 10). The virtual chain that models the top platform movement is analogous to dim $\$_P^C$ in Equation (13).

The intended purpose of the platform is to maintain the top platform in a horizontal position in an ocean surface environment. Using the information gathered by the IMU in the bottom platform, it is possible to compute the length of the prismatic actuators that, for a given orientation of the bottom platform, the top platform tends towards a horizontal position. A position controller is used for this task. The encoders on the prismatic actuators give the current lengths of the limbs, and the data provided by the IMU and the equations described in Section 3 give the desired lengths of the limbs. The adopted control system takes the desired position and orientation of the top platform as input, and the output gives the limb's desired length, with a feedback PID controller to improve the stability and accuracy of the outputs. The PID controller was manually tuned for the simulator. The values of $k_p = 5000.0$, $k_i = 0.5$, and $k_d = 10.0$ will likely have to be adjusted for the real mechanism. A diagram of the control system can be seen in Figure 11.

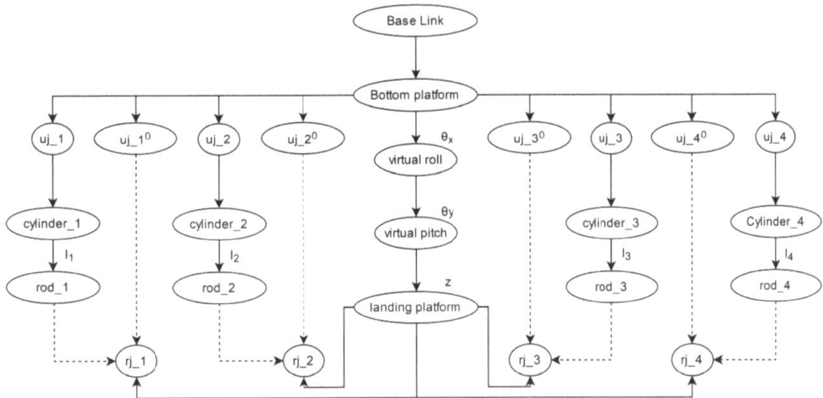

Figure 10. Tree structure of the platform links. Dashed lines—scripted connections.

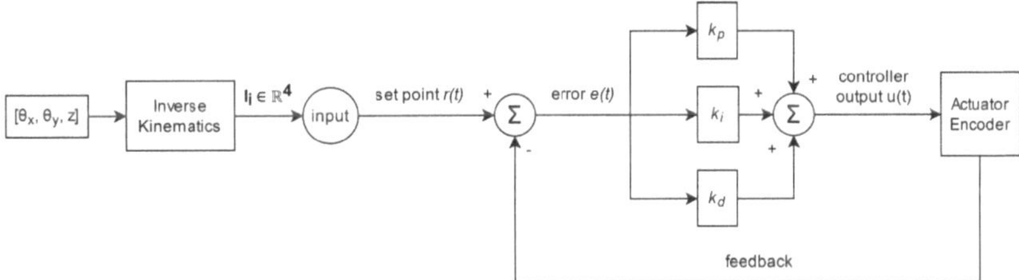

Figure 11. Control system of the 4-UPR platform.

The operational conditions of the platform are determined by several factors, including the capabilities of the UAV, the joint limits, and the sea state. The technical aiding equipment limitations, such as the support boat, also contribute to these conditions. A wave height of 1 m is defined as the the platform's operational limit, which corresponds to wave state 3 on the Douglas sea scale. Additionally, the UAV cooperating with this platform should not be operated in wind speeds exceeding 20 knots as its control starts becoming unreliable.

The maritime environment is physically simulated using the Gazebo plug-ins from Virtual Robot X [35]. The wave motions follow the model of Gerstner Waves [36] to approximate the influence of ocean waves in a simplified manner intended to balance physical fidelity and visual realism with real-time execution requirements. The motion of the bottom platform can be obtained with an IMU, which provides the orientation, angular velocity, and linear acceleration. In order to compare the movements of the top and bottom platform, another IMU is placed in the top platform. Since the control of the mechanism is executed with a position controller, the efficacy (η) of the mechanism for a given wave state and cylinder velocity is given by Equation (29), where α_{bottom} denotes the angle of the Z-axis to the Z'-axis in the global frame O, provided by the IMU. The same logic applies to α_{bottom}. Considering that the prismatic actuators have a linear speed of 0.1 m/s, which, depending on the current orientation of the platform, translates to an angular velocity up to 1.7 rad/s of the upper platform, the results for different wave states are shown in

Table 2. The selected wave parameters were based on realistic values that can be measured in the different wave states. While there is generally a correlation between wave height and period, the shorter periods for some large waves reflect how the platform may respond to rapid changes in steepness. All the tests were conducted over 30 s. The resulting plots are shown in Figure 12. The machine conducting the simulations has an i5-10600K CPU @ 4.10 GHz, a GeForce RTX 2060, and 15.5 GiB memory, running ROS Noetic and Gazebo 11.

$$\eta = \% \int_0^t \frac{\alpha_{bottom} - \alpha_{top}}{\alpha_{bottom}} dt \tag{29}$$

Table 2. Efficacy of the mechanism for different wave states.

Wave State	Number of Waves	Amplitude [m]	Period [s]	Direction [x,y]	η (%)
1	1	0.2	3	[1,1]	98.23
2	2	0.2;0.3	5;3	[1,0];[0,1]	97.16
3	3	0.2;0.3;0.4	7;5;3	[1,0];[0,1];[1,1]	94.84

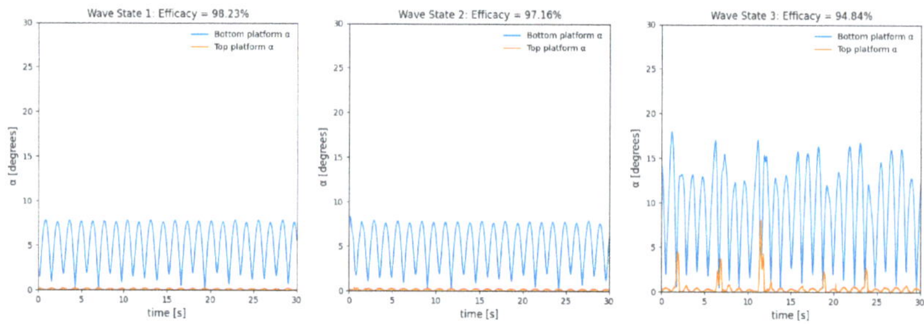

Figure 12. Resulting plots for the different wave states presented in Table 2.

Overall, the mechanism has a very good η for calmer (lower amplitudes, longer periods) wave states. However, as the waves grow more aggressive, the mechanism tends to have a much lower η. This can be attributed to three factors:

- The speed capacity of the linear actuators: with faster cylinders, the position correction can also occur faster, which will lead to less motion of the top platform. To improve the stability, one can acquire a more expensive, albeit faster, cylinder, which can be harder to control properly.
- The adopted control system: the simplicity of the adopted control system leads to poor adaptation to quick rates of change induced by waves with high amplitudes and short periods (evidenced in Figure 12). Due to tidal wave predictability and stability, and since UAVs usually do not operate well in strong winds (which cause highly irregular waves), the implementation of a predictive control system is recommended. Other studies such as that of Halvorsen et al. [37] have shown the success of this type of prediction algorithm for wave synchronization for rougher sea states.
- The small size of the vessel also implies that its roll and pitch orientation will be more affected by the wave's motion. Thus, a bigger vessel could also provide better stability.

6. Discussion

This paper presented a surface vessel with a landing platform for UAVs consisting of a four-UPR parallel manipulator with three DoFs for wave synchronization. The estimation of the platform's DoFs was achieved using the well-established reciprocal screw theory. The mechanism can perform roll and pitch rotations as well as translations in the Z-axis. An inverse kinematics approach allowed a better understanding of how the variation of the parameters of the platform affects the mechanism properties. The workspace representation was achieved with a numerical sweep that checks all the possible orientations for a given z of the upper platform. For $z = 0.5$ m, the platform pitch and roll combinations are limited to the area of a circle, given by Equation (28). The mechanism concept was further tested in a simulated scenario, resorting to ROS and Gazebo to emulate the physical model and environment. The inverse kinematics analysis and the use of a virtual chain were able to overcome the limitation of closed-loop chains in URDF files and achieve satisfying results in terms of the expected behaviour in these types of mechanisms. The control system applied in this work showed satisfying results for lower wave amplitudes (up to 0.4 m), with an η of 94.84%. This metric shows how well the top platform was able to sustain the horizontal position compared to the bottom one. However, for more aggressive sea states, the η starts dropping considerably, implying that a more robust control system is needed. Overall, the results achieved with this work show how wave synchronization in small surface vessels can be achieved with the methods demonstrated and the limitations imposed by the sea state.

Author Contributions: P.P., conceptualisation, validation, investigation, writing—original draft preparation; R.C., writing—review and editing, supervision; A.P., writing—review and editing, supervision, funding acquisition, project administration. All authors have read and agreed to the published version of the manuscript.

Funding: This work is funded by the European Commission under the European Union's Horizon 2020—The EU Framework Programme for Research and Innovation 2014–2020 under grant agreement No. 871571 (ATLANTIS).

Institutional Review Board Statement: Not applicable.

Informed Consent Statement: Not applicable.

Data Availability Statement: Not applicable.

Conflicts of Interest: The authors declare no conflict of interest.

Abbreviations

The following abbreviations are used in this manuscript:

UAV	Unmanned aerial vehicle
ASV	Autonomous surface vehicle
ROS	Robot operating system
R-joint	Revolute joint
U-joint	Universal joint
IMU	Inertial measurement unit
DoF	Degrees of freedom
URDF	Unified robot description format
PID	Proportional–integral–derivative

References

1. Lynch, K.M.; Park, F.C. *Modern Robotics: Mechanics, Planning, and Control*, 1st ed.; Cambridge University Press: Cambridge, MA, USA, 2017.
2. Ghobakhloo, A.; Eghtesad, M.; Azadi, M. Position control of a Stewart-Gough platform using inverse dynamics method with full dynamics. In Proceedings of the 9th IEEE International Workshop on Advanced Motion Control, Istanbul, Turkey, 27–29 March 2006; pp. 50–55.
3. Meng, X.; Gao, F.; Wu, S.; Ge, Q.J. Type synthesis of parallel robotic mechanisms: Framework and brief review. *Mech. Mach. Theory* **2014**, *78*, 177–186. [CrossRef]
4. Stewart, D. A platform with six degrees of freedom. *Proc. Inst. Mech. Eng.* **1965**, *180*, 371–386. [CrossRef]
5. Zhao, J.; Feng, Z.; Chu, F.; Ma, N. *Advanced Theory of Constraint and Motion Analysis for Robot Mechanisms*; Academic Press: Cambridge, MA, USA, 2013.
6. Merlet, J.P. *Parallel Robots*; Springer Science & Business Media: Berlin/Heidelberg, Germany, 2005; Volume 128.
7. Kong, X.; Gosselin, C. *Virtual-Chain Approach for the Type Synthesis of Parallel Mechanisms*; Springer: Berlin/Heidelberg, Germany, 2007.
8. Zhao, J.S.; Zhou, K.; Mao, D.Z.; Gao, Y.F.; Fang, Y. A new method to study the degree of freedom of spatial parallel mechanisms. *Int. J. Adv. Manuf. Technol.* **2004**, *23*, 288–294. [CrossRef]

9. Guan, L.W.; Wang, J.S.; Wang, L.P. Mobility analysis of the 3-UPU parallel mechanism based on screw theory. In Proceedings of the 2004 International Conference on Intelligent Mechatronics and Automation, Chengdu, China, 26–31 August 2004; pp. 309–314.

10. Wang, L.; Xu, H.; Guan, L. Mobility analysis of parallel mechanisms based on screw theory and mechanism topology. *Adv. Mech. Eng.* **2015**, *7*, 1687814015610467. [CrossRef]

11. Cai, J.; Deng, X.; Feng, J.; Xu, Y. Mobility analysis of generalized angulated scissor-like elements with the reciprocal screw theory. *Mech. Mach. Theory* **2014**, *82*, 256–265. [CrossRef]

12. Fang, H.; Fang, Y.; Zhang, K. Kinematics and workspace analysis of a novel 3-DOF parallel manipulator with virtual symmetric plane. *Proc. Inst. Mech. Eng. Part C J. Mech. Eng. Sci.* **2013**, *227*, 620–629. [CrossRef]

13. Zhang, H.; Fang, H.; Zhang, D.; Luo, X.; Zou, Q. Forward kinematics and workspace determination of a novel redundantly actuated parallel manipulator. *Int. J. Aerosp. Eng.* **2019**, *2019*, 4769174 . [CrossRef]

14. Pashkevich, A.; Chablat, D.; Wenger, P. Kinematics and workspace analysis of a three-axis parallel manipulator: The Orthoglide. *Robotica* **2006**, *24*, 39–49. [CrossRef]

15. Desai, R.; Muthuswamy, S. A forward, inverse kinematics and workspace analysis of 3RPS and 3RPS-R parallel manipulators. *Iran. J. Sci. Technol. Trans. Mech. Eng.* **2021**, *45*, 115–131. [CrossRef]

16. Laryushkin, P.; Antonov, A.; Fomin, A.; Essomba, T. Velocity and Singularity Analysis of a 5-DOF (3T2R) Parallel-Serial (Hybrid) Manipulator. *Machines* **2022**, *10*, 276. [CrossRef]

17. Du, X.; Wang, B.; Zheng, J. Geometric Error Analysis of a 2UPR-RPU Over-Constrained Parallel Manipulator. *Machines* **2022**, *10*, 990. [CrossRef]

18. Jones, T.; Dunlop, G. Analysis of rigid-body dynamics for closed-loop mechanisms—Its application to a novel satellite tracking device. *Proc. Inst. Mech. Eng. Part I J. Syst. Control. Eng.* **2003**, *217*, 285–298.

19. Wapler, M.; Urban, V.; Weisener, T.; Stallkamp, J.; Dürr, M.; Hiller, A. A Stewart platform for precision surgery. *Trans. Inst. Meas. Control* **2003**, *25*, 329–334. [CrossRef]

20. Ladenburg, J.; Hevia-Koch, P.; Petrović, S.; Knapp, L. The offshore-onshore conundrum: Preferences for wind energy considering spatial data in Denmark. *Renew. Sustain. Energy Rev.* **2020**, *121*, 109711. [CrossRef]

21. Campos, D.F.; Pereira, M.; Matos, A.; Pinto, A.M. DIIUS-Distributed Perception for Inspection of Aquatic Structures. In Proceedings of the OCEANS 2021: San Diego—Porto, San Diego, CA, USA, 20–23 September 2021; pp. 1–5.

22. Pinto, A.M.; Matos, A.C. MARESye: A hybrid imaging system for underwater robotic applications. *Inf. Fusion* **2020**, *55*, 16–29. [CrossRef]

23. Silva, R.; Matos, A.; Pinto, A.M. Multi-criteria metric to evaluate motion planners for underwater intervention. *Auton. Robots* **2022**, *46*, 971–983. [CrossRef]

24. Shafiee, M.; Zhou, Z.; Mei, L.; Dinmohammadi, F.; Karama, J.; Flynn, D. Unmanned aerial drones for inspection of offshore wind turbines: A mission-critical failure analysis. *Robotics* **2021**, *10*, 26. [CrossRef]

25. Pereira, P.N.D.A.A.D.S.; Campilho, R.D.S.G.; Pinto, A.M.G. Application of a Design for Excellence Methodology for a Wireless Charger Housing in Underwater Environments. *Machines* **2022**, *10*, 232. [CrossRef]

26. Aissi, M.; Moumen, Y.; Berrich, J.; Bouchentouf, T.; Bourhaleb, M.; Rahmoun, M. Autonomous solar USV with an automated launch and recovery system for UAV: State of the art and Design. In Proceedings of the 2020 IEEE 2nd International Conference on Electronics, Control, Optimization and Computer Science (ICECOCS), Kenitra, Morocco, 2–3 December 2020; pp. 1–6.

27. Li, W.; Ge, Y.; Guan, Z.; Ye, G. Synchronized Motion-Based UAV–USV Cooperative Autonomous Landing. *J. Mar. Sci. Eng.* **2022**, *10*, 1214. [CrossRef]

28. Liu, J.; Zhang, D.; Chen, Y.; Xia, Z.; Wu, C. Design of a class of generalized parallel mechanisms for adaptive landing and aerial manipulation. *Mech. Mach. Theory* **2022**, *170*, 104692. [CrossRef]

29. Guo, J.; Li, G.; Li, B.; Wang, S. A ship active vibration isolation system based on a novel 5-DOF parallel mechanism. In Proceedings of the 2014 IEEE International Conference on Information and Automation (ICIA), Hailar, China, 28–30 July 2014; pp. 800–805. [CrossRef]

30. Chen, B.Y.; Chiang, M.H. Simulation and experiment of a turbine access system with three-axial active motion compensation. *Ocean Eng.* **2019**, *176*, 8–19. [CrossRef]

31. Cai, Y.; Zheng, S.; Liu, W.; Qu, Z.; Zhu, J.; Han, J. Sliding-mode control of ship-mounted Stewart platforms for wave compensation using velocity feedforward. *Ocean Eng.* **2021**, *236*, 109477. [CrossRef]

32. Pinto, A.M.; Marques, J.V.A.; Campos, D.F.; Abreu, N.; Matos, A.; Jussi, M.; Berglund, R.; Halme, J.; Tikka, P.; Formiga, J.; et al. ATLANTIS—The Atlantic Testing Platform for Maritime Robotics. In Proceedings of the OCEANS 2021, San Diego—Porto, San Diego, CA, USA, 20–23 September 2021; pp. 1–5. [CrossRef]

33. Zhao, T.S.; Li, Y.W.; Chen, J.; Wang, J.C. A Novel Four-DOF Parallel Manipulator Mechanism and Its Kinematics. In Proceedings of the 2006 IEEE Conference on Robotics, Automation and Mechatronics, Bangkok, Thailand, 7–9 June 2006; pp. 1–5. [CrossRef]

34. Neves, F.S.; Claro, R.M.; Pinto, A.M. End-to-End Detection of a Landing Platform for Offshore UAVs Based on a Multimodal Early Fusion Approach. *Sensors* **2023**, *23*, 2434. [CrossRef]

35. Bingham, B.; Agüero, C.; McCarrin, M.; Klamo, J.; Malia, J.; Allen, K.; Lum, T.; Rawson, M.; Waqar, R. Toward maritime robotic simulation in gazebo. In Proceedings of the OCEANS 2019, Seattle, WA, 27–31 October USA, 2019; pp. 1–10.

36. Thon, S.; Dischler, J.M.; Ghazanfarpour, D. Ocean waves synthesis using a spectrum-based turbulence function. In Proceedings of the Computer Graphics International 2000, Geneva, Switzerland, 19–24 June 2000; pp. 65–72.
37. Halvorsen, H.S.; Øveraas, H.; Landstad, O.; Smines, V.; Fossen, T.I.; Johansen, T.A. Wave motion compensation in dynamic positioning of small autonomous vessels. *J. Mar. Sci. Technol.* **2021**, *26*, 693–712. [CrossRef]

Article

The Viability of a Grid of Autonomous Ground-Tethered UAV Platforms in Agricultural Pest Bird Control

Joshua Trethowan, Zihao Wang and K. C. Wong *

School of Aerospace, Mechanical and Mechatronic Engineering, The University of Sydney, Sydney, NSW 2006, Australia
* Correspondence: kc.wong@sydney.edu.au; Tel.: +61-2-9351-2347

Abstract: Pest birds are a salient problem in agriculture all around the world due to the damage they can cause to commercial or high-value crops. Recent advancements in Unmanned Aerial Vehicles (UAVs) have motivated the use of drones in pest bird deterrence, with promising success already being demonstrated over traditional bird control techniques. This paper presents a novel bird deterrence solution in the form of tethered UAVs, which are attached and arranged in a grid-like fashion across a vineyard property. This strategy aims to bypass the power and endurance limitations of untethered drones while still utilising their dynamism and scaring potential. A simulation model has been designed and developed to assess the feasibility of different UAV arrangements, configurations, and strategies against expected behavioural responses of incoming bird flocks, despite operational and spatial constraints imposed by a tether. Attempts at quantifying bird persistence and relative effort following UAV-induced deterrence are also introduced through a novel bird energy expenditure model. This aims to serve as a proxy for selecting control techniques that reduce future foraging missions. The simulation model successfully isolated candidate configurations, which were able to deter both single and multiple incoming bird flocks using a centralised multi-UAV control strategy. Overall, this study indicates that a grid of autonomous ground-tethered UAV platforms is viable as a bird deterrence solution in agriculture, a novel solution not seen nor dealt with elsewhere to the authors' knowledge.

Keywords: unmanned aerial vehicle; UAV; tethered UAV; pest bird control; drone; intelligent system; path planning

Citation: Trethowan, J.; Wang, Z.; Wong, K.C. The Viability of a Grid of Autonomous Ground-Tethered UAV Platforms in Agricultural Pest Bird Control. *Machines* **2023**, *11*, 377. https://doi.org/10.3390/machines11030377

Academic Editor: Dan Zhang

Received: 27 December 2022
Revised: 26 February 2023
Accepted: 3 March 2023
Published: 11 March 2023

1. Introduction

Damage caused by birds is a long-standing and costly problem in agricultural sectors across the world, especially in high-value crops such as blueberries, cherries and wine grapes [1–4]. The United States—considered one of the top five agriculture-producing countries globally [4]—experienced more than USD 800 million in losses in 2004 solely due to European Starlings, with total losses from all bird species estimated in excess of USD 4.7 billion [5]. The heavy damage is attributed to the fact that most of the existing bird damage control methods have not proven to be effective [2].

With improvements in UAV platforms and autonomous technologies, several studies have explored the use of UAVs, in particular multirotor type UAVs, as an effective form of bird deterrence through pursuit and shepherding manoeuvres [6–9]. Wang et al. in [7] utilised a novel predator multirotor UAV system and demonstrated successful bird deterrence up to a 50 m radius centred on the UAV itself. Similarly, a study in [10] used a quadcopter system to deter birds from a vineyard in the United States and concluded that random patterns and distress calls could significantly reduce bird foraging activity in the region, with up to a 50% reduction during flight cycles compared to controlled no-flight cycles over a 14-day period. Anecdotally, many farmers have also started using commercial off-the-shelf UAVs for bird deterrence with some success. However, these

UAVs still require human operators throughout their mission to be successful. Evidently, for a bird deterrence system to be cost-effective, it must aim to be autonomous and mitigate any additional manual labour. If the user is required to dedicate a large portion of their time using the bird control system, then they lose productive time that could be used elsewhere on the property.

A major limitation of current multirotor UAV technologies is their limited flight time and endurance. Most typical small multirotor UAVs have an endurance of approximately 10 to 30 min before they need to recharge [11]. This directly conflicts with the established aim of mitigating additional manpower since an operator would be required at frequent intervals throughout the day to recharge the UAV or replace its battery.

A potential solution to this problem is using a power-over-tether multirotor system. Recent advancements in tether technology have enabled capabilities such as the ground stations in [12–15], with the one in [12] autonomously varying a 12 m long tether, using a winch system to ensure that the tether is kept taut without excessively pulling on the multirotor and impacting its dynamics. Compared to a free-flying multirotor with limited endurance, such a system would have indefinite endurance through power delivered via their physical tethers from a ground station. The tether would also improve safety by physically restricting the flying vehicles' movements, preventing collision accidents, and removing fly away concerns, which would make the system more likely to be approved by regulators.

The objective of this paper is to explore the feasibility of using multiple multirotor UAVs that are tethered to the ground in a grid-like manner to deter bird flocks from a given property. The project builds on the prior study in [7,8,16] to develop a simulation model and investigate the suitable UAV arrangements, configurations, and strategies against expected behavioural responses of incoming bird flocks.

2. Simulation Model and Methodology

This work extends and adapts the trajectory planning algorithm based on probability maps as well as the bird behaviour model based on field trials used by Wang [17]. The focus of this paper is placed on the multi-UAV arrangement and coordination due to dynamic and spatial constraints imposed by finite tether lengths. Modelling and predicting bird behaviour is notoriously difficult; thus, simplifications and assumptions must be made on the three key components which define this problem: the property or 'world' model, the tethered UAV model (hereon 'agent'), and the bird flock model (hereon 'target' and 'target flock').

2.1. Property World Set-up

Observations of typical vineyard plot sizes in Australia have been found to be approximately 100 m by 100 m in size. Hence, to simplify this problem, it will be assumed throughout the simulation that the property 'world' is controlled as a square of these dimensions.

The property itself is modelled as a probability map or occupancy grid, similar to those described in [16,18]. This method models the search area as a grid of cells wherein each cell is given a single value representing the probability of a given target occupying it based on external sensor information. In the problem of UAVs deterring target flocks, this is an appropriate scheme to adopt as the precise location of the target is unknown representative of search-and-capture or pursuit–evasion missions. Hence, UAV agents chase targets by pursuing regions of high probability despite lacking knowledge of the true state of the target.

Though the real-life problem involving UAVs and birds is inherently 3D, from field trials in [7], birds that pass well above the ground are of no concern to the crop being protected, while UAV altitude ranges from ground level to approximately 15 m have shown comparable success in initiating flight responses or evasive manoeuvres in targets within the same height range [7]. Hence, the occupancy grid map is modelled in 2D to reduce computational complexity. Despite this, aspects of 3D dynamics are still implemented elsewhere within the agent and target models.

Each cell within the simulation property lies within the horizontal dimensions $[x_{min}, x_{max}]$ in x and $[y_{min}, y_{max}]$ in y to form the spatial domain M, defined as:

$$M = \left\{ \bar{c} \,\middle|\, \begin{array}{l} \bar{c}_x \in [x_{min}, x_{max}] \\ \bar{c}_y \in [y_{min}, y_{max}] \end{array} \right\}, \tag{1}$$

where $\bar{c}$ is a vector representing the centre coordinates of each individual grid cell.

From the square property assumption earlier, it is intuitive to use square-shaped cells of edge length L to populate a discrete $N \times N$ cell set grid map. The modified spatial domain and centre coordinates of each cell is therefore defined as:

$$\tilde{M} = \left\{ \bar{c} \,\middle|\, \begin{array}{l} \bar{c}_x = x_{min} + L(i - 0.5) \\ \bar{c}_y = y_{min} + L(i - 0.5) \end{array}, i = 1, 2, \ldots, N \right\}. \tag{2}$$

For this project, grid cells with an edge length of $L = 5$ m were chosen to populate the 100 m by 100 m property (forming a 20 by 20 cell set), as shown in Figure 1, since a higher fidelity than this in localization is not expected with current technologies. The probability score k of a cell ranges between 0 and 1, where $k = 1$ represents a 100% likelihood of a target occupying that cell while $k = 0$ is the opposite. Hence, observing Figure 1c, regions of higher probability correlate to dark grey cells, while the converse is true with light grey to white cells.

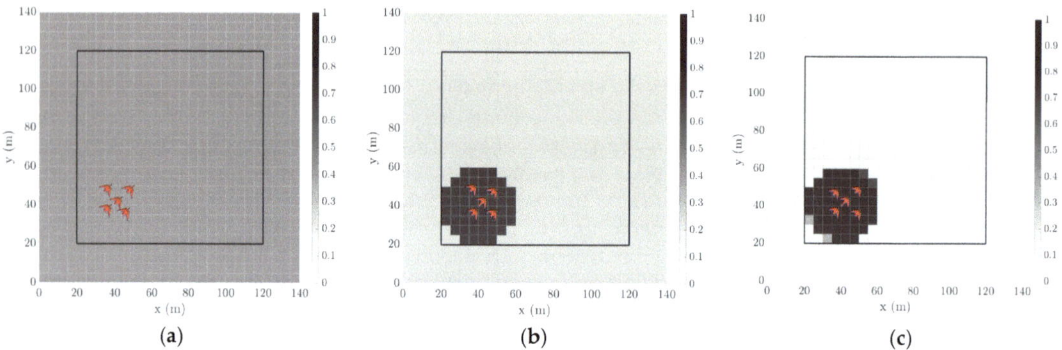

Figure 1. Target probability map over incremental time steps (darker cells represent higher probability of target occupancy based off sensor estimates while the red markers represent the true target flock position). Note flocks are not drawn to scale. (**a**) Probability grid at $t = 0$ s; (**b**) Probability grid at $t = 1$ s; (**c**) Probability grid at $t = 2$ s.

Importantly, this probability score is updated at every time step so the system can have continued confidence regarding the location of the target. However, if there is no sensor information, the probability score of that cell will begin approaching a non-zero nominal value (in other words, away from the certainties of $k = 0$ and $k = 1$). The nominal probability value $k_{nominal}$ is approached as:

$$k(t+1, \bar{c}) = \tau_{probability} k(t, \bar{c}) + \left(1 - \tau_{probability}\right) k_{nominal}, \tag{3}$$

where $\tau_{probability} \in [0, 1]$ is a time constant determining the rate that k approaches $k_{nominal}$. The probability nominalisation model and the parameters are chosen to reflect the limited bird behaviour observed by the authors. Further data are needed to verify the accuracy of the model.

Each agent is equipped with a sensor model that detects targets within a circular sector of radius r_{sensor} and angle θ_{sensor}, as illustrated in Figure 2. All of the cells within the sector are designated a probability score k_{high} if a target is detected by the sensor. Conversely, all of the cells are designated k_{low} upon zero detection. In addition, virtual ground cameras are positioned at the corners of the property to provide sensor information across the whole protected area (designated by the black box in Figure 1) that each agent has access to at every time step. Bird detection and localisation using an arrangement of ground cameras in tandem with mobile ones mounted on UAVs can be found in [19]. A camera network like this improves localisation accuracy in detecting target flocks.

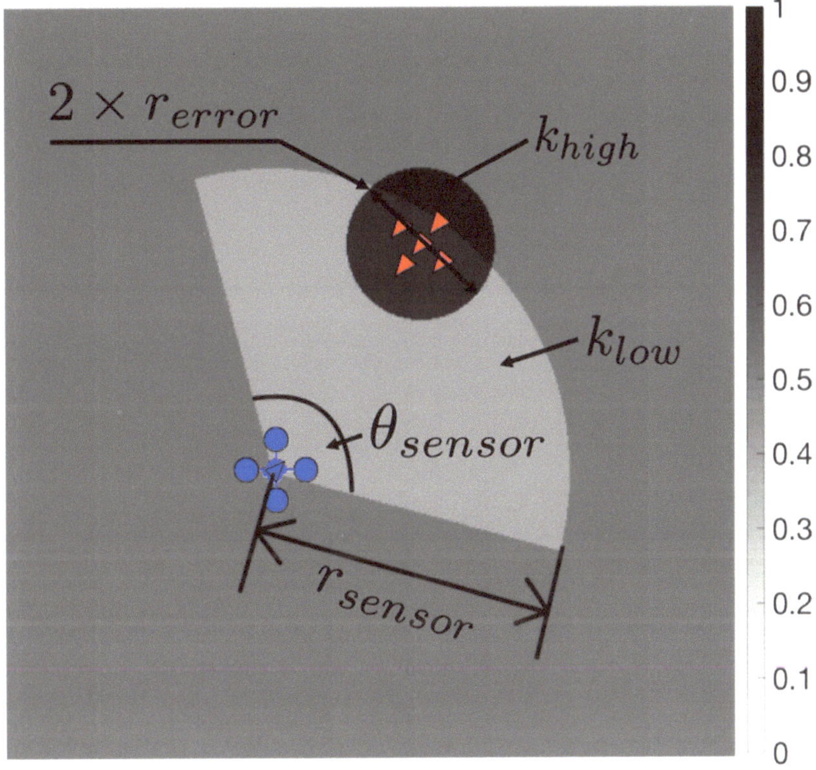

Figure 2. Illustration of the sensor model. The greyscale bar indicates the probability of a target occupying the cells, darker indicates higher probability.

At the initialisation of the simulation, all of the grid cells are initialised with the same nominal value before any sensor data are processed, as can be seen in Figure 1a, where each cell has a probability score of $k = 0.4$. In the next time step, shown in Figure 1b, the global probability values are updated, such that the cells near the true target position are now darker ($k = 0.7$), while the regions further away become lighter ($k = 0.2$). Finally, by the next time step, shown in Figure 1c, the colour contrast becomes more salient, indicating increased confidence in the system as to where the target flock both is and is not.

2.2. Tethered UAV Model

2.2.1. UAV Dynamics

All of the agents in this problem are assumed to be tethered multirotor UAVs. The agents are modelled using simple second-order dynamics with the following state vector X_{UAV}:

$$X_{UAV} = (x, y, z, v_x, v_y, v_z, \psi), \tag{4}$$

where (x, y, z) represent the position; (v_x, v_y, v_z) represent the velocity; and ψ represents the heading. The vertical dynamics (z-axis) have been decoupled from the horizontal dynamics (xy-plane) due to the aforementioned assumptions, where positional accuracy and modelling in height are of limited concern to the results. This similarly assumes the pitch and roll to be zero and are thus not included in the state vector as a further simplification. The horizontal dynamics are governed by the relative error between the current position of the agent and the desired waypoint and are constrained by maximum speed (v_{max}), acceleration (a_{max}), yaw rate (ψ_{max}), and tether length (L_T).

In this simulation, the effective tether length is defined as the horizontal projection along the ground, as demonstrated in Figure 3, as opposed to the true length in a 3D space, since the horizontal distance has a significantly greater impact on UAV dynamics and stability. Hence, this tether length defines a maximum radius in which an agent can move, mapped to a circle in 2D (Figure 3a) and a cylinder in 3D (Figure 3b) and will thus be hereon defined as the tether radius (R_T) for this simulation.

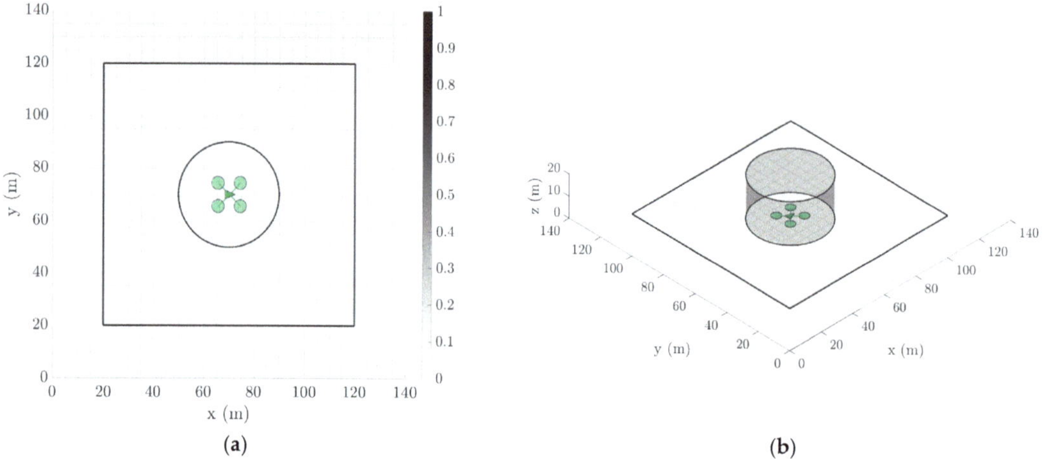

(a) **(b)**

Figure 3. Agent UAV radius of operation sample case $R_T = 20$ m. The maximum range maps to a circle in 2D (**a**) and a cylinder in 3D (**b**). Note the UAV is not drawn to scale.

The vertical dynamics of the agents are governed by the relative error between the ground tether points when they are undeployed and a set nominal height value when deployed. These set heights can be different for each agent to minimize the likelihood of collisions for overlapping sample arrangements, similar to that in Figure 4b. However, this is unnecessary for configurations where the tether radii of the agents do not overlap at all, as in Figure 4a.

2.2.2. Trajectory Planning

The optimal agent waypoint is chosen using a cost function strategy that evaluates every cell of the occupancy grid map, as defined in Section 2.1. This is an appropriate method as many existing commercial-off-the-shelf (COTS) autopilots, such as Pixhawk, accept GPS coordinates as waypoints when generating UAV trajectories [20].

A receding horizon control strategy described by [16] is used. In this strategy, at every time step t, a subset of reachable or possible waypoints in the next n_{RHC} steps for each agent is selected, as defined by their maximum velocity and turning rate (but still constrained by tether radius). Within this subset, the optimality of each cell is considered using a cost function that incorporates the target probability score first and foremost and then the difficulty for the agent to reach that position in terms of distance, heading change, tether limits, as well as the trajectory or proximity of a target or another agent. Therefore, the optimal waypoint is defined as the cell $\bar{c}$ with a minimum cost C :

$$\min_{C} C = \alpha \cdot f_{k'}(\bar{c}) + \beta \cdot f_d(\bar{c}) + \gamma \cdot f_h(\bar{c}) + \delta \cdot f_R(\bar{c}) + \varepsilon \cdot (f_{beyond}(\bar{c}) + f_T(\bar{c}) + f_{other}(\bar{c})) \quad (5)$$

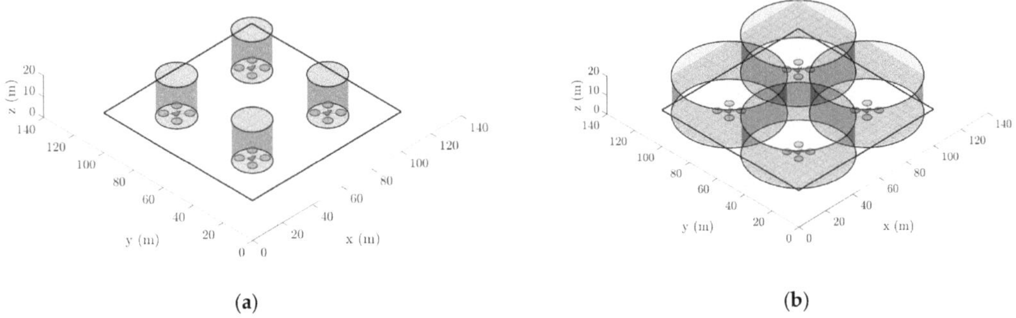

(a)   (b)

Figure 4. Sample arrangement involving four UAVs with different spatial constraints and operating regions imposed by their tether lengths. (**a**) No height constraints are necessary for this configuration. (**b**) The set heights should be different in this configuration to avoid collisions.

The first term in Equation (5) correlates to the complement of the target probability score, as shown in Equation (6) (the higher the probability, the lower the associated cost of that cell):

$$f_{k'}\left(\bar{c}\right) = 1 - k\left(t + n, \bar{c}\right). \quad (6)$$

The second term in Equation (5), f_d, corresponds to the distance cost based on a normalized horizontal distance between the centre coordinates of every reachable cell in the receding horizon and the agent's current position:

$$f_d\left(\bar{c}\right) = \frac{\sqrt{\left(\bar{c}_x - \bar{c}_{UAV_x}\right)^2 + \left(\bar{c}_y - \bar{c}_{UAV_y}\right)^2}}{n \cdot t \cdot v_{max}}. \quad (7)$$

The third term, f_h, corresponds to the normalized heading change cost:

$$f_h\left(\bar{c}\right) = \frac{f_\psi\left(\tan^{-1}\left(\frac{\bar{c}_y - \bar{c}_{UAV_x}}{\bar{c}_y - \bar{c}_{UAV_y}}\right), \psi_{UAV}\right)}{\pi}, \quad (8)$$

where f_ψ is a function that returns the smallest angle between its two input arguments.

The next introduced term, f_R, correlates to a repulsion field or potential field. In this potential field, the cells that have increased in target probability score over a time step are given a higher cost, with the converse effect for cells with decreasing probability score. In other words, cells with $+\Delta k$ map to the forward direction of a moving target. This motivates agents to approach a target flock from behind or from the side in preference to head-on, which may otherwise result in a collision or affect the shepherding smoothness and cohesive flock assumptions made. This resulting repulsion–attraction field cost (mapped to [0,1]) is defined as:

$$f_R\left(\bar{c}\right) = \frac{1 + k\left(t, \bar{c}\right) - k\left(t - 1, \bar{c}\right)}{2k_{max}}, \quad (9)$$

where k_{max} is the maximum probability score possible for a cell.

The last term in Equation (5) is composed of three parts and correlates to the cost associated with the cells that cannot be reached within the receding horizon $\left(f_{beyond}\right)$; the cells located outside the allowable tether radius (f_T); and cells that are proximal to other agents (f_{other}). The values for these functions are binary; there is a zero cost for all reachable cells but a maximum cost of 1 for unreachable cells. Note that the tether limit cost here is implemented to further discourage agent UAVs from approaching their radial limits at speed or overshooting and over-tugging the tether.

The parameters of $\alpha, \beta, \gamma, \delta$ serve as relative weights in terms of importance placed on each of the components in Equation (5), and the sum to unity is as shown:

$$\alpha + \beta + \gamma + \delta = 1. \tag{10}$$

Note that the weight parameter of ε from Equation (5) is instead given a nominal value $\gg 1$ to ensure that the unreachable grid cells are never optimal after applying the rest of the cost function. The different components of the cost function can be visualized from the perspective of the left agent in the sample scenario shown in Figure 5; the cell with minimum cost after applying every term in the cost function is then designated as the optimal waypoint for that specific agent, as indicated in Figure 6. The individual weight is tuned based on the emphasis of the mission. As an example, if a smoother path is required, the heading cost weight γ should be increased. Whereas a larger α will make sure the agents focus on chasing the target flocks at the expense of more unnecessary accelerations and decelerations.

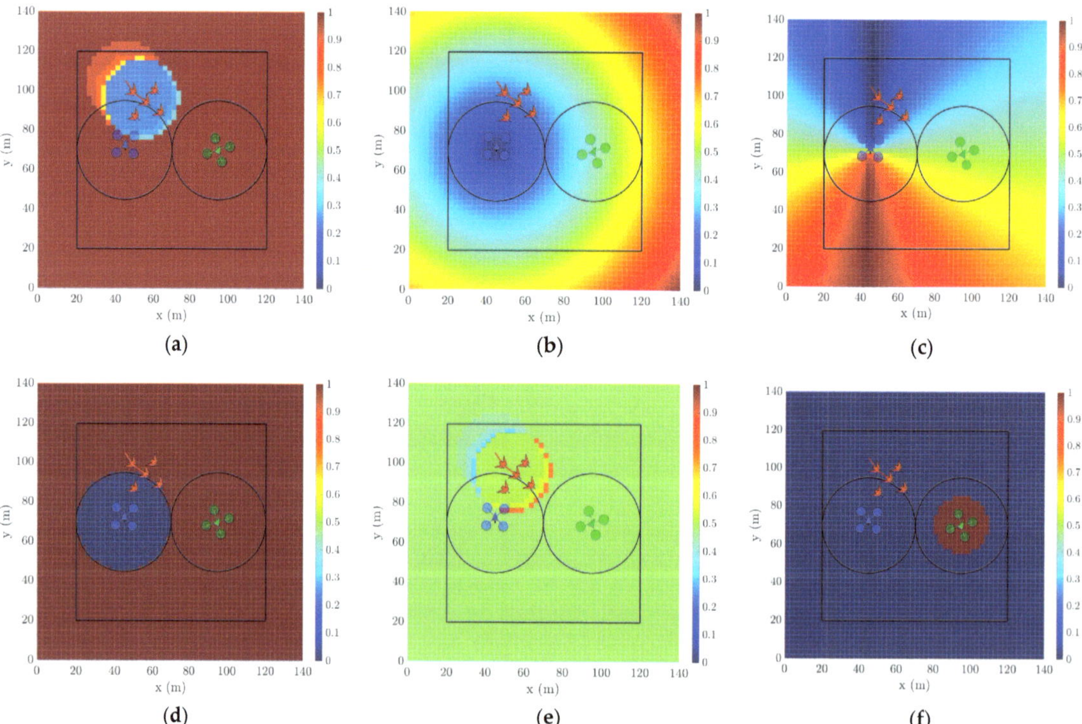

Figure 5. Snapshot visualization of cost function components at $t = 5$ s from perspective of the left agent. (**a**) Target probability cost; (**b**) distance cost; (**c**) heading cost; (**d**) tether limit cost; (**e**) repulsive field cost; (**f**) other agent cost.

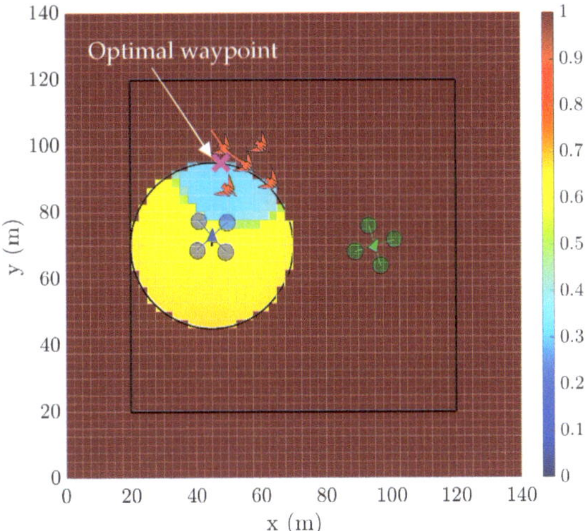

Figure 6. Combined cost function with weightings applied at $t = 5$ s. Optimal waypoint for left agent is shown and represents the grid cell with minimum cost.

2.2.3. Agent Grid Arrangement

With the dynamics and trajectory planning now defined for the agents, the next step is to determine the relative arrangement or coordinates of the agents within the property world; in other words, the ground tether points that define the centres of the pursuit region of each agent. The agent grid arrangement scheme aims to distribute n points as evenly as possible in a $b \times h$ sized grid, where the n points represent the number of agents. Centroidal Voronoi Tessellation (CVT) was chosen as a common and fairly computationally quick method to evenly partition a given plane into n convex polytopes [21]. CVT is a special form of Voronoi Tessellation in which the 'generating' point or node of a given partition is constrained to be coincident with its own centroid, resulting in neighbouring nodes being spaced as far apart from each other as possible.

Hence, for a set of Voronoi regions $\{V_i\}_{i=1}^{k}$, the mass centroid c_i over a region with probability density $\rho(y)$ is defined as:

$$c_i = \frac{\int_{V_i} y \rho(y) dy}{\int_{V_i} \rho(y) dy}.$$

(11)

Lloyd's Algorithm was deemed to be appropriate for implementing and generating a CVT based on a desired number of nodes (agent number in this case) as opposed to redistributing existing nodes or data points that might be randomly distributed on a plane [21]. Note that CVTs are only defined for three or more generator points when using Delaunay triangles (the dual tessellation of the Voronoi diagram), and thus, the special degenerate cases of 1 and 2 agents were manually defined (see [21,22] for more information on CVT methods and Delaunay triangulation). The CVT method proved effective for generating a desired number of the agents' coordinates as evenly as possible within a given grid size, as shown in Figure 7.

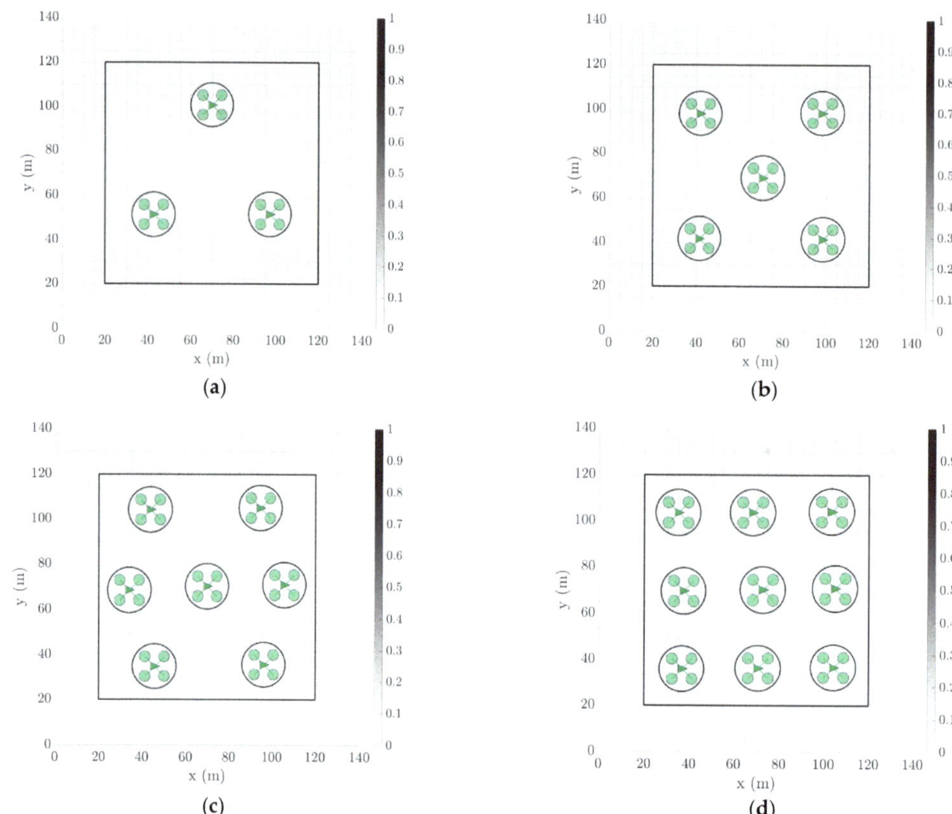

Figure 7. Sample grids with differing agent numbers generated using CVT method (all with $R_T = 10$ m). (**a**) $n = 3$ agents; (**b**) $n = 5$ agents; (**c**) $n = 7$ agents; (**d**) $n = 9$ agents.

2.2.4. Chasing Strategy and Multiple UAV Coordination

The remaining consideration for the agent UAV model is the chasing strategy and the coordination of multiple agents. A single agent can pursue its target waypoint using the cost function strategy outlined previously; however, this method becomes problematic when there are multiple agents and multiple targets since there is no hierarchy or control scheme that dictates which flock is chased by which agent.

A centralized control strategy is used to coordinate the multiple UAVs on the property. In this strategy, the coordinates of each agent at a given time step are relayed to a central computer, which then determines which agents should deploy or remain undeployed based on their proximity to high target probability areas. An agent simply needs to be within the 'scaring range', as defined by Equation (12), to be deployed:

$$r_{scare} \leq R_T + r_{no-go}, \tag{12}$$

where R_T is the maximum tether length, as previously defined, while r_{no-go} is the maximum radius around an agent UAV that elicits escape manoeuvres by the target flocks, as observed in experimental field trials in [7].

For the simulation, each agent can have one of the following states at any given time (colour-coded for clarity to match the simulation snapshots provided in Section 3):

- **Chase**: the agent is deployed to pursue a target flock which is in range.

- **Home**: the agent is undeployed and stationed at the ground tether point–this agent state does not scare away target flocks, irrespective of distance.
- **Return**: the agent is deployed but returning to the ground tether point since the target flock is no longer within scaring range.

In this control strategy, any in-range agents are deployed, not just the closest agent. As the position of a target is estimated using a shared probability map, an agent may pursue a midpoint position among multiple targets since it has found the geometric centre of the highest probability scores. Hence, a k-means clustering algorithm is proposed to demarcate the targets (distinct target flocks) and to help inform the agents as to which target should be pursued. The k-means algorithm is an iterative process that aims to partition a data set into k pre-defined distinct clusters through the following procedure:

1. Choose the number of clusters k.
2. Initialize centroids through random selection of k data points without replacement.
3. Compute the sum of the squared distance between all data points and every centroid.
4. Assign each data point to the closest centroid (to form distinct clusters).
5. Compute new centroids by taking the average value of data points within each respective cluster.
6. Keep iterating on steps 3–5 until converged.

In this simulation, the k value (cluster number) correlates to the number of targets and is assumed to always be known. In reality, sensor information, such as the camera detection and localisation network discussed in Section 2.1, would need to deduce how many targets are truly present. It is important to note that this procedure works for data sets with values that are spatially distributed. In a target probability map, the 'data points' or probability scores are evenly distributed since every grid cell has a value; in other words, there are no distinct clusters. Hence, to map probability score into clusters, when performing this procedure, scores below a certain threshold are removed to isolate the higher probability cells into distinct spatial clusters. Through this k-means clustering, any number of k target flocks can enter the property, and the control strategy is such that an agent will pursue the geometric centre of the closest cluster if it exists and is in range.

2.3. Target Flock Model

This simulation assumes that all targets are Common Starlings (Sturnus vulgaris) to reduce the number of variables and simplify the problem.

2.3.1. Target Flock Dynamics and Trajectory Planning

Identical to the agent UAV model, the target flocks are modelled using simple second-order dynamics with the following state X_{bird}:

$$X_{bird} = (x, y, z, v_x, v_z, \psi). \tag{13}$$

Furthermore, all of the targets that make up a single target flock are assumed to remain in flock formations throughout the simulation based on field observations made in [7]. Hence, the behavioural responses and trajectory planning of the whole flock are simplified to be modelled on an individual target positioned at its geometric centre. The field observations also found that no birds were seen or remained within 50 m of a UAV. This value corresponded to the 'no-go radius' r_{no-go} of Equation (12) and was grossly estimated using GPS coordinates of the UAV and observations made by the eye of the flock positions on relatively large plots of land. Hence, for this simulation with the smaller plot size of 100 m by 100 m, a conservative, reduced 'no-go' radius of 20 m has been implemented into the target flock model.

The trajectory planning of the targets is modelled using the same 'interests' system used by Wang [17]. This interests system for target flocks parallels the probability map system whereby each cell within the grid map is allocated an interest score i between 0 (minimum interest) and 1 (maximum interest). The cells close to the agents are given reduced interest scores, and a target will migrate to a cell with the highest interest once its current position falls below a certain threshold $i_{threshold}$ (to satisfy the observation that stationary birds limit movement to conserve energy). Furthermore, the cells outside of the protected area are given a very-low-interest score to ensure that the targets continuously aim to remain within the property where possible.

2.3.2. Target Flock Energy Expenditure System

The estimation of target flock energy expenditure in activities such as flight and foraging can serve the purpose of characterizing the perceived relative effort or energy that a target flock requires to forage in a field that is being protected by bird-deterring UAVs. An ideal bird deterrence system would aim to elicit maximum energy expenditure in a target flock to decrease the likelihood of future foraging attempts or increase the time until the given targeted flock returns to the protected property, as purported by optimal foraging theory [23,24].

Hence, a novel target flock energy expenditure system is introduced here. This model estimates an initial energy reserve dedicated for a single foraging event and makes deductions from this reserve as the target flock undergoes different states or flight modes. The important parameters that are needed to define this energy expenditure system are:

- Initial energy expenditure reserve.
- Body mass.
- Basal metabolic rate (BMR).
- Power costs for regular long flights and short burst flights.

In the literature on Common Starlings, the number of flights per day (FPD) is approximately 135 flights [25]. The average flight speed V is assumed to be close to the minimum power speed as a conservative assumption; a value of approximately 10 m/s [26]. Most Starlings nests are also usually within 500 m of desirable foraging zones. Hence, the total flight time between foraging and nesting sites can be estimated by:

$$t_{flight,total} = \frac{FPD \cdot d_{travel}}{V}, \tag{14}$$

where d_{travel} is the distance between a foraging zone and a nesting site. The average flight cost $P_{flight,avg}$ for the Common Starling has been estimated to be 20.4 W [27], which allows for the calculation of the allowable energy expenditure per trip for a single target flock (in J):

$$E_{exp.,single} = \frac{P_{flight,avg} \cdot t_{flight,total}}{FPD}. \tag{15}$$

The value from Equation (15) is then fed as the initial energy condition in the simulation. Note that this is a conservative case since the target flocks in the simulation initialize along the boundary of the protected property rather than at their more distant nesting sites. A summary of these parameters for the Common Starling is summarized in Table 1.

Table 1. Energy and mass data for the Common Starling.

Parameter	Mass (kg)	BMR (W)	$P_{flight,avg}$ (W)	P_{short} (W)	P_{long} (W)
Value	0.079	0.877	20.4	27.2	10.0

2.4. Simulation Testing Plan

Variables of Interest

The dependent variables for this simulation should indicate the performance of the bird-deterring system, and they are:

- Time taken to repel a target (if successful).
- The energy expended by a given target in a single mission. (Unsuccessful missions are capped at $t = 150$ s.)

The independent variables are determined after considering the cause and effect between different parameters:

- Number of agents.
- Tether length.
- Maximum agent speed.
- Number of targets.

At the initialization of each simulation, the target is initialized to enter at any random point along the property boundary with a heading pointing to the centre of the grid map. The simulation is then run for all the combinations of the parameters outlined in Table 2. Following this, the entire simulation process is repeated but with a different target initial position to measure the average performance of the system. Similarly, this entire simulation process is also repeated using multiple targets.

Table 2. Simulation parameter definitions and values.

Variable Type	Simulation Parameters
Independent	Number of agents ($n = 1$–10), tether radius ($R_T = 0$–20 m), agent maximum speed ($v_{max} = 8$, 12, 16, 20 m/s), number of targets ($n_t = 1$–3)
Dependent	Time taken to deter target (t); energy expended by an individual target flock per mission ($E_{expended}$)
Nuisance variables and sources of randomness	Target flock entry positions, generated agent grid positions
Held-fixed	Property size, target flock type (Common Starling), agent controller policy, agent distribution function, agent sensor range, distribution of 'interests' (corresponding to crop type and positioning), climatic conditions, individual target experience and variation

3. Simulation Results

3.1. Simulation Convergence Study

A simulation convergence study was performed to verify that the number of simulation repetitions is sufficient to provide a fair representation of average performance that is no longer sensitive to randomness in target entry position as well as the non-uniqueness in agent grid formations for $n \geq 3$ agents. Similarly, it is important to avoid unnecessarily large simulation repeats if results have already converged to reduce computation costs. It is assumed that the system is stable and that by the Law of Large Numbers [28], after a sufficient number of simulation repetitions, the performance of the system will converge to its true mean. Hence, a test case to observe system convergence with different configurations between $N = 1 - 500$ simulation repeats was performed (see Appendix A Figure A1). Note that the tether radius has been kept constant at a mid-range value of 14 m throughout each simulation.

This tether radius value was chosen since simulation results at the tether radius extremes were expected to converge considerably faster as the non-uniqueness issues of agent configuration and the randomness in the target entry point are diminished or less sensitive when the agents can either reach zero targets ($R_T = 0$ m) or they can reach many targets regardless of their ground tether point ($R_T = 20$ m). It was observed that there is no considerable variation in both the time performance and energy expenditure running averages after 300 simulations. Hence, it is sufficient to perform $N = 300$ simulation repeats to fairly represent the problem, and this number will be the assumed value used in the following scenarios.

3.2. Single Target Flock Scenarios

The scenarios in this section all involve single target flocks tested against the various agent configurations outlined in the simulation testing plan. The full averaged results are presented in Figures 8 and 9.

Observing the time performance results in Figure 8, as the tether radius R_T increases, the average time taken to dispel a target flock decreases; however, this is dependent on the number of agents in the configuration as well. A successful configuration is defined as one that deters a target flock in over 90% of test cases across all the independent simulation runs, and as can be seen by the red markers, there is a 'critical' minimum tether radius for when this occurs with each agent number. For one or two agents ($n = 1 - 2$), no tether radius value allows for a successful deterrence event. Performance improves marginally for $n = 3$ agents at higher tether radius values but is still limited to a 20% success rate (at $R_T = 20$ m), while a much larger increase in performance can be seen for $n \geq 4$ agents and tether radii of $R_T \approx 13 - 14$ m—tether values above this provide diminishing improvements in performance. For this property size, the tether radius appears to have no significant impact on time performance across all values when there are $n \geq 9$ agents. The configuration that minimises the agent number and the tether radius while still remaining successful in 98% of cases at an average deterrence time of $t = 64$ s, is $n = 4$ agents and $R_T = 14$ m.

Observing the target flock energy expenditure results in Figure 9, as the tether radius R_T increases, the energy expenditure in a single target bird also increases from $n = 1 - 4$ agents, after which energy expenditure decreases with increasing agent number ($n > 4$). The agent configuration that elicits maximum target energy expenditure while still satisfying the 90% success rate threshold can be seen as $n = 4$ agents with tether radii of 13 m ($E = 628$ J). Note that all the averaged energy expenditure values are still below the initialised reserve value of 1020 J.

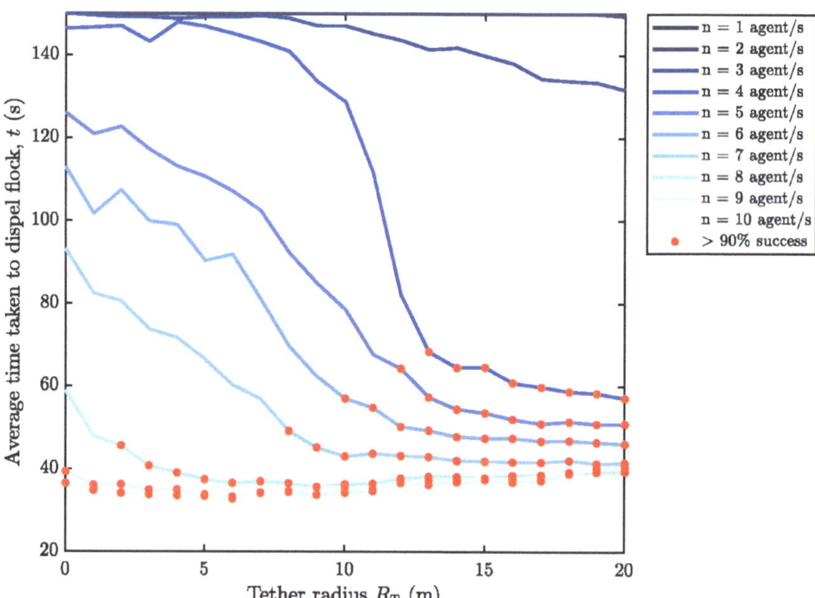

Figure 8. Single target flock scenarios–time performance (darker lines indicate fewer agent numbers while red markers indicate data points whose average success rate is above the given threshold across the 300 independent simulation runs). Maximum agent speed: $v_{max} = 12$ m/s.

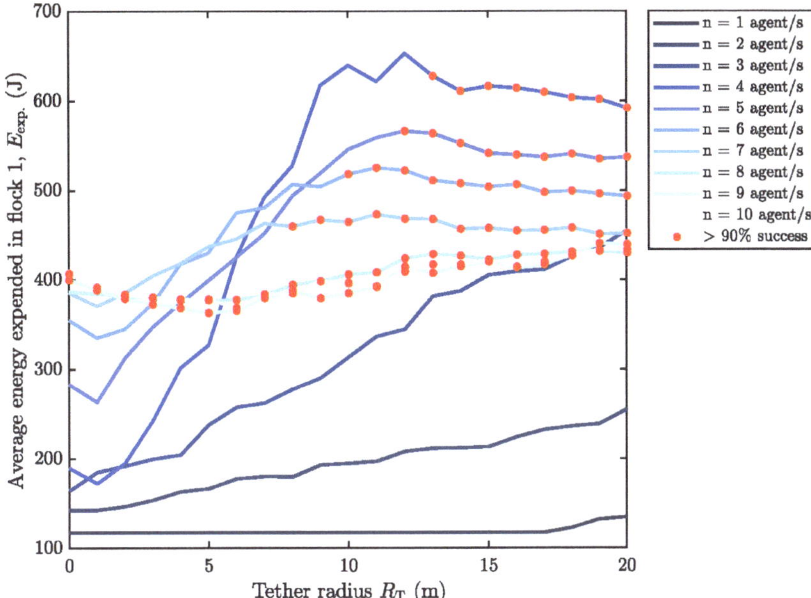

Figure 9. Single target flock scenarios–energy performance (darker lines indicate fewer agent numbers while red markers indicate data points whose average success rate is above the given threshold across the 300 independent simulation runs). Maximum agent speed $v_{max} = 12$ m/s.

The configuration of $n = 4$ agents and $R_T \approx 13 - 14$ m was shown to be salient; thus, snapshots for a conservative sample single target flock scenario with $n = 4$ agents and $R_T = 14$ m is shown as a reference demonstration in Figure 10 due to its marginally higher overall success rate.

In this sample simulation, the target flock is initialised with its full energy reserve at the lower left-hand side of the property (Figure 10a). After entering the property and the closest agent is deployed in its chase state (blue), the target flock seeks a waypoint on the other side of the property (Figure 10b). Once the target nears the next agent, it is deterred once again and seeks the next best interest location on the property, and this process continues to repeat (Figure 10c), with the historical agent flight paths and states shown in their respective colours. At $t = 86$ s, the target is successfully dispelled from the property and is consequently greyed out (Figure 10d). Note that its original energy reserve has been quartered by the end of the simulation. The full 3D trajectory of this simulation can be found in Appendix A Figure A2 for reference and visually captures the vertical ascents and descents that correlate to higher energy expenditure manoeuvres.

3.3. Multiple Target Scenarios

The scenarios in this section involve multiple target flocks tested against the same UAV configurations outlined in the single target flock scenarios above and in the testing plan. The full averaged results, run for 300 simulations on each configuration, are presented in Figures 11 and 12 ($n_T = 2$ targets). The results for $n_t = 3$ targets provided similar results, and for brevity, are omitted here and included in Appendix A Figures A3 and A4 for reference. The 2D snapshots and 3D trajectories for the triple target simulation are included in Appendix A Figures A5 and A6.

The time required to deter two target flocks is shown in Figure 11. Note that a given simulation run is not successfully complete in these scenarios until both target flocks have been dispelled. Similar trends to the single target flock scenarios can be seen here for double target flock deterrence time. The successful configuration that minimises agent number and tether radius is evidently $n = 4$ agents and $R_T = 14$ m. This configuration provides an average deterrence time of $t = 71$ s at a success rate of 91%.

The target flock energy expenditure results are shown in Figure 12. Note that these results only constitute the values for one out of the two target flocks; conducted for the sake of brevity since the second flock results are near-identical, as expected. The best configuration eliciting maximum target flock energy expenditure ($E_{\text{exp.}} = 512$ J) is similarly $n = 4$ agents and $R_T = 14$ m, with the same success rate of 91%.

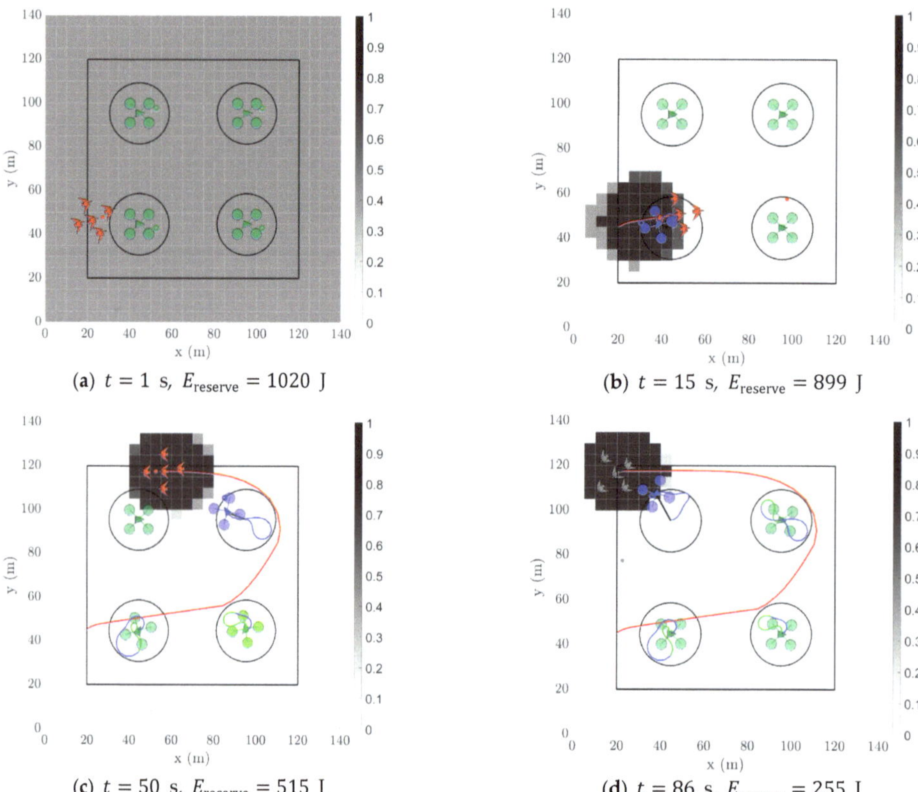

(**a**) $t = 1$ s, $E_{\text{reserve}} = 1020$ J

(**b**) $t = 15$ s, $E_{\text{reserve}} = 899$ J

(**c**) $t = 50$ s, $E_{\text{reserve}} = 515$ J

(**d**) $t = 86$ s, $E_{\text{reserve}} = 255$ J

Figure 10. Snapshots of a sample single target flock simulation with the following configuration: $n = 4$ agents, $R_T = 14$ m, $v_{max} = 12$ m/s. Simulation begins in (**a**) with each target in the target flock at the full energy reserve of 1020 J, all agents undeployed in their Home state, and each cell in the probability map initialised with the same value. As the simulation progresses from (**b**,**c**), the target flock seeks to find grid cells of maximum interest indicated by the red dot but is repelled by nearby deployed agents, shown in their Chase state, while agents that were deployed but no longer in range are shown in their Return state. Simulation is completed in (**d**) with the target centre (greyed out) no longer in the protected area by $t = 86$ s and the original energy reserve quartered.

The snapshots for a sample double target flock simulation using this configuration are presented in Figure 13. The two target flocks are initialised independently with full energy reserves along the property boundary: target 1 (red) at the bottom left-hand side and target 2 (magenta) at the bottom right-hand side (Figure 13a). After a period of time, the targets are each deterred by the closest agents in range and seek new waypoints away from the scaring stimuli (Figure 13b). Once the target flocks reach the new area in the property, they are deterred by the next pair of agents (Figure 13c). Finally, by $t = 79$ s, both targets are dispelled from the property at the same time (Figure 13d). Note that the energy reserves of both targets are approximately halved from their initial value. The full 3D trajectory for this sample simulation is shown in Appendix A Figure A7 for reference.

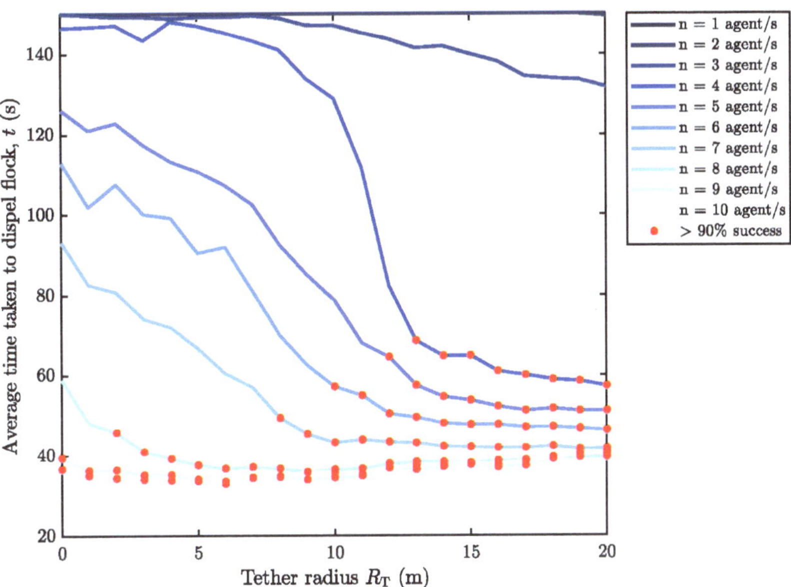

Figure 11. Double target flock scenarios–time performance (darker lines indicate fewer agent numbers while red markers indicate data points whose average success rate is above the given threshold across the 300 independent simulation runs). Maximum agent speed $v_{max} = 12$ m/s.

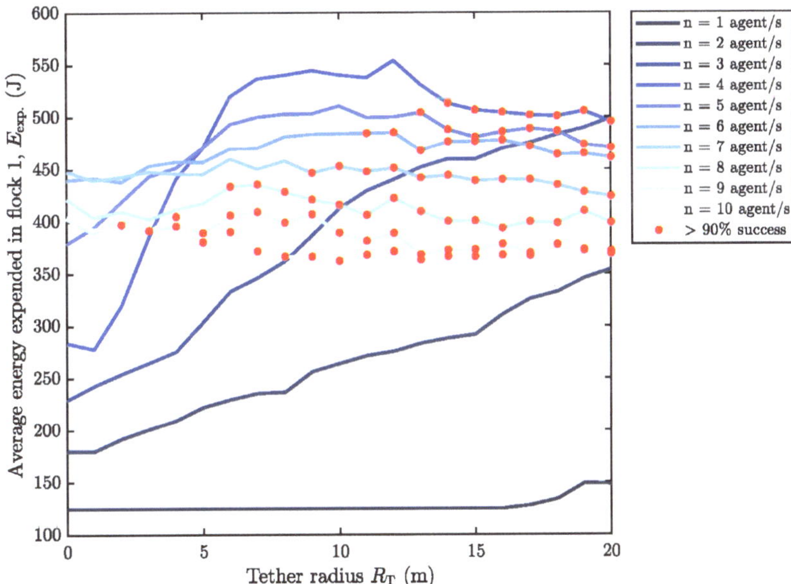

Figure 12. Double target flock scenarios–energy performance (darker lines indicate fewer agent numbers while red markers indicate data points whose average success rate is above the given threshold across the 300 independent simulation runs). Maximum agent speed $v_{max} = 12$ m/s.

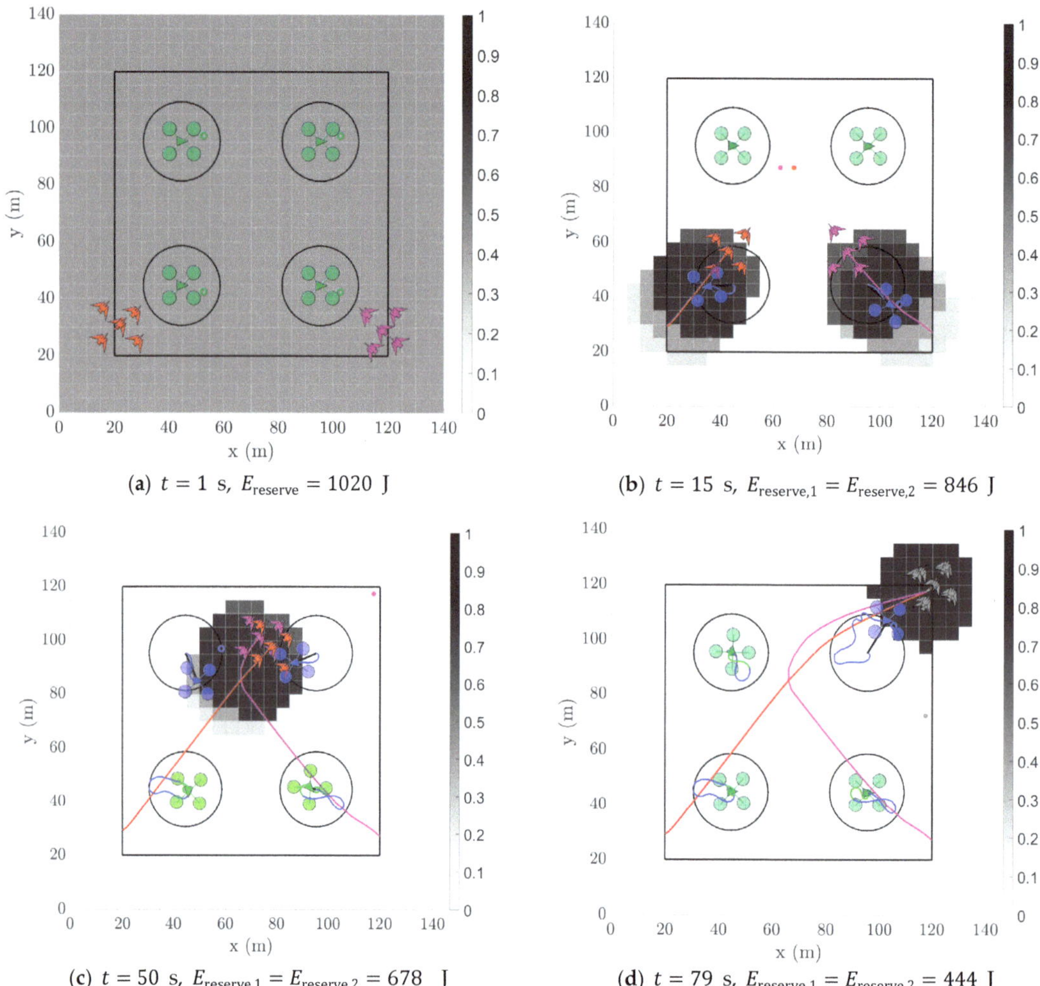

(**a**) $t = 1$ s, $E_{\text{reserve}} = 1020$ J

(**b**) $t = 15$ s, $E_{\text{reserve},1} = E_{\text{reserve},2} = 846$ J

(**c**) $t = 50$ s, $E_{\text{reserve},1} = E_{\text{reserve},2} = 678$ J

(**d**) $t = 79$ s, $E_{\text{reserve},1} = E_{\text{reserve},2} = 444$ J

Figure 13. Snapshots of a sample double target flock simulation with the following configuration: $n = 4$ agents, $R_T = 14$ m, $v_{max} = 12$ m/s. Simulation begins in (**a**) with each target in their respective target flock at the full energy reserve of 1020 J, all agents undeployed in their Home state, and each cell in the probability map initialised with the same value. As the simulation progresses from (**b**,**c**), both target flocks seek to find grid cells of maximum interest indicated by their respective red (Flock 1) and magenta (Flock 2) dots but are repelled by nearby deployed agents, shown in their Chase state, while agents that were deployed but no longer in range are shown in their Return state. Simulation is completed in (**d**) with both target centres (greyed out) no longer in the protected area at the same time of $t = 79$ s and the original energy reserve approximately halved.

3.4. Testing Agent Maximum Speed

The deterrence time and energy expenditure results for $n = 4$ agents and a single target flock at various tether radii and maximum flight speeds are shown in Figures 14 and 15, respectively. Similar performance can be seen for both metrics across all agents' maximum speeds, as was seen in previous results: as the tether radius increases, deterrence time decreases with notable improvement around the $R_T = 10 - 12$ m range, while energy expenditure increases with increasing tether radius, though this does appear to plateau after the same $R_T = 10 - 12$ m range. Overall, there do not appear to be any salient trends or points of interest that depend on agent maximum speed, with all speed cases here tracing out near-identical performance plots.

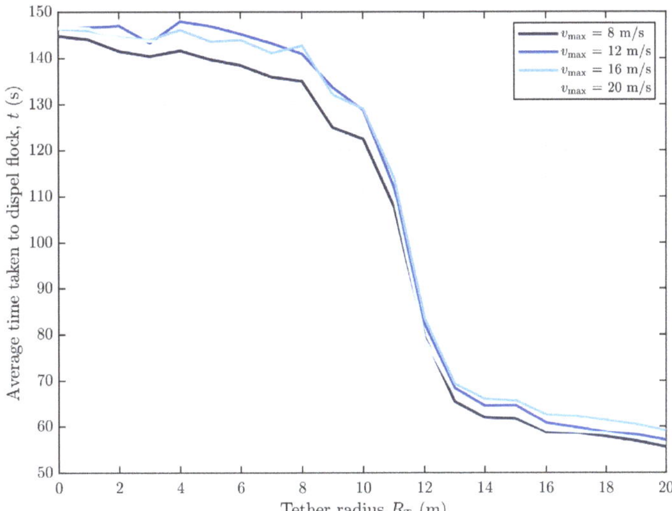

Figure 14. Impact of UAV maximum speed on time deterrence for a configuration of $n = 4$ agents. Darker lines indicate lower maximum speed.

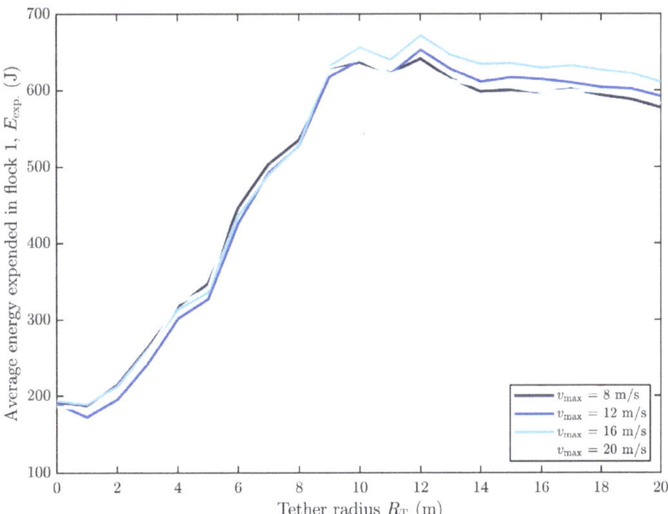

Figure 15. Impact of agent maximum speed on energy expenditure for a configuration of $n = 4$ agents. Darker lines indicate maximum speed.

4. Discussion

4.1. Summary and Analysis of Results

Across the single target flock and both multi-target flock scenarios, the proposed tethered UAV strategy successfully deterred the target flocks from the property area. In each scenario, the average time to deter a target flock reduced with increasing tether radius and agent number. As was shown, every given agent number had a 'critical' tether radius value, which represented the minimum length required for at least a 90% average success rate across the full 300 simulation runs. This critical value only appears for $n \geq 4$ agents and decreases with increasing agent number, while tether radii values above the critical point provide diminishing or marginal improvements in performance. The trajectories for both the target flocks modelled via 'interests' maps and informed by field trials–as well as the pursuing agents modelled via cost functions and constrained by tether radius, are shown to be smooth. The candidate configurations that exhibited the best performance across the different target scenarios are summarised in Table 3.

Table 3. Simulation parameter definitions and values. Note the outlier for the case of $n_t = 2$, $R_T = 13$ m where success rate was markedly lower than other cases.

No. Targets, n_t	Tether Radius, R_T (m)	Success Rate (%)
1	13	97
	14	98
2	13	81
	14	91
3	13	91
	14	96

Considering the combination of $R_T = 13$ and $n_t = 2$, the targets only had a success rate of 81%, it is therefore ideal to opt for the 14 m version, which maintained a success rate of at least 91%; crops only need to be lightly damaged once to be downgraded in terms of quality and value; hence, the more consistently effective system should be chosen. This provides further justification for excluding the three agent configurations that only achieved a maximum success rate of 20% at $R_T = 20$ m, despite having an average deterrence time below the pseudo-infinity of $t = 150$ s in each of the time performance plots. Intriguingly, the success rate for the three target flock scenarios was *higher* than the two target flock scenarios; however, this is most likely due to the coarse modelling and the fact that having more targets means there is a higher chance to activate a nearby agent; once an agent is active, it may be more effective in chasing and funnelling the remaining targets. Note that varying the maximum agent speed for the optimal configuration, as shown in Figures 14 and 15, did not provide any salient features—all speeds provided comparable results. This is presumably because the operating regions constrained by the tether radius for each agent are quite small, and the different maximum flight speeds neither help nor hinder the overall deterrence capabilities since they are not reached in most cases.

Analysis of time performance across the different target flock number scenarios showed that the average time taken to deter all of the intruding target flocks intuitively increased with an increasing target number, seen by the general shift towards the top right in the critical minimum tether value and overall deterrence time in the plots. This indicates that longer tether radii are required to dispel more targets in an equivalent time frame when there are more targets. Though improvements in deterrence time are certainly beneficial, it is ultimately much more important that the system be functional and economically viable. Hence, though the configuration chosen provides the slowest deterrence time among those with success rates above 90%, it is preferable as the cheapest viable option.

The energy results across the different target number scenarios showed that the average target flock energy expenditure reduced and plateaued with increasing numbers of target flocks. This is presumably because when there are more target flocks, there is a higher chance that more agent UAVs are triggered for deployment, which can, in turn, have the effect of funnelling the target flocks into similar routes and positions and forcing them to take shorter flight paths across the property (compare the flight path lengths of the sample single target flock and double target flock scenarios in Figures 10 and 13, respectively). Additionally, note that the average energy expended across all scenarios and all configurations fell well below the maximum reserve value of 1020 J, indicating that all 'successful' scenarios were purely by physically repelling the targets off the property while none were through full depletion of the energy reserve, forcing the given target to stop its foraging mission in that property. Similarly, the minimum target flock energy expenditure was approximately 125 J across all scenarios, which closely matches that seen in the zero lateral movement benchmark case; this value is non-zero since the target flocks are assumed to fly in at altitude and at speed and, therefore, must consume energy to land at their first foraging location, even if it is acquired within the first few seconds upon entering.

A comparison of the optimal configuration against the free-moving agent benchmark case is shown in Table 4 and helps serve as a reference, considering that the field trial observations that informed the construction of this simulation model were on a drone. Note that a free-moving agent here is analogous to an infinite tether.

Table 4. Comparison between free-moving, 'infinite'-tether benchmark case ($n = 1$) with chosen tethered UAV strategy ($n = 4$). The free agent slightly outperformed in average energy expenditure elicited while the tethered team system had a better time performance metric.

No. Targets (n_t)	Tether Radius R_T (m)	Deterrence Time t (s)	Energy Expenditure $E_{exp.}$ (J)	Success Rate (%)
1	-	65	677	100
	14	65	611	98
2	-	103	662	94
	14	71	512	91
3	-	113	599	89
	14	78	491	96

For a single target flock, the tethered team configuration and free agent are on par in average deterrence time, while the free agent slightly outperforms in average energy expenditure elicited in each target and overall success rate. For the multi-target flock scenarios, the tethered system performed 31% faster than the free-moving agent in both scenarios but elicited 23% ($n_t = 2$) and 18% ($n_t = 3$) smaller energy expenditures. Importantly, however, the tethered system maintains a high success rate through to three targets.

4.2. Limitations of the Simulation Model

Though the tethered UAV strategy shows promise, it is important to recognise the limitations of this simulation model to ascertain the degree of confidence in how accurately it reflects reality, particularly in the following components:

- Target flock behaviour modelling limitations: as previously discussed, this factor represents the largest limitation. Within the simulation, it is assumed that targets remain within cohesive flocks the entire duration and will predictably flee from a scaring stimulus in the same manner each time and always flee to the most 'optimal' cell or spot on the property. Though some of these behaviours are modelled on true bird responses to UAVs in vineyards, it omits other behavioural factors and does not account for the fact that there can be many different bird types and species being deterred on the property at the same time, which in turn can influence decision making on where to forage if target flocks of different species do not want to be near each other. This is a fine approximation for flight initiation distance (FID) considering all bird types in the field experiments of [7] fled at distances of 50 m or greater (except for Silvereyes, which exhibit different foraging behaviours). Instead, it challenges foraging assumptions and the universality of the energy expenditure results since they are solely based on data and estimations for the Common Starling. As discussed in the energy expenditure modelling setup, this is a highly crude model due to the number of assumptions made and the scope in which each metric can vary. Ultimately, the energy model is more useful and accurate as a relative measure for identifying configurations that elicit higher expenditure costs rather than a source of absolute truth for what these expenditure values actually are. Though this simulation does not account for returning target flocks, it is assumed by principles such as optimal foraging theory that target flocks become dissuaded the more energy they expend in searching for food on that property and will thus seek food elsewhere. This can also be beneficial and conservative for the simulation since every target flock enters with full energy reserves as opposed to depleted reserves from previous foraging attempts. A final big limitation of the target flock behaviour modelling is that it does not account for habituation; it assumes that target flocks will respond at the same distance to the scaring stimuli every time, irrespective of exposure history or perceived learning. Though this assumption may be valid for a single foraging mission, it certainly does not hold when the same target flocks are exposed to the same stimuli over periods of time.

- Agent modelling limitations: in terms of agent modelling, there were simplifications in dynamics with the hybrid 2D and 3D approach. A 3D probability map was avoided to reduce significantly higher computational costs for only marginal gains, indicating that the vertical trajectory planning for the agents was constrained to operate between two pre-set heights (the ground and 15 m), with the optimal waypoint calculated by the cost function only carrying relevance in the horizontal plane. Similarly, no dynamic considerations due to tether were modelled; it was assumed that, for the configurations under study, the horizontal offsets from the tether point were sufficiently small to not impact overall dynamics and stability to an appreciable degree, as long as there is an appropriate winch system to maintain sufficient tautness such as seen in the near-horizontal 12 m powered tether used in [13]. This adds greater pressure and motivation for a shorter tether radius, where this assumption becomes increasingly more valid. Another simplification is the assumption that agent arrangement is optimal when they are spaced as far apart from each other as possible with the motivation of maximising property area coverage. This may not be the case since most birds in field trials are observed to enter the property from one of its four sides (as opposed to directly top-down from flight), as confirmed by the higher proportion of bird crop damage normally located along the peripheral zones of the property. Hence, a more optimal configuration may account for this factor and redistribute agents closer to the property boundary accordingly. A further assumption is that the number of target flocks on the property is always known based on ground camera sensor information. The verification of this assumption lies beyond the scope of this paper. It is also important to note that the bird behavioural responses are based on a UAV that is equipped with specialised scaring stimuli (auditory and visual) and not just a regular multirotor. Hence, prototyping of the physical tethered system must incorporate and account for existing scaring stimuli.
- Property modelling limitations: in reality, most crops are not perfect squares in shape, nor are they only 100 m by 100 m in size. This simulation model further reduces the problem to an idealised scenario and assumes that all sections within the property are equally attractive to target flocks, not accounting for topographical variations nor existing pockets of potentially higher or lower foraging interest through higher local crop yields or pre-existing damage, respectively. In reality, the optimal tether radius will also need to be modified and scaled with fields of different sizes since the strategy chosen in this paper will presumably fail for much larger property dimensions.
- Probability map modelling limitations: The probability map model does not evolve in the temporal domain, nor does it vary according to environmental light and weather. At the current stage, there is not enough bird behavioural study related to the foraging activity in vineyards under different weather conditions to enable a more dynamic model. Further investigation is needed to construct a more accurate probability map model.

Nevertheless, the simulated multi-UAV tethered strategy has shown to be able to deter multiple target flocks and is a novel approach and technique not seen anywhere in the literature. The use of a team of tethered drones dedicated to bird deterrence and simulation of target energy expenditure in response to pursuing drones are introduced here as firsts of their kind to the authors' knowledge. The simulation model has been constructed using established methods and techniques used in the literature for UAV path planning via cost function and probability map and has been benchmarked against a free-moving single agent case as performed through simulation and field trials by Wang in [7], [17]. Attempts at increasing confidence in the system have focused on the imposition of dynamic constraints based on typical multirotor capabilities as well as spatial constraints due to the tether. A configuration of four agents for a plot that is only 100 m by 100 m may seem overly costly upon first inspection, particularly if the intent is to scale up the system for larger property areas; however, it is important to remember that the scaring distances were greatly underestimated as a conservative assumption in the simulation model and that the reality may be quite different with some target flocks certainly fleeing at further distances across and within different studies. Furthermore, it is not certain that this configuration will perform as expected in a field trial, but it at least remains feasible in size and scope for testing. Hence, this returns to the original question of this paper of whether a tethered system is feasible. This simulation model has shown that it could work, even if not fully optimised, though it must be validated through physical testing.

5. Conclusions and Future Work

5.1. Future Testing

The next most important step is certainly a validation of the simulation model and the chosen configuration. This requires physical testing of sub-component features such as the multi-UAV coordination strategy as well as verification of stable flight under a tether. Ultimately, true verification, once sub-component features have been tested, is through in situ field trials with live birds. Unfortunately, restrictions imposed on the Greater Sydney region between July and October 2021 due to the COVID-19 pandemic limited all access to university laboratories as well as crops in regional areas where field trials would be performed. Hence, the scope of this research paper remains strictly theoretical via simulation modelling since all forms of physical testing were not possible. Nevertheless, various proposed tests and prototype possibilities for a future study when restrictions are eased are outlined below. Note that although features such as power-over-tether serve as driving factors for the tethered design in the first place, it is well-established that power-over-tether technologies already exist and, therefore, verification or feasibility tests for this lie outside the scope of this paper. Rather, the simulation results presented depend heavily on the success of the multi-UAV coordination strategy, the impact of the tether on dynamics being minimal, and that the target flocks respond to the tethered drone system in a similar manner to free-flying drones recorded in existing field trials. Hence, these areas must be tested first and foremost to truly evaluate whether a team of tethered drones can coordinate in a grid-confined arrangement to scare incoming target flocks away. Planned future testings include:

- Autonomous Multi-UAV Coordination Test: evaluate the coordination strategy between multiple drones using real hardware.
- Tethered Drone Dynamics Test: considering that the influence of a tether on the dynamics and stability of a drone was not explicitly modelled in the simulation, it is important to verify that its impact is near-minimal in the desired testing range of $R_T = 0 - 14$ m and up to a target altitude of 15 m.
- Bird Response to Tethered UAV Test: after the multi-UAV coordination strategy and tethered dynamics have been independently tested and analysed (with consequent changes made wherever needed), the system can be tested as a full prototype in situ on a vineyard.

5.2. Future Research

Target Flock Modelling

Modelling target flock behaviour predictably and accurately continues to be a challenge and will, therefore, always need to be verified through physical tests with live birds. Nevertheless, research into foraging styles and preferences could assist in nuancing which areas of a property are more at risk of bird damage. For instance, flocks of European Starlings have been shown to concentrate their feeding to habitually visited areas and completely ignore other viable spots [29], leaving some parts of a crop experiencing disproportionate amounts of damage. Similarly, the target flock energy expenditure model introduced in this paper should be further refined. Though the energy model serves more as a proxy for relative effort and target flock persistence, a catalogue of the energy and flight data for more target species seen on food crops would allow for a better comparison of different target responses to UAVs.

Furthermore, it was assumed that all target flocks were the same size and caused the same amount of damage. However, not all targets will enter and forage as part of larger flocks, and differently sized crops can influence the amount of damage or show varied reception to scaring stimuli. Hence, research should be undertaken to understand the impact of flock size on bird damage and the efficacy of UAVs in deterring them to further inform the chasing strategy of the agents. The potential research areas include:

- Multi-UAV coordination optimisation: research and development into a decentralised or distributed control architecture where each agent has the capacity to make more 'informed' decisions in chasing.
- Multi-UAV arrangement optimisation: the current simulation model assumes that every spot along the property boundary is an equally likely entry point; however, further research should explore different optimised arrangements and grid-generating algorithms which do account for local geography and areas with historically greater recorded bird damage.
- Property size and shape: importantly, further work should analyse the impact of different property sizes and shapes and determine whether there is a relationship that determines the number of drones required per square metre and the scalability of different configurations. Similarly, it should be determined whether there is a relationship between tether length and property size (for a given number of agents or property 'coverage' ratio).
- Other applications: the strategy could be analysed in other specialised or smaller contexts, such as hangars or airports where birds can similarly cause disruption, but deterrence regions may only need to be in very specific zones.

5.3. Conclusions

The objective of this research paper was to determine the feasibility of deterring pest birds using multiple UAVs that are individually tethered to the ground in a grid. This objective was achieved; however, the research used simulation and remains theoretical due to the COVID-19 restrictions imposed on movement and university access which prohibited physical testing and validation. The simulation model was constructed as an extension of the novel autonomous trajectory planning algorithm based on a probability map and bird field trial observations by Wang in [17]. The simulation model was then tested and benchmarked to find suitable tethered agent configurations for the bird deterrence problem that would have then been presented for physical tests. An analysis highlighted the optimal candidate configuration for the 100 m by 100 m plot as requiring four evenly distributed agent UAVs with tether radii of 14 m.

This research made the following explicit contributions to the tethered UAV bird deterrence problem:

- Development of a simulation model that can operate on a wider testing range, including number of UAVs, number of target flocks and maximum speeds.

- Introduction of tethered UAVs as a potential solution to the bird deterrence problem—implemented both as a spatial constraint and in the cost function strategy within the simulation model.
- Addition of a grid-generation algorithm to arrange any number of UAVs evenly within a given rectangular area.
- Expansion of the UAV bird deterrence problem to a pseudo-3D model involving a 2D probability map representing target flock occupancy but 3D dynamics for the UAVs and target flocks.
- Introduction of a centralised multi-UAV control strategy that can cope and coordinate the chasing of multiple incoming target flocks.
- Introduction of a novel target flock energy expenditure model based on the Common Starling used to relativise effort required by a bird in a given foraging mission due to a particular agent configuration and which can be scaled up to include a catalogue of other target species.

The tethered multi-UAV strategy for bird deterrence is a novel solution proposed in this paper for agriculture and not seen elsewhere to the authors' knowledge. Although future field experiments are essential in ultimately confirming the viability of such a system, simulations of this novel bird deterrence method still show great promise that this solution is, in fact, feasible.

Author Contributions: Conceptualisation, K.C.W. and Z.W.; Methodology, J.T. and Z.W.; Software: J.T. and Z.W.; Validation: J.T. and Z.W.; Formal Analysis: J.T.; Investigation: J.T.; Resources: J.T., Z.W. and K.C.W.; Data curation: J.T.; Writing—original draft preparation: J.T. and Z.W.; writing—review and editing: J.T., Z.W. and K.C.W.; visualisation: J.T.; supervision: K.C.W. and Z.W.; Project administration: K.C.W. and Z.W. All authors have read and agreed to the published version of the manuscript.

Funding: This research received no external funding.

Institutional Review Board Statement: Not applicable.

Informed Consent Statement: Not applicable.

Data Availability Statement: The source code of the simulation and the videos are open source and available at https://github.com/ZihaoUSYD/TetheredBirdDeterringDrone (Accessed on 9 March 2023).

Acknowledgments: We would like to thank Agent Oriented Software and CEO Andrew Lucas for the idea and support that made this project possible.

Conflicts of Interest: The authors declare no conflict of interest.

Appendix A

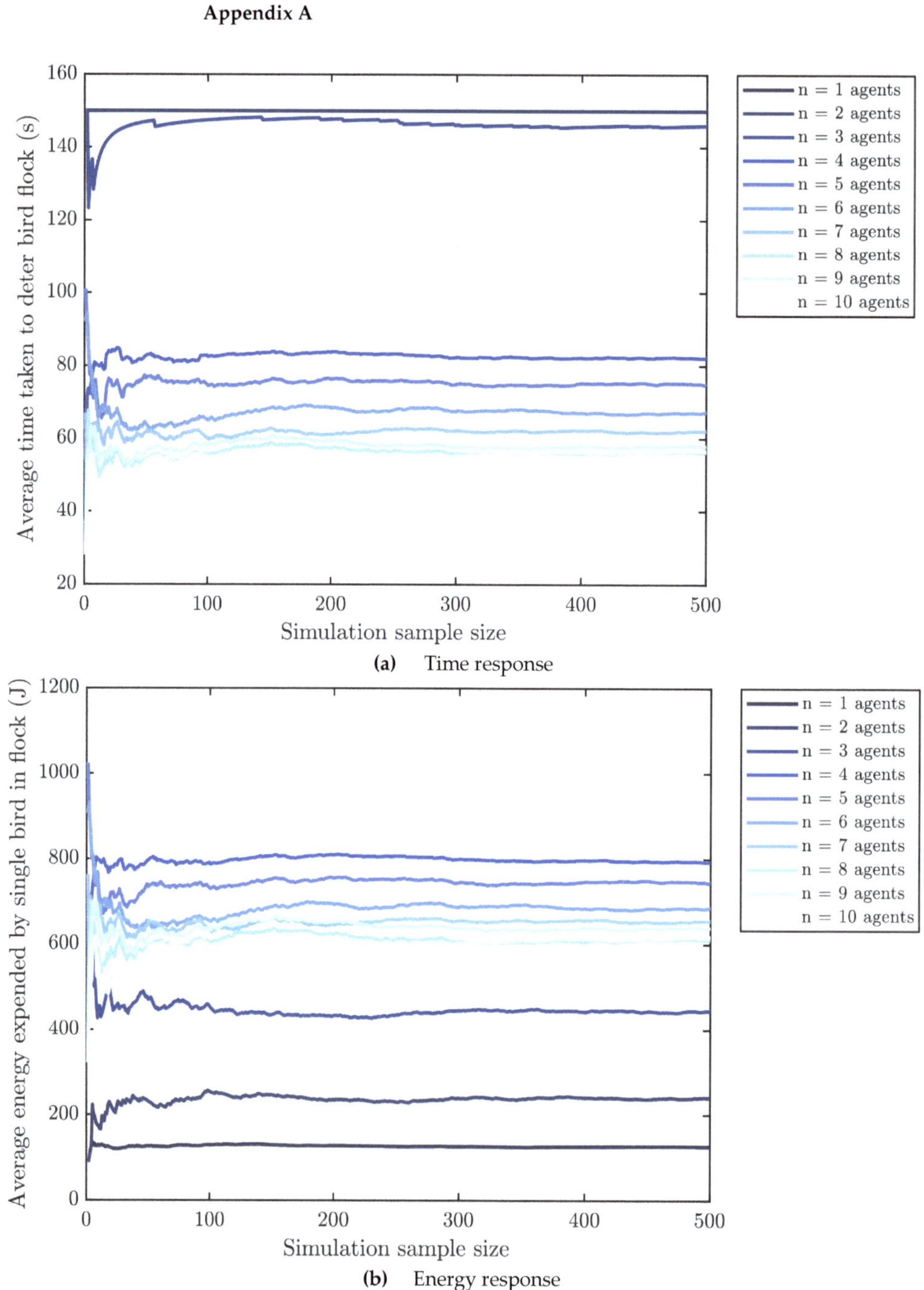

(a) Time response

(b) Energy response

Figure A1. Simulation convergence study for (**a**) time response and (**b**) energy expenditure, with varied agent number but fixed tether radius at $R_T = 14$ m. As simulation sample size increases, more values are being used to calculate the mean; every data point represents the running average of all preceding simulations from a minimum of 1 simulation up to a maximum of 500 simulations.

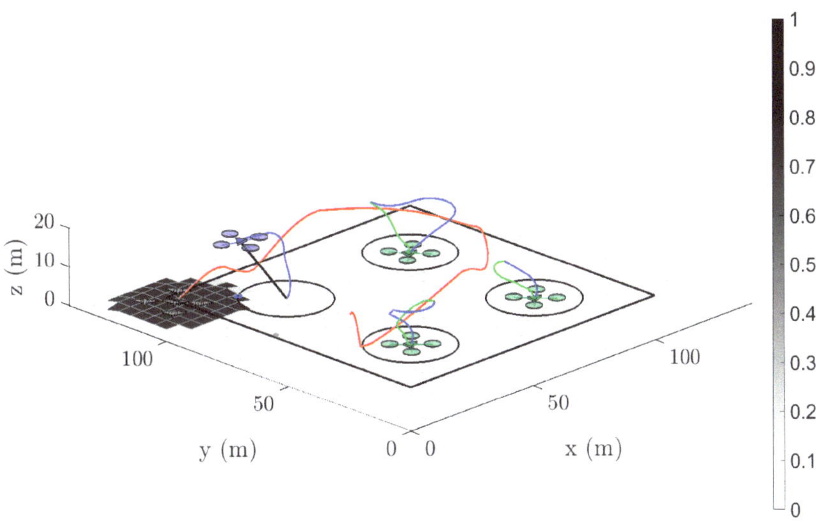

Figure A2. Single-target scenarios—sample simulation 3D trajectory by t = 86 s, $E_{reserve}$ = 255 J. Regions of maximum energy expenditure correlate to high acceleration areas such as take-offs and landings.

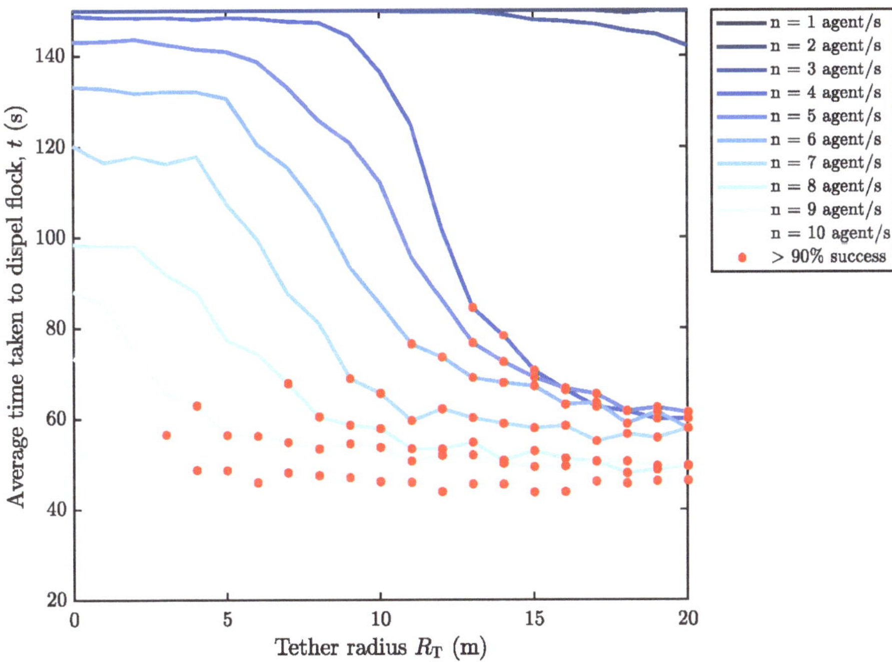

Figure A3. Triple-target scenarios—time performance (darker lines indicate fewer agent numbers while red markers indicate data points whose average success rate is above the given threshold). Maximum agent speed v_{max} = 12 m/s.

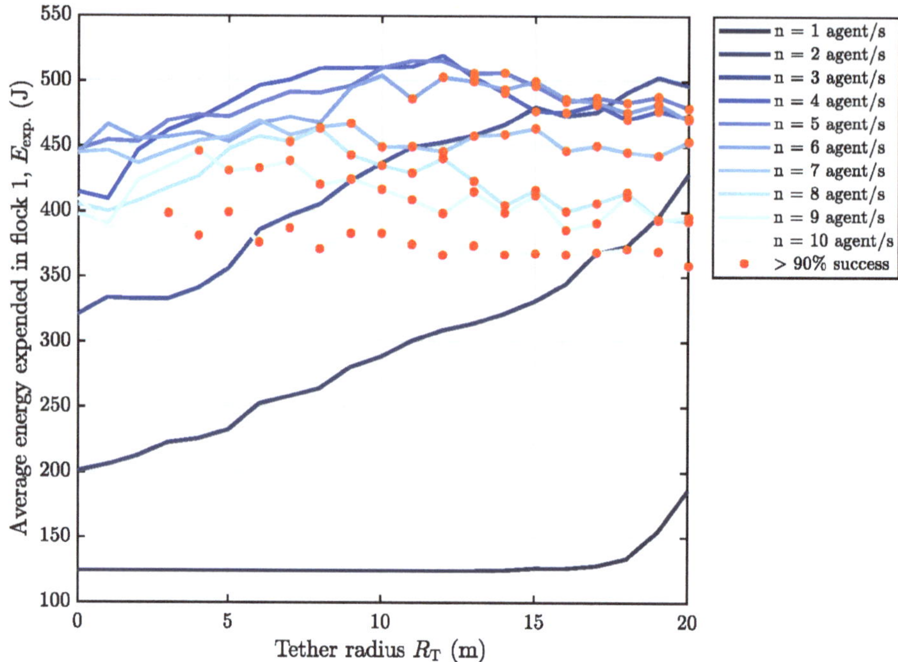

Figure A4. Triple-target scenarios—time performance (darker lines indicate fewer agent numbers while red markers indicate data points whose average success rate is above the given threshold across the 300 independent simulation runs). Maximum agent speed $v_{max} = 12$ m/s.

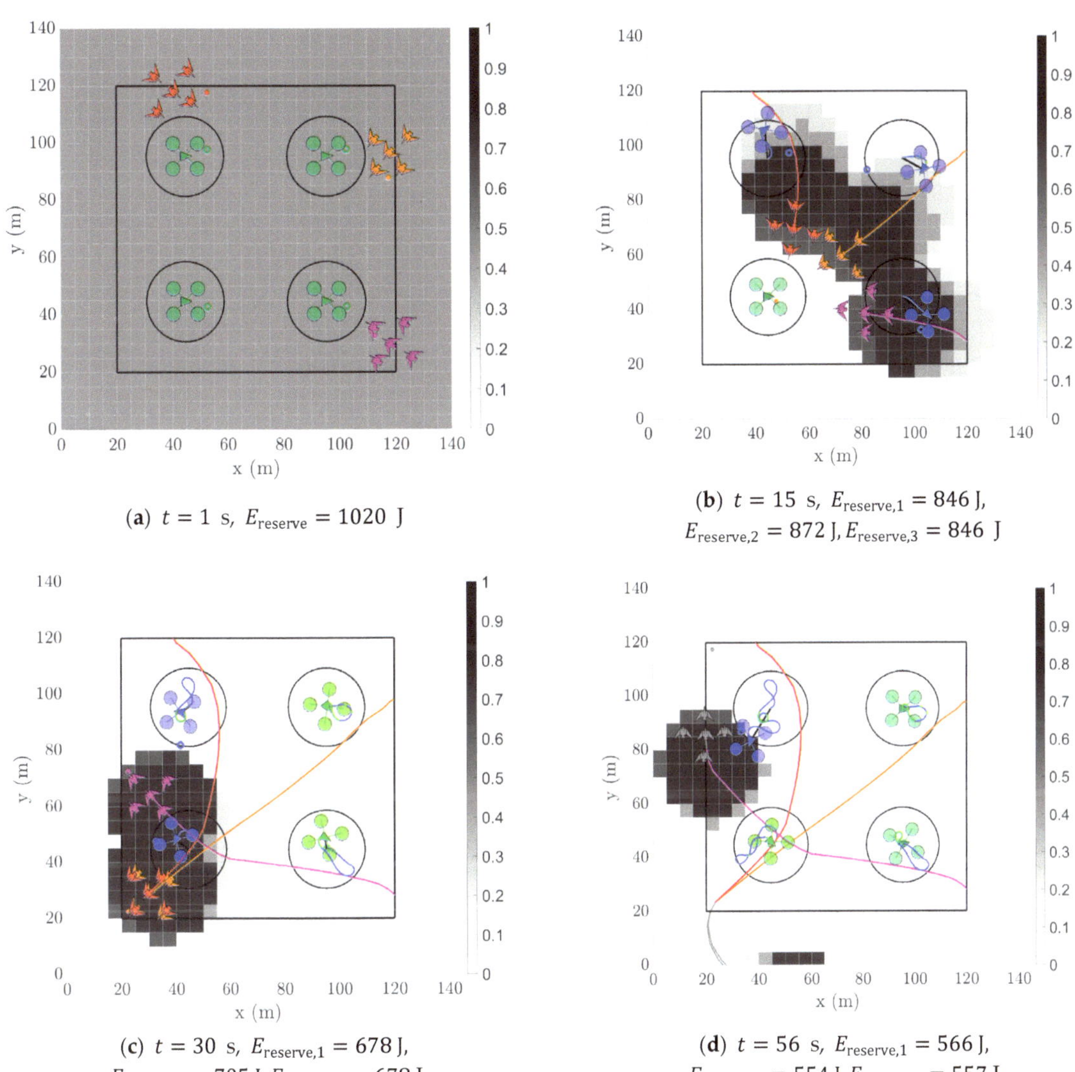

(**a**) $t = 1$ s, $E_{\text{reserve}} = 1020$ J

(**b**) $t = 15$ s, $E_{\text{reserve},1} = 846$ J, $E_{\text{reserve},2} = 872$ J, $E_{\text{reserve},3} = 846$ J

(**c**) $t = 30$ s, $E_{\text{reserve},1} = 678$ J, $E_{\text{reserve},2} = 705$ J, $E_{\text{reserve},3} = 678$ J

(**d**) $t = 56$ s, $E_{\text{reserve},1} = 566$ J, $E_{\text{reserve},2} = 554$ J, $E_{\text{reserve},3} = 557$ J

Figure A5. Snapshots of a sample triple target simulation with the following configuration: $n = 4$ agents, $R_T = 14$ m, $v_{max} = 12$ m/s. Simulation begins in (**a**) with each target in the flock at the full energy reserve of 1020 J, all agentjer I u s undeployed in their Home state, and each cell in the probability map initialised with the same value. As the simulation progresses from (**b**,**c**), both target flocks seek to find grid cells of maximum interest indicated by their respective red (Flock 1), magenta (Flock 2) and orange (Flock 3) dots but are repelled by nearby deployed agents, shown in their Chase state, while agents that were deployed but no longer in range are shown in their Return state. Simulation is completed in (**d**) with both target flocks centre (greyed out) no longer in the protected area at the same time of $t = 56$ s and the original energy reserve approximately halved.

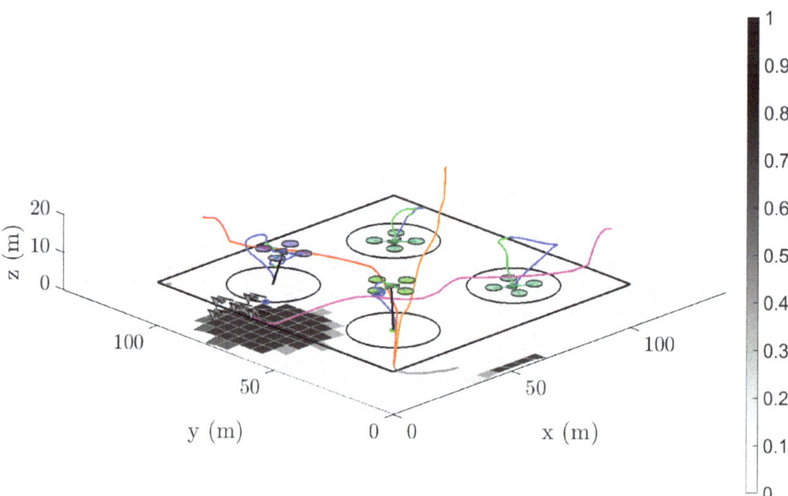

Figure A6. Triple-target scenarios—sample simulation 3D trajectory by $t = 56$ s, for all three target flocks. Regions of maximum energy expenditure correlate to high acceleration areas such as take-offs and landings.

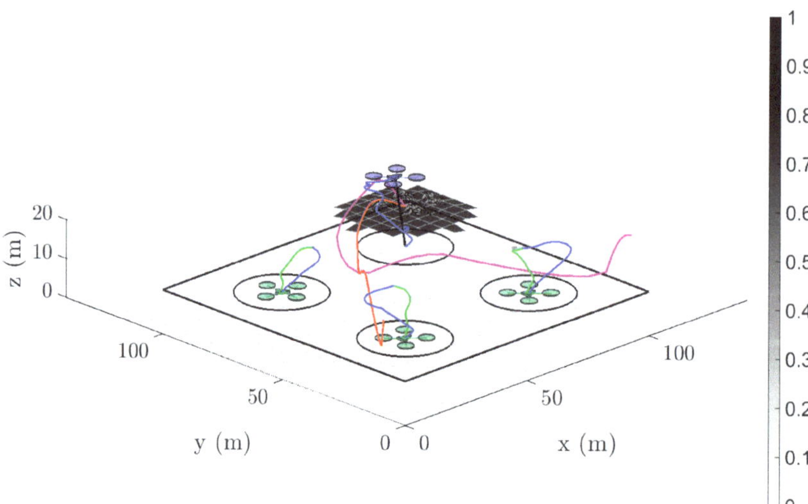

Figure A7. Double-target scenarios—sample simulation 3D trajectory by $t = 79$ s, $E_{reserve} = 444$ J for both target flocks. Regions of maximum energy expenditure correlate to high acceleration areas such as take-offs and landings.

References

1. Tracey, J.; Saunders, G. *Bird Damage to the Wine Grape Industry*; Australian Government Bureau of Rural Sciences: Canberra, ACT, Australia, 2003.
2. Tracey, J.; Bomford, M.; Hart, Q.; Saunders, G.; Sinclair, R. *Managing Bird Damage to Fruit and Other Horticultural Crops*; Bureau of Rural Science: Canberra, ACT, Australia, 2007.
3. Anderson, A.; Lindell, C.; Moxcey, K.; Siemer, W.; Linz, G.; Curtis, P.; Carroll, J.; Burrows, C.; Boulanger, J.; Steensma, K.; et al. Bird damage to select fruit crops: The cost of damage and the benefits of control in five states. *Crop Prot.* **2013**, *52*, 103–109. [CrossRef]

4.   Elser, J.L.; Lindell, C.A.; Steensma, K.M.M.; Curtis, P.D.; Leigh, D.K.; Siemer, W.F.; Boulanger, J.R.; Shwiff, S.A.; Lindell, J.L.; Steensma, C.A.; et al. Measuring bird damage to three fruit crops: A comparison of grower and field estimates. *Crop Prot.* **2019**, *123*, 1–4. [CrossRef]
5.   Pimentel, D.; Zuniga, R.; Morrison, D. Update on the environmental and economic costs associated with alien-invasive species in the United States. *Ecol. Econ.* **2005**, *52*, 273–288. [CrossRef]
6.   Storms, R.F.; Carere, C.; Musters, R.; van Gasteren, H.; Verhulst, S.; Hemelrijk, C.K. Deterrence of birds with an artificial predator, the RobotFalcon. *J. R. Soc. Interface* **2022**, *19*. [CrossRef] [PubMed]
7.   Wang, Z.; Griffin, A.S.; Lucas, A.; Wong, K.C. Psychological warfare in vineyard: Using drones and bird psychology to control bird damage to wine grapes. *Crop Prot.* **2019**, *120*, 163–170. [CrossRef]
8.   Wang, Z.; Fahey, D.; Lucas, A.; Griffin, A.S.; Chamitoff, G.; Wong, K.C. Bird damage management in vineyards: Comparing efficacy of a bird psychology-incorporated unmanned aerial vehicle system with netting and visual scaring. *Crop Prot.* **2020**, *137*, 105260. [CrossRef]
9.   Wandrie, L.J.; Klug, P.E.; Clark, M.E. Evaluation of two unmanned aircraft systems as tools for protecting crops from blackbird damage. *Crop Prot.* **2019**, *117*, 15–19. [CrossRef]
10.   Bhusal, S.; Khanal, K.; Karkee, M.; Steensam, K.; Taylor, M.E. Unmanned aerial systems (uas) for mitigating bird damage in wine grapes. In Proceedings of the 14th International Conference on Precision Agriculture, Montreal, QC, Canada, 24–27 June 2018.
11.   Schacht-Rodríguez, R.; Ponsart, J.-C.; García-Beltrán, C.D.; Astorga-Zaragoza, C.M. Prognosis & Health Management for the prediction of UAV flight endurance. *IFAC-PapersOnLine* **2018**, *51*, 983–990. [CrossRef]
12.   Zikou, L.; Papachristos, C.; Tzes, A. The Power-over-Tether system for powering small UAVs: Tethering-line tension control synthesis. In Proceedings of the 2015 23rd Mediterranean Conference on Control and Automation (MED), Torremolinos, Spain, 16-19 June 2015; pp. 681–687. [CrossRef]
13.   Kiribayashi, S.; Yakushigawa, K.; Nagatani, K. Design and Development of Tether-Powered Multirotor Micro Unmanned Aerial Vehicle System for Remote-Controlled Construction Machine. In Proceedings of the 11th Conference on Field and Service Robotics (FSR), Zurich, Switzerland, 12–15 September 2017; pp. 637–648.
14.   Nicotra, M.M.; Naldi, R.; Garone, E. Taut Cable Control of a Tethered UAV. *IFAC Proc.* **2014**, *47*, 3190–3195. [CrossRef]
15.   Lupashin, S.; D'Andrea, R. Stabilization of a flying vehicle on a taut tether using inertial sensing. In Proceedings of the 2013 IEEE/RSJ International Conference on Intelligent Robots and Systems, Tokyo, Japan, 3–7 November 2013; pp. 2432–2438. [CrossRef]
16.   Furukawa, T.; Bourgault, F.; Lavis, B.; Durrant-Whyte, H.F. Recursive Bayesian search-and-tracking using coordinated UAVs for lost targets. In Proceedings of the IEEE International Conference on Robotics and Automation, Orlando, FL, USA, 15–19 May 2006; Volume 2006, pp. 2521–2526. [CrossRef]
17.   Wang, Z.; Wong, K.C. Autonomous Pest Bird Deterring for Agricultural Crops Using Teams of Unmanned Aerial Vehicles. In Proceedings of the 2019 12th Asian Control Conference (ASCC), Kitakyushu, Japan, 9–12 June 2019; pp. 108–113.
18.   Lum, C.W.; Rysdyk, R.T.; Pongpunwattana, A. Occupancy based map searching using heterogeneous teams of autonomous vehicles. In Proceedings of the AIAA Guidance, Navigation, and Control Conference and Exhibit, Keystone, CO, USA, 21–24 August 2006; Available online: http://arc.aiaa.org/doi/pdf/10.2514/6.2006-6196 (accessed on 9 March 2023).
19.   Wang, Z. *Intelligent UAVs for Pest Bird Control in Vineyards*; The University of Sydney: Sydney, NSW, Australia, 2021.
20.   mRobotics. mRo Pixhawk 2.4.6 Cool Kit (Limited Edition). 2016. Available online: https://store.mrobotics.io/product-p/mro-pixhawk1-fullkit-mr.htm (accessed on 12 August 2018).
21.   Du, Q.; Faber, V.; Gunzburger, M. Centroidal Voronoi tessellations: Applications and algorithms. *SIAM Rev.* **1999**, *41*, 637–676. [CrossRef]
22.   Ju, L.; Ringler, T.; Gunzburger, M. *Voronoi Tessellations and Their Application to Climate and Global Modeling*; Springer: Berlin/Heidelberg, Germany,, 2011; pp. 313–342.
23.   Pyke, G.H. Optimal Foraging Theory: A Critical Review. *Annu. Rev. Ecol. Syst.* **1984**, *15*, 523–575. [CrossRef]
24.   Kie, J.G. Optimal Foraging and Risk of Predation: Effects on Behavior and Social Structure in Ungulates. *J. Mammal.* **1999**, *80*, 1114–1129. [CrossRef]
25.   Coleman, J.D. The foods and feeding of starlings in Canterbury. *Proc. N. Z. Ecol. Soc.* **1977**, *24*, 94–109.
26.   Heldbjerg, H.; Fox, A.D.; Thellesen, P.V.; Dalby, L.; Sunde, P. Common Starlings (*Sturnus vulgaris*) increasingly select for grazed areas with increasing distance-to-nest. *PLoS ONE* **2017**, *12*, e0182504. [CrossRef]
27.   Wiersma, P. *Working for a Living: Physiological and Behavioural Trade-Offs in Birds Facing Hard Work*; University of Groningen: Groningen, The Netherlands, 2003.
28.   Dinov, I.D.; Christou, N.; Gould, R. Law of Large Numbers: The Theory, Applications and Technology-Based Education. *J. Stat. Educ.* **2009**, *17*, 1–9. [CrossRef]
29.   Whitehead, S.C.; Wright, J.; Cotton, P.A. Winter Field Use by the European Starling Sturnus vulgaris: Habitat Preferences and the Availability of Prey. *J. Avian Biol.* **1995**, *26*, 193. [CrossRef]

Article

Use of Unmanned Aerial Vehicles for Building a House Risk Index of Mosquito-Borne Viral Diseases

Víctor Muñiz-Sánchez [1], Kenia Mayela Valdez-Delgado [2], Francisco J. Hernandez-Lopez [3], David A. Moo-Llanes [2], Graciela González-Farías [1] and Rogelio Danis-Lozano [2,*]

1 Monterrey Campus, Centro de Investigación en Matemáticas, A.C., Km. 10 Autopista al Aeropuerto, Parque de Investigación e Innovación Tecnológica (PIIT), Av. Alianza Centro 502, Apodaca 66628, Mexico
2 Centro Regional de Investigación en Salud Pública (CRISP), Instituto Nacional de Salud Pública (INSP), 4a Av. Norte esquina 19 Calle Poniente s/n, Tapachula 30700, Mexico
3 CONACYT—Centro de Investigación en Matemáticas, A.C., CIMAT Unidad Mérida, Parque Científico y Tecnológico de Yucatán (PCTY), Sierra Papacal, Mérida 97302, Mexico
* Correspondence: rdanis@insp.mx

Abstract: The Vector Control Program in Mexico has developed operational research strategies to identify entomological and sociodemographic parameters associated with dengue transmission in order to direct targeted actions and reduce transmission. However, these strategies have limitations in establishing their relationship with landscape analysis and dengue transmission. This study provides a proof of concept of the use of unmanned aerial vehicle technology as a possible way to collect spatial information of the landscape in real time through multispectral images for the generation of a multivariate predictive model that allows for the establishment of a risk index relating sociodemographic variables with the presence of the vector in its different larval, pupal, and adult stages. With flight times of less than 30 min, RGB orthomosaics were built, where houses, roads, highways, rivers, and trails are observed in detail, as well as in areas with a strong influence of vegetation, detailing the location of the roofs or the infrastructure of the house, grass, bushes, and trees of different dimensions, with a pixel resolution level of 5 centimeters. For the risk index, we developed a methodology based on partial least squares (PLS), which takes into account the different type of variables are involved and the geographic distribution of the houses as well. Results show the spatial pattern of downtown low-risk housing, which increases as we approach the outskirts of the town. The predictive model of dengue transmission risk developed through orthomosaics can help decision makers to plan control and public health activities.

Keywords: *Aedes aegypti*; UAV; risk model; FAMD; PLS

Citation: Muñiz-Sánchez, V.; Valdez-Delgado, K.M.; Hernandez-Lopez, F.J.; Moo-Llanes, D.A.; González-Farías, G.; Danis-Lozano, R. Use of Unmanned Aerial Vehicles for Building a House Risk Index of Mosquito-Borne Viral Diseases. *Machines* **2022**, *10*, 1161. https://doi.org/10.3390/machines10121161

Academic Editor: Dimitrios E. Manolakos

Received: 5 November 2022
Accepted: 1 December 2022
Published: 4 December 2022

Publisher's Note: MDPI stays neutral with regard to jurisdictional claims in published maps and institutional affiliations.

1. Introduction

Unmanned aerial vehicle (UAV) technology has developed in recent years; its applications avoid many of the limitations associated with satellite data, such as long repetition times, cloud contamination, low spatial resolution, and lack of homogeneity in camera angle or shooting time [1]. It also offers the possibility of collecting detailed spatial information in real time at a relatively low cost [2,3]. They carry out highly detailed imaging work with essential high spatial resolution, spectral, and temporal characteristics [4–7]. Likewise, unmanned aerial vehicles (UAVs) are developing technology for remote-sensing applications based on passive sensors, such as RGB, multispectral, hyperspectral, thermal cameras, and active sensors, such as LiDAR or RADAR [8,9].

The UAVs can provide precise spatial and temporal data to understand the links between disease transmission and environmental factors [2,3,10]. Moreover, it is a flexible and low-cost solution for mapping mosquito breeding sites and the operational dissemination of strategies in mosquito-elimination campaigns. It can be used in communication materials to demonstrate the specific risk conditions for a given area [5,11]. Through aerial surveillance, possible mosquito breeding sites could be identified in domestic areas inaccessible to traditional ground surveillance, which allows them to be proposed as a useful and complementary tool in surveillance programs and mosquito control [12]. Likewise, it can help operators of vector-control actions to optimize treatments by accurately identifying and mapping the coverage of breeding sites by using standard and multispectral images [13,14]. Orthomosaics provide vital information for other public health planning activities [11].

The physical landscape characteristics influence the distribution of adult female mosquitoes [15] as interrelated socioeconomic, environmental, and behavioral factors can explain the presence and abundance of *Aedes aegypti* [16]. Anthropogenic environmental changes can create an overabundance of resources responsible for sustaining vector mosquito species' invasion, spread, and colonization of urban areas [17]. Vegetation cover is an important environmental factor that allows optimal conditions for mosquito life-cycle development [18,19]. The organization of the backyard, knowledge about vector biology, and cleaning containers are identified as the main topics for future prevention strategies for the transmission of dengue in the local and national context [20].

New technologies for mosquito surveillance are growing; drones and specialized cartography, machine learning, deep learning, big data, and citizen science can be added as tools for better understanding the mosquito population dynamics and determining the areas at risk to better conditions for spreading mosquito populations [21]. Predictive models can be used for decision making in preventing various arboviruses transmitted by *Ae. aegypti* in endemic urban and semiurban areas. Building a model that determines the power and association of different variables is a necessary challenge because the *Ae. aegypti* mosquito exists in complex environments with complex dynamics, where numerous factors intervene in different dimensions, and each one can have a different role of lesser or greater importance. However, it does not stop influencing the response of the vector to the differences in the landscape.

In this study, by UAV cartography, field surveillance, and algorithm development, we design a risk index to determine which houses have the environmental, demographic, and socioeconomic characteristics to be classified as "aedic" houses for immature and adult *Ae. aegypti* mosquitoes in Tapachula, Chiapas, a city on the Mexican southern border. This proposed methodology can adapt the model to the different regions of our country, and this flexibility is allowed mainly by the availability of real-time information provided by UAV, one of its most important contributions.

Nonetheless, for the best performance of the UAV, it is necessary to consider the meteorological conditions and know the flight area [22,23] and the terrain's characteristics, such as tall trees and telecommunication antennae, to avoid electromagnetic interference [22]. As a user of UAVs, the national regulations must be observed [24]. Moreover, it should be considered that one of the factors for UAV flight coverage depends on the battery time [3,23]. Indeed, before the flights, the field teams should visit and talk with the health authorities, principal neighborhood actors, and community leaders, who are necessary for developing these new tools [25]. The costs of this technology must be taken into account and considered as open-source software for affordable technology. A multidisciplinary team is also required to address the entire strategy.

Therefore, this study answers the following. (1) How to take advantage of UAV technology by obtaining landscape elements to be included in an index that responds to the complexity of scenarios in which the mosquito develops. (2) How to combine the information obtained from different sources, both from field data and the variables derived from high-definition cartography, that have a practical utility, sense of reality, fidelity, and opportunity to strengthen mosquito-surveillance and control programs.

2. Related Work

Studies on UAV applications in health are scarce. They addressed simulations on using drones for out-of-hospital care in cardiac arrest emergencies, identification of people after accidents, transport of blood samples, and improvement of surgical procedures in war zones [26]. UAVs can transport medical supplies, specimens, biological materials, and vaccines to hard-to-reach areas at a reduced travel time. Moreover, the research on the use of drones in health is almost limited to simulation scenarios [26,27].

In mosquito-surveillance and control programs, UAVs have been designed to control mosquitos by spraying larvicides, dropping larvicide tablets, spreading larvicide granules, the application of adulticides in ultra-low volume, sampling in bodies of water, and surveillance of adult mosquitoes [28,29]. Likewise, drones can help operators of vector-control actions to optimize treatments by accurately identifying and mapping the coverage of breeding sites by using standard and multispectral images [14] and as tools for the release of sterile mosquitoes [30,31].

Indeed, three years ago, different studies were conducted to evaluate the use of UAVs to identify the larval breeding sites of different mosquito species. In Peru, multispectral mapping was used with random forest analysis to identify *Anopheles darling* larval breeding sites [32]. In Brazil, it was used to detect *Ae. aegypti* breeding sites [33]. In Mexico, the first study evaluating UAVs was carried out to identify *Ae. aegypti* breeding sites [12].

In studies of spatiotemporal distribution models of *Ae. aegypti* vectors, the limited availability of high-resolution data for physical variables stands out as part of the inconsistencies in the number and influencing factors [34]. Few studies comprehensively highlight the interaction between environmental, socioeconomic, meteorological, and topographical systems [34]. Maximum entropy species-distribution models have been used, with human population density, distance to vegetation, water channels [35], population density, and poverty [36] being the main predictor variables of the vector suitability of a given area.

There are few works reported in the literature about the use of machine learning (ML) and nonparametric techniques for mosquito surveillance. In [37], authors used three ML models (nonlinear discriminant analysis, random forest, and generalized linear models) to analyze the environmental suitability of three indigenous mosquito species in the Netherlands: *Culiseta annulata*, *Anopheles claviger*, and *Ochlerotatus punctor*, where the response variable is mosquito abundance sampled at 778 locations by using CO_2-baited mosquito magnet traps. As covariates, they used environmental variables obtained from satellite images and meteorological data. The best results were achieved with random forests, and they showed the most discriminative variables for each species. In [38], the authors used classification methods based on k–nearest neighbor, support vector machines, neural networks, and random forest to predict mosquito abundance in 90 sampling sites from Charlotte, NC, USA. The data consisted of mosquito samples collected with gravid traps and socioeconomic and landscape factors. The dependent variable to predict mosquito abundance was binarized as low (0) and high (1) according to the median value, and the independent variables or predictors were divided into seven socioeconomic variables and seven landscape variables obtained from remote-sensing images. Two versions of the input variables were used, one dichotomized into binary values according to the median of each variable, and one with the raw continuous values scaled. The best performance was achieved with the k–nearest neighbor algorithm in both cases; however, there are no further details about the distance metric used in the binary input variables case, because, except for random forest, all classification methods use Euclidean distance as the default option, even when a nonlinear kernel or activation function is used; i.e., they are originally formulated for continuous input variables. In [39], 1066 females adult *Ae. aegypti* were collected by using battery-powered Prokopack aspirators from 128 indoor and outdoor houses in northeastern Thailand to map and predict the female adult *Ae. aegypti* abundance. They used five popular supervised learning models (logistic regression, support vector machine, k-nearest neighbor, artificial neural networks, and random forest) based on socioeconomic, climate change, dengue knowledge, attitude, and practices (KAP), and/or landscape factors to predict the abundance considering only two classes (high and low). In this work, the random forest method had the highest prediction model accuracy, and the study demonstrated that climate change, KAP, and landscape factors were more important than socioeconomic factors in explaining mosquito abundance. Similar to [38], it is not clear how the mixed input data (categorical and continuous) are managed, because the classification models they used (except random forest) cannot naturally handle mixed-type data. This may be the reason why random forest showed the best results.

Those approaches are interesting in terms of prediction, and the use of ML models provides flexibility because most of them are nonparametric and do not assume a priori any probability distribution of data. However, for us it is very important to gain some insight into the local characteristics of the area of study, which makes it more prone to mosquito-borne diseases. A study of the importance of the variables can be done based on the parameters of the models [37,39], when it is possible, correlation analysis, or by a specific hypothesis tested on the models [38], based, for example, on manually adding or excluding individual variables or sets of variables [39]. All those approaches must be carried out carefully when our variables are mixed-type data, particularly for ML models based on Euclidean distance between observations. For this reason, in our approach we pay special attention to obtaining an adequate distance metric for mixed-type data, which allows us to obtain an index representing the local characteristics of houses that are related in some way to dengue disease.

Despite the scientific evidence available, the objectives of each previous study—to analyze the available entomological, epidemiological, and environmental data in defined study areas and to consider different types of variables—have been limited in terms of technology, resolution, and frequency of images available. Having high-resolution, real-time, and more accessible images taken by UAVs dictates the need to evaluate their use with predictive power within the prevention and control of vector-borne diseases. Prior knowledge is a good foundation, but current technology allows us to analyze the dynamics of different changes at different spatial scales, such as overpopulation, urbanization, and climate change.

3. Materials and Methods

3.1. Study Area

Tapachula is the second-most important city in Chiapas, the economic capital of the state and the main populated center of the Soconusco region. In 2018, it was first, epidemiologically, in cases of dengue, with 876 cases registered, an increase of 188% in relation to 2019. Tapachula city has 217,550 inhabitants and 78,039 houses, and is located on the southern Pacific coast of Chiapas. It has an average annual temperature of 21.7 °C and rainfall of 1000–5000 mm [40]. The climate is warm, humid, and temperate, with a heavy rainy season from May to November. The area of study is El Vergel, a colony in the northwest of Tapachula, with an extension of 433 km^2 (Figure 1). It has an estimated population of 2590 people (residents and workers) in 592 houses. There were 597 inhabitants per km^2, with an average age of 11 years and an average of seven years of schooling.

3.2. Field Surveys

Surveys were conducted in 216 houses in the El Vergel neighborhood in Tapachula, Chiapas, from 19 November to 5 December 2020 (Figure 1). The surveys were carried out among inhabitants of legal age.

Four household surveys were conducted in El Vergel neighborhood, Tapachula, Chiapas: (a) the Premise Condition Index (PCI, [41]) and the Modified Premise Condition Index (PCI$_2$, [42]), (b) the entomological survey for adults of *Ae. aegypti*, (c) the entomological survey for larval breeding sites of *Ae. aegypti*, and (d) a demographic survey. These surveys are described below.

Figure 1. Orthomosaic of the El Vergel neighborhood, Tapachula, Chiapas, pointing out the houses surveyed (**left**), more resolution details for landscape elements recognition (**right**).

3.2.1. Premise Condition Index (PCI)

This survey consists of nine questions related to the house appearance or house condition, the yard condition, and the degree of shade; the value of this index oscillates in three levels: low, medium, and high [41]. The modified premise condition index (PCI_2) was also obtained from this survey, including the yard conditions and the degree of shade, without house conditions [42].

3.2.2. Entomological Survey for Adults of *Ae. aegypti*

These surveys were carried out once in all rooms of each house for the manual collection of *Ae. aegypti* at rest by using backpack aspirators [43]; this tool also can collect mosquitoes while flying at 10 or 15 cm from the mouth of the catch container. The mosquito count was done manually in the field and recorded per house. Then, the adult mosquitoes were transported in small cages covered with mash and labeled, one per house, into a cooler to the laboratory. The adults were then identified morphologically, according to [44].

3.2.3. Entomological Survey for *Ae. aegypti* Breeding Sites

In the same manner, this survey was carried out once and consisted of a visual inspection of each house, looking for all containers that could be holding water to look for eggs, larvae, or pupae of *Ae. aegypti*, indoor, outdoor, and on roofs of houses when it was possible. A container had eggs, larvae, or pupae of *Ae. aegypti* was identified as a positive breeding site. The number of eggs, larvae per stage, and pupae per container were counted manually and recorded per house [45].

Ladles, mesh nets, mosquito larvae dippers, and plastic pipettes were used to check each container. The small and medium containers were emptied for larvae and pupae counted, and for the large containers, the personnel used small containers for which all the immature stages of *Ae. aegypti* were counted by depth until no larvae or pupal activity was observed in the container [45]. The species were identified in the field according to [44]. Previous training was necessary for field personnel to strengthen their knowledge about the vector and fill out the entomological surveys.

3.2.4. Overcrowding

This index was constructed from the demographic and socioeconomic survey applied once to each house. It consists of the number of inhabitants/number of house rooms of 216 houses surveyed.

3.3. Drone Photography and Cartography Construction

We use a Matrice 600 DJI® RPAS equipment [22] with a high-resolution 16-megapixel X5 optical camera [46] and a MicaSense Rededge® multispectral camera [47] in order to create cartographic products such as orthomosaics, contour lines, housing classification, and vegetation indices (NDVI, GNDVI, NDVIRedEdge, and CIgreen) [10]. The drone flights were made in collaboration with the National Centre for Disaster Prevention of the Mexican Government (CENAPRED) by using the CENAPRED protocols [48] and following the national normative law [49] and manufacturers' recommendations [22]. The flight plan was carried out by using the PIX4D software [50] which allows flight speed and height variables, camera tilt angle, and overlap between images to be kept constant. All flights were made at 100 m of altitude with a resolution of 2.59 cm/pixel by noon, trying to avoid the shadows [12,51,52], and on the same week of the field surveillance. At the same time, we take drone photographs on RGB by X5 optical camera and on five spectral bands (R, G, B, RedEdge, and NIR) by Micasense® camera.

Once the flight plan is carried out over the study area, photographs captured by the sensor are obtained in JPG format and transferred to the computer for processing. All aerial photographs were processed by using PiX4D [50] software to obtain the georeferenced orthomosaic. These models are created from the process on the point cloud generated by the captured photographs to obtain products such as the orthophoto, digital surface model (DSM), and digital terrain model (DTM) with a spatial resolution of 5 cm. These cartographic products are obtained for both cameras, performing separate processing (Pix4Dmapper version 4.3.31 [53]). Raster data of all structures and vegetation represents the DSM and the development of the geodesic surface represents the DTM. For the classification of vegetation and houses, the RGB orthomosaic was subjected to radiometric calibration by using the ENVI spectral signature library [54] with the empirical line compute (ELC) factors algorithm, and thus it can perform postprocessing on the image, such as supervised classification with the maximum likelihood estimation (MLE) method. The RGB orthomosaic was obtained of the El Vergel neighborhood, where we can see houses, roads, highways, rivers, and trails in detail. Areas with a strong influence of vegetation can be observed, detailing the location of the houses' roofs or infrastructure, as well as the vegetation types, grass, bushes, and trees of different dimensions (see Figure 1). Once the raster formats were obtained, the processing of polygons that cover the coverage of buildings and vegetation continued. The 216 houses were located in the aforementioned raster. Two types of polygons were digitized in the total area of each house: house coverage by roofs and vegetation coverage by trees. Both were divided into three strata: 1. less than 25%; 2. from 26 to 50%; and 3. more than 50%. Shade cover (ShadeDrone) was based on the building cover (roofs) plus the tree cover per house. The cartography of different vegetation indices (Normalized Difference Vegetation Index (NDVI), Green Normalized Difference Vegetation Index (GNDVI), Normalized Difference Red Edge Index (NDVIre), Green Chlorophyll Index (CIgreen) [55], and tree height cartography) were generated by using the ArcMap software from the raster calculator tool [56] that allows the application of different formulas already estimated previously to obtain them, using as inputs the bands of our multispectral sensor (R-G - B- IR- RE).

Finally, the shapefiles of the houses ($N = 216$) of El Vergel, Tapachula, Chiapas were individually separated house by house. Subsequently, 100 points were randomly created on the ArcMap® 10.3 platform within each house. Each variable derived from the vegetation indices (NDVIre, GNDVI, NDVI, and CIgreen), DSM, and DTM consisted of an average of 100 points per house. A summary of the variables we consider in our analysis is shown in Table 1. Time photo capture and flight height were allowed the resolution necessary for recording data of the landscape variables for *Aedes* risk index (Table 1). The UAV photos taken simultaneously with the data recorded through field surveillance reflect the elements found in the field.

Table 1. Description of the variables obtained from field surveys and specialized cartography of El Vergel neighborhood, Tapachula, Chiapas, Mexico.

Variable	Type	Description
Premise Condition Index (PCI)	Discrete	A house score about the house condition, the yard condition, and the degree of shade assessment of houses surveyed. Score 3 to 9.
Premise Condition Index (PCI) weighted	Categorical	Qualification criteria based on the house score of PCI. 1 (Low) = 3, 2 (Medium) = $[4, 6]$ and 3 (High) = $[7, 9]$.
Modified Premise Condition Index (PCI_2)	Discrete	A house score about the yard condition and the degree of shade assessment of houses surveyed. Score 3 to 6.
Modified Premise Condition Index (PCI_2) weighted	Categorical	Category criteria are based on the house score of PCI_2. 1 (Low) = 2, 2 (Medium) = $[3, 4]$ y 3 (High) = $[5, 6]$.
Shade conditions	Categorical	Category of the degree of shadow per house (PCI or PCI_2) 1 : $[0, 25\%)$, 2 : $[25\%, 50\%)$, 3 : $[50\%, 100\%]$. Based on personnel criteria.
Male mosquitoes	Continuous	Number of *Ae. aegypti* male mosquitoes per house.
Female mosquitoes	Continuous	Number of *Ae. aegypti* female mosquitoes per house.
Mosquitoes	Continuous	Number of *Ae. aegypti* mosquitoes per house.
Pupae	Continuous	Number of *Ae. aegypti* pupae per house.
1st instar larvae	Continuous	Number of *Ae. aegypti* 1st instar larvae per house.
2nd instar larvae	Continuous	Number of *Ae. aegypti* 2nd instar larvae per house.
3rd instar larvae	Continuous	Number of *Ae. aegypti* 3rd instar larvae per house.
4th instar larvae	Continuous	Number of *Ae. aegypti* 4th instar larvae per house.
Larvae	Continuous	Number of *Ae. aegypti* all instar larvae per house.
Breeding sites	Continuous	Number of breeding sites of *Ae. aegypti* with eggs, larvae, or pupae per house.
Overcrowding	Continuous	The number of inhabitants/Number of house rooms per house.
ShadeDrone	Categorical	Category of the degree of shadow per house, delimited on aereal images made by drone: 1 : $[0, 25\%)$, 2 : $[25\%, 50\%)$ and 3 : $[50\%, 100\%]$.
TreeCover	Categorical	Category of the tree cover per house, delimited on aereal images made by drone: 1 : $[0, 25\%)$, 2 : $[25\%, 50\%)$ and 3 : $[50\%, 100\%]$.
TreeHeight	Continuous	Tree high average per block, based on the houses surveyed and trees identified on drone cartography per block.
NDVIRe	Continuous	Average of 100 random points per house of Normalized Difference Red Edge Index (NDRE).
GNDVI	Continuous	Average of 100 random points per house of Green Normalized Difference Vegetation Index (GNDVI).
NDVI	Continuous	Average of 100 random points per house of Normalized Difference Vegetation Index (NDVI).
Digital Surface Model	Continuous	Digital Surface Model (DSM) per house on masl.
CIgreen	Continuous	Average of 100 random points per house of Green Chlorophyll Index (CIgreen).
Digital Terrain Model	Continuous	Digital Terrain Model (DTM) per house on masl.

3.4. Methodology

Given the data gathered according to Section 3.2, our proposed index is based on two steps (see Figure 2). First, we obtain a vectorial representation of each house included in our study in such a way that we can obtain similarities among them. This step involves an encoding procedure to the raw data, which takes into account the different types of variables we have. In the second step, we obtain the index as a latent variable derived from a particular regression procedure by introducing a dependent variable related to the *Ae. aegypti*. In the following sections, we describe in detail those steps.

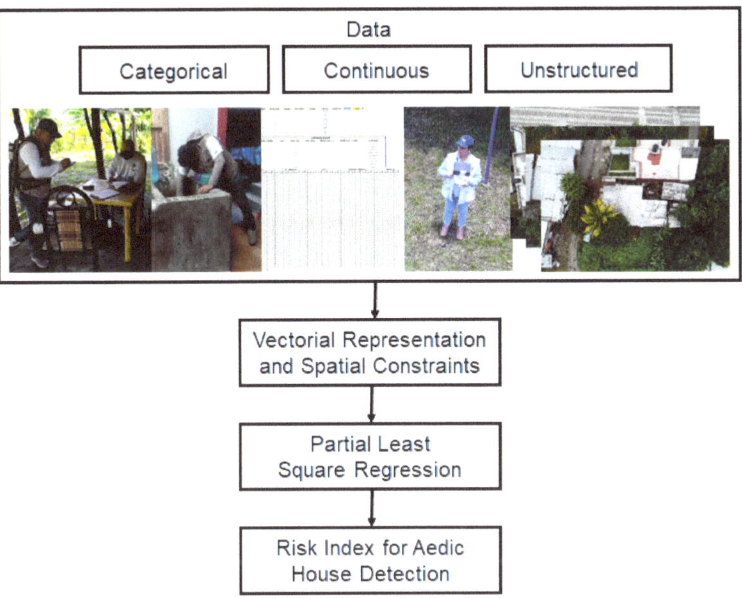

Figure 2. Block diagram of the proposed methodology.

3.4.1. Vectorial Representation of the Data

The methodology we propose to obtain the risk index requires a vectorial representation of the data. In the case of medium- to high-dimensional data, it is recommended that this representation be low dimensional, and therefore, a dimensionality reduction procedure is envolved.

Principal Component Analysis (PCA [57]) and Multiple Correspondence Analysis (MCA [58,59]) are well-known techniques to obtain low-dimensional representations of quantitative and qualitative data, respectively. When we have mixed-type data, i.e., quantitative and qualitative, as in our case, there are some proposals in the literature where, prior to the dimensionality reduction, an encoding procedure is done to one or both types of variables. The simplest encoding procedure is the discretization of the quantitative features by mapping continuous values into predefined intervals (e.g., low, medium, high), in order to obtain a complete categorical dataset; however, valuable information can be ignored, and defining the appropriate intervals can be tricky. It is also possible to obtain a quantitative dataset by introducing dummy variables, where qualitative features are transformed via a one-hot encoding procedure, whereby each category is represented as a binary variable. Although it is a common practice in regression models, several drawbacks have been noted, such as the increase in size of the dataset, and mainly, the multicollinearity problems that can be found [60]. Another approach is to construct a metric space with a similarity measure between pairs of objects defined in such a way that it can handle mixed-type data, such as multidimensional scaling (MDS, [61]) with Gower distance ([62]). Another sophisticated method for encoding mixed-type data based on Shannon entropy has been proposed in [63], where it is shown that the obtained representation can preserve the information in the original data with a memory-efficient procedure, which is particularly useful for big data scenarios (large and high-dimensional datasets). Another popular alternative for low to medium-size datasets, is factor analysis for mixed data (FAMD, [64,65]), which can be viewed as a mix between PCA and MCA. As is noted in [64], FAMD can be obtained by PCA including quantitative variables with a particular metric, which includes the contribution of quantitative and qualitative variables (previously encoded) to the distance between pairs of observations. This is the approach we use in our proposal, and in the next paragraph, we will describe the FAMD methodology following [66].

Factor Analysis for Mixed Data

Let $\mathbf{X}_{n \times d}$ be our data matrix, with n observations and $d = c + q$ variables, of which c are quantitative and q are qualitative. Now, let us define the encoded data matrices $\mathbf{Z}^C$ and $\mathbf{Z}^Q$ for quantitative and qualitative variables, respectively. For quantitative variables, we use the centered-scaled version: $Z_{ij}^C = (X_{ij} - \bar{x}_j)/\sigma_j$ for $j \in C$, where C is the set of quantitative variables. For qualitative variables, $\mathbf{Z}^Q$ is the so-called complete disjunctive table, which is a multivariate indicator matrix, the rows of which are the concatenation of q one-hot encoding vectors for each $q \in Q$, the set of qualitative variables. Suppose that the qth variable has K_q categories and $K = \sum_q K_q$ is the total number of categories for all qualitative variables. Then, $\mathbf{Z}^Q$ is a matrix of size $n \times K$, with entries $Z_{ik}^Q = 1$ if observation i belongs to the kth category and 0 otherwise, for $i = 1, 2, \ldots, n$ and $k = 1, 2, \ldots, K$. Generally, we use the scaled version of $\mathbf{Z}^Q$, where each entry is set to $Z_{ik}^Q/p_k - 1$, where p_k is the proportion of observations possessing category k. Those transformations give us a new data matrix $\mathbf{Z} = [\mathbf{Z}^C, \mathbf{Z}^Q]$, with observations $\mathbf{z} \in \mathbb{R}^s$, where the dimension of this metric space is $s = c + K$. For any two observations $\mathbf{u}, \mathbf{v} \in \mathbb{R}^s$, a weighted norm is defined as $\|\mathbf{u}\|_W^2 = \mathbf{u}'\mathbf{W}\mathbf{u}$, and its corrreponding weighted distance will be $d_W^2(\mathbf{u}, \mathbf{v}) = \|\mathbf{u} - \mathbf{v}\|_W^2 = (\mathbf{u} - \mathbf{v})'\mathbf{W}(\mathbf{u} - \mathbf{v})$.

In order to take into account the two types of variables in the distance defined before, in the standard version of FAMD, the weight matrix is defined as $\mathbf{W} = \operatorname{diag}([\mathbf{1}_c, \mathbf{p}])$, where $\mathbf{1}_c$ is the vector of ones with length c, and $\mathbf{p}$ is the vector of category proportions of length K. The latter implies a weighted scheme on the columns of $\mathbf{Z}^Q$ when the distance between encoded observations is computed, which together with the scaling defined before, emphasizes the differences between observations with rare categories. However, another weighting scheme is possible, as is noted in [66]. The fitting of the model is similar to PCA on the transformed variables, based on singular value decomposition (SVD), where an additional weighting scheme is applied for the row space of $\mathbf{Z}$ by means of $\mathbf{B}_{n \times n} = \operatorname{diag}(\mathbf{1}n^{-1})$; then, the solution is based on $\operatorname{SVD}(\mathbf{B}^{1/2}\mathbf{Z}\mathbf{W}^{1/2})$. See [64] for details.

Similar to PCA and MCA, FAMD gives us low-dimensional representations for variables and observations. In our case, we will only use the representations for observations, which allow us to measure similarities in houses according to their characteristics.

3.4.2. Spatial Constraints

The spatial relationships between the houses in our study are based on their geographic distribution. We used a ML methodology to generate a categorical variable for each house, which we add to the original variables, corresponding to the cluster they belong to, according to a particular clustering structure obtained with a clustering algorithm with spatial constraints. The constraint we used is the contiguity in space, specifying in this way which observations (houses) are considered connected. When we include these connectivities in a clustering algorithm, the observations in a cluster are required not only to be similar to one other according to their covariates, but also to comprise a contiguous set of observations. To this end, we define a connectivity matrix $\mathbf{C}_{n \times n}$ with entries $C_{ij} = 1$ if observations i and j are considered contiguous and 0 otherwise. A cluster is then considered contiguous if there is a path between every pair of observations in that cluster, i.e., the subgraph is fully connected. Several classical ML clustering algorithms have been modified to take into account this type of constraint [67,68]. We decided to use an agglomerative hierarchical clustering method adapted to handle connectivity constraints implemented in the module `cluster` from `sklearn` package in Python [69]. We chose this method mainly because of its simplicity and the good results that we obtained.

3.4.3. Risk Index with Partial Least Squares

Our proposed index is based on partial least squares regression (PLS, [70,71]). PLS is a class of techniques for modeling the association between blocks of observed variables through latent variables.

Let $\widetilde{\mathbf{X}}_{n \times p}$ and $\mathbf{Y}_{n \times t}$, be the matrices representing a set of independent and dependent variables, respectively, which we assume, are linearly related. PLS aims to find a set of vectors of latent scores $\{\alpha_1 \dots, \alpha_R\}$ and $\{\omega_1 \dots, \omega_R\}$, which are orthogonal and related to the column space of $\widetilde{\mathbf{X}}$ and $\mathbf{Y}$, in such a way that, for any value between 1 and R, this set of vectors span the "most interesting" subspaces of the ranges of both matrices [72]. Taken separately, α_r and ω_r for $r = 1, \dots, R$, can be the PCA solution for $\widetilde{\mathbf{X}}$ and $\mathbf{Y}$; however, the solution of PLS is not accounting for $\widetilde{\mathbf{X}}'\widetilde{\mathbf{X}}$ or $\mathbf{Y}'\mathbf{Y}$ only, but for the cross-covariance $\widetilde{\mathbf{X}}'\mathbf{Y}$. Let us focus on the first pair of latent scores α_1 and ω_1. In this case, PLS can be stated as a method to find vectors $\mathbf{a}$ and $\mathbf{b}$ such that

$$\mathrm{Cov}(\alpha_1, \omega_1) = \max_{\|\mathbf{a}\| = \|\mathbf{b}\| = 1} \mathrm{Cov}(\widetilde{\mathbf{X}}\mathbf{a}, \mathbf{Y}\mathbf{b}). \tag{1}$$

If $\mathbf{a}_1$ and $\mathbf{b}_1$ are the solution to (1), it can be shown that $\mathrm{Cov}(\alpha_1, \omega_1)\mathbf{a}_1\mathbf{b}_1'$ is the best rank-one approximation of $\widetilde{\mathbf{X}}'\mathbf{Y}$ in the least-squares sense [72].

In its classical form, the fitting procedure of PLS is based on the nonlinear iterative partial least squares (NIPALS) algorithm [73], which give us the whole set of solutions (α_r, ω_r) and $(\mathbf{a}_r, \mathbf{b}_r)$, for $r = 1, \dots, R$.

In our study, the data matrix $\widetilde{\mathbf{X}}_{n \times p}$ is the p–dimensional representation of the original mixed-type dataset obtained with the methodology described in Section 3.4.1. The dependent variable $\mathbf{Y}$ is a $n \times 1$ matrix (i.e., a vector). In this case, the solution is given by a particular case of PLS known as PLS1 [71,72].

Our particular interest is the solution given by the first pair of PLS scores, because our proposed index is given by $\widetilde{\mathbf{X}}\mathbf{a}_1$.

4. Results

We performed an extensive set of experiments in order to explore interesting patterns and to select a good model to obtain our proposed index. Those experiments where carried out based on a grid of parameters, which we consider important, and are described on Table 2.

Table 2. Main parameters considered in our model. The possible values of those parameters were chosen arbitrarily based on the results we observed.

Parameter	Description	Values
$PC_{DroneImage}$	The number of principal components (PCA) used to represent the vegetation indices (NDVI, GNDVI, NDVIre and CIgreen), and cartographic information (DSM and DTM)	$\{1, 2\}$
$nclust$	The number of clusters considered to model the spatial relationships between the houses.	$\{2, 3, 4, 5, 6\}$
k_{nn}	The number of k nearest neighboring houses for defining the connectivities in the agglomerative hierarchical clustering with spatial constraints	$\{3, 4, 5\}$
PC_{FAMD}	The number of principal components to be used, obtained with FAMD	$\{2, 3, 4, 5, 6\}$
PC_{PLS}	The maximum number of principal components to obtain in PLS to generate our proposed index. Observe that, altough we use just the first score of PLS ($\widetilde{\mathbf{X}}\mathbf{a}_1$), results may vary for different values of PC_{PLS}. Also, this parameter must satisfy $PC_{PLS} \leq PC_{FAMD}$	$\{2, 3, 4, 5, 6\}$

As in any regression-based model, in PLS we can obtain some metrics accounting for its predictive performance; however, it is not our objective (it is very easy to show that a performance metric such as the coefficient of determination, can be artificially "inflated", resulting in an overfitted model). In our case, we want to take advantage of a unique characteristic of PLS: to construct the best set of latent variables (components) based on the set of factors we obtained with FAMD, which are correlated to our response variable of interest, as we explained in Section 3.4.3. In this sense, we want to exploit the explanatory habilities of PLS, which will allow us to obtain a set of components that can represent all covariates we used (from different sources and types) in a useful way to be considered as a risk index.

Then, our main evaluation criteria was the explained inertia (covariance) obtained with the FAMD procedure according to the PC_{FAMD} variable, and the residuals on log counts of the response variables. Because we have many possible combinations of the parameters, we will not show here those experiments, but many of them can be found as supplementary material to this paper in https://github.com/victorm0202/mosquito_borne_viral_disease_paper.

We observe that the patterns of risk we found are consistent in most of the parameterizations we used, except on those where the model overfits, and we observed also that this tends to occur when PC_{PLS} is high (5 or 6, mainly). In this case, the risk index does not show a particular spatial trend, even when we vary the other parameters.

Based on those experiments, we decided to use the following parameters of our proposed model for the results we will show:

- $PC_{DroneImage}$: 1
- $nclust$: 4
- k_{nn}: 3
- PC_{FAMD}: 2
- PC_{PLS}: 2

obtaining an explained inertia of 0.661.

Based on this set of parameters, and the geographic distribution of the houses included in our study, we generate a cluster configuration by using agglomerative hierarchical clustering with spatial constraints, as is explained in Section 3.4.2, where we included all independent variables related to the houses of our dataset.

In Figure 3, we show the connectivities induced by **C** by using the centroids of the polygons defining the houses and the cluster configuration of houses according to four regions. We add a categorical variable to our dataset which represents the region assigned to each house.

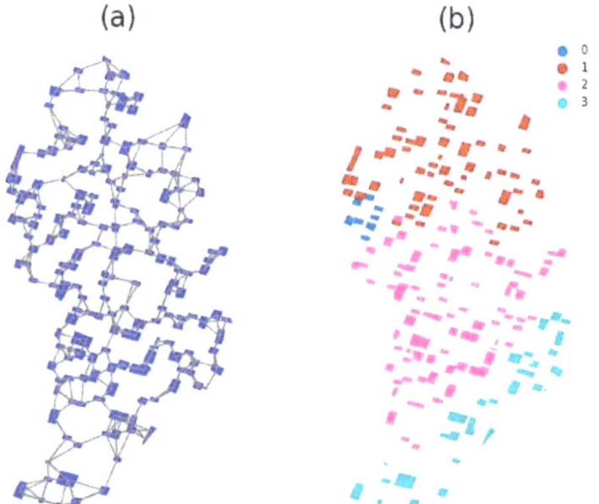

Figure 3. Hierarchical clustering with connectivity constraints. (**a**) Spatial distribution of houses and their connectivities with $k = 3$ NN. (**b**) Cluster structure with 4 regions or clusters.

We apply FAMD to the complete dataset of independent variables, including that for the region. As was explained in Section 3.4.1, we can obtain low-dimensional representations for variables and observations, but it will be restricted to observations only, because our interest is to have a useful representation for the houses according to all their covariates. Because FAMD is based on MCA and PCA, we can obtain representations based on each type of variable, as we show in Figure 4. For visualization purposes, we show two-dimensional representations of observations based on qualitative variables (Figure 4a), quantitative variables (Figure 4b), and the complete set of variables (Figure 4c), which is our main interest.

Figure 4. Two-dimensional representations of observations (houses) obtained with FAMD. (**a**) Based on qualitative variables. (**b**) Based on quantitative variables. (**c**) Based on both quantitative and qualitative variables

We can see that some patterns emerge when we use categorical and continuous variables only, but perhaps it is more clear in the case of continuous variables (Figure 4b), where we can observe a main cluster of houses, and three smaller clusters. Moreover, we can see at least two clusters when we consider the whole set of variables, as is shown in Figure 4c. We tried to explore a little more on this fact by applying k–means clustering considering the 3 PC we get with FAMD and $k = 4$ clusters. The results are shown in Figure 5. Initially, we wonder if those patterns corresponded to the spatial configuration we found in Figure 3. This was not the case, but we will use the clusters we found with k–means to compare with the results we obtained with the proposed index.

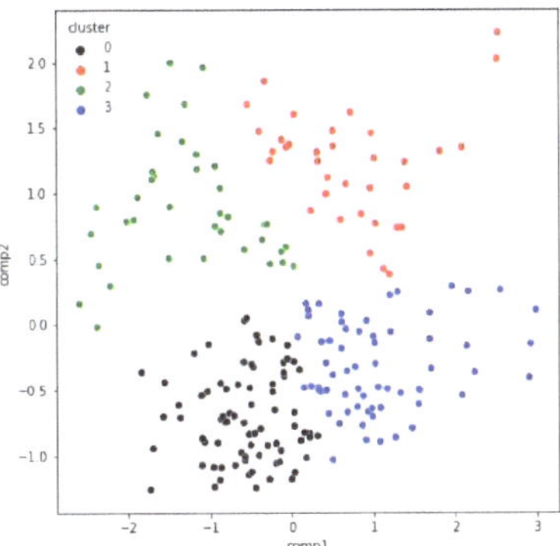

Figure 5. k–means clustering for observations (houses) obtained with FAMD. We used $k = 4$ clusters considering 3 PC as input.

To construct our univariate index with PLS, as we explained in Section 3.4.3, we use different dependent variables $\mathbf{Y}$. In all cases, our index is defined as $\gamma = \tilde{\mathbf{X}}\mathbf{a}_1$, a vector of length n, and it corresponds to the first PLS score.

We use univariate and multivariate $\mathbf{Y}$ as dependent variables for PLS. The univariate dependent variables are denoted by $\mathbf{Y}_{\text{pupae}}$, $\mathbf{Y}_{\text{mosquitoes}}$ and $\mathbf{Y}_{\text{stages}}$. Meanwhile, the multivariate version includes all the previous variables concatenated by columns, resulting in a matrix $\mathbf{Y}_{\text{all}}$ of size $n \times 3$.

For visualization purposes, and in order to compare our proposed indices, we show in Figure 6 the bivariate risk index, which corresponds to the first two scores obtained with PLS for each **Y**. The points represent the houses, and the cluster assigned to each point is the same that we obtained with k–means in the FAMD representation (Figure 5). We observe similar patterns in the FAMD representation and the index obtained; furthermore, we can see that all bivariate indexes in Figure 6 are very similar.

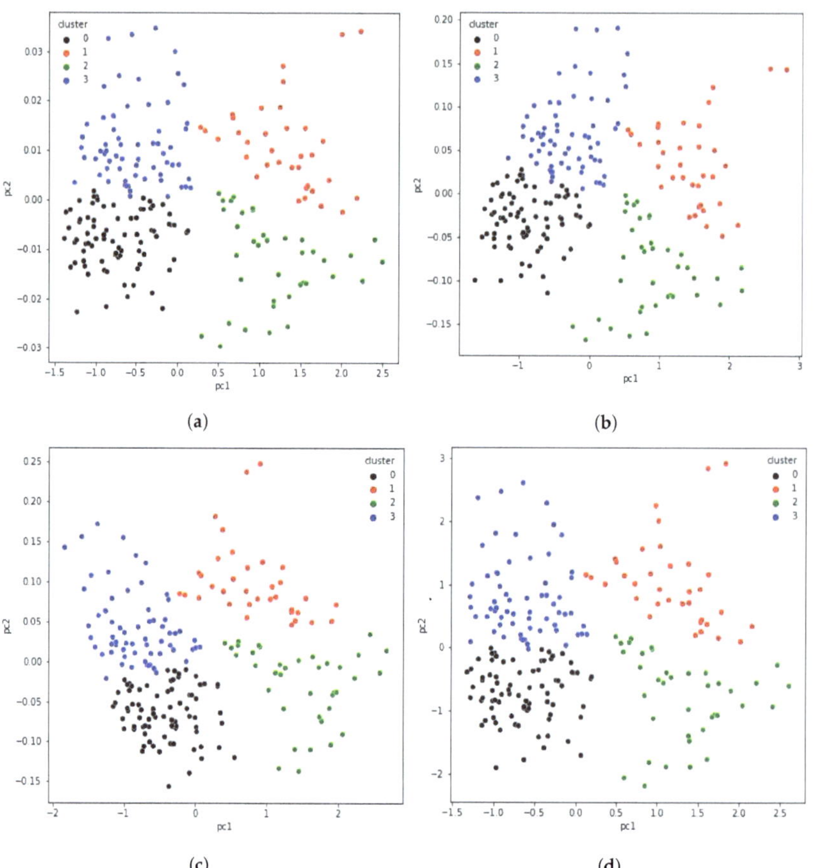

Figure 6. Bivariate risk index obtained with PLS for different response variables **Y**. (a) $\mathbf{Y}_{\text{stages}}$. (b) $\mathbf{Y}_{\text{mosquitoes}}$. (c) $\mathbf{Y}_{\text{pupae}}$. (d) $\mathbf{Y}_{\text{all}}$. Points in the scatterplots represent the houses.

For practical purposes, our proposed index is univariate and, as we said before, corresponds to the first score obtained with PLS, which was scaled between zero and one. In Figure 7 we show the house-level risk index we propose based on PLS for each dependent variable **Y** defined before. This figure confirms the similarity between all indices.

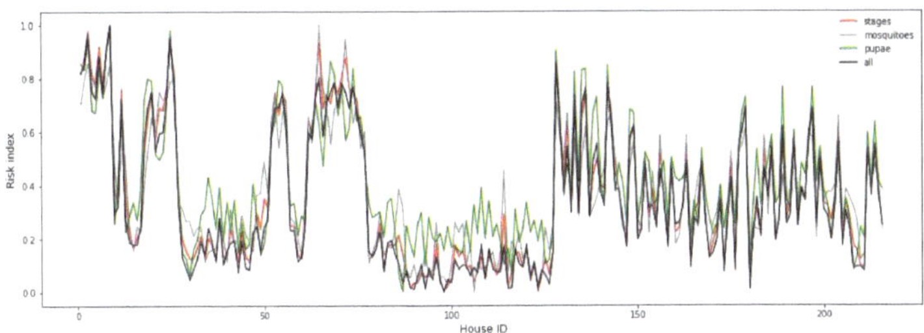

Figure 7. House-level risk index based on PLS for different response variable **Y**.

In addition, we show in Figure 8 the geographic distribution of houses, depicted with polygons, and the risk index for each house. Very interesting patterns can be seen in both Figures 7 and 8. There are some evident peaks of the risk index, which mainly correspond to houses on the periphery of the region of study. This spatial pattern of high-risk houses corresponds to region 3 in Figure 3, and some houses of regions 0 and 1. Houses with low risk can be found in the center of the study region and correspond to the region 2 of Figure 3. The risk increases as we approach the periphery of the region, as can be seen in Figure 8.

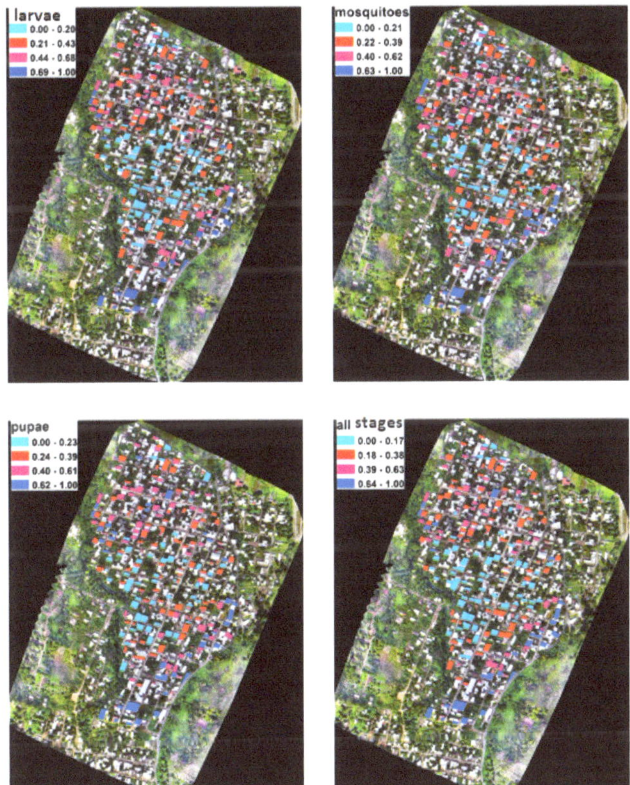

Figure 8. House-level risk index obtained with PLS for different response variable **Y**. Maps shows the geographic distribution of houses.

The house risk index of less than 0.5 for larvae, pupae, adults, and all stages was in 69.4% (150), 70.4% (152), 71.8% (155), and 70.4% (152) of houses, respectively. The major part of the houses with low risk (less than 0.5) was located in the downtown of the El Vergel neighborhood. The characteristics of this area are different from the rest of the dwellings in several aspects: they have concrete slab roofs and tile floors, better-quality services, and less tree cover, among other things, as opposed to the peripheral houses, which are characterized by more vegetation cover, sheet metal roofs, deficiency in the provision of services and constant population movement. This may represent a greater risk for dengue transmission in these particular areas. For all of the above, the generated maps can be added to the layers of geographic information that strengthen decision making on mosquito prevention and control measures in the area. They can permit direct mosquito control activities for those houses with high risk for *Ae. aegypti* in different life-cycle stages with adequate activities can involve community participation.

5. Discussion

The use of new technologies such as UAV have proven useful in supporting the implementation of entomological surveillance strategies for emerging and reemerging diseases. For malaria, the use of drones has allowed vegetation to be associated with the abundance and presence of the vector. On the other hand, methodology for monitoring and identifying breeding sites of *Ae. aegypti* mosquitoes has recently been reported.

The traditional entomological surveillance was carried out in homes by the Vector Control Program in Mexico and used entomological indices. The house (premises), index (percentage of houses infested with *Ae. aegypti*, larvae and/or pupae), container index (percentage of water-holding containers infested with *Ae. aegypti* larvae and/or pupae), and the Breteau index (number of positive containers per 100 houses recorded) [74], showed little relationship to adult densities. In some situations, their values were not associated with cases in the locality [75–77]. Likewise, the national dengue program uses some indices to determine risk areas where the vector develops. One of them is the Premise Condition Index (PCI). This index considers the house's appearance, the shade, and the disorder in the patio of the dwellings to determine the degree of risk in the dwelling in the presence or absence of the vector [42]. In [42], the potential of aerial surveillance for identifying breeding sites of *Ae. aegypti* was demonstrated two decades ago. Satellite aerial photographs intended to identify high-altitude breeding and resting sites of *Ae. aegypti* in residential areas were compared with the modified Housing Condition Index (PCI_2) application [41] as a rapid assessment technique within the dwellings. However, that was limited to the resolution of the images and could not identify small elements of backyards. Nevertheless, the satellite images are taken at different seasons, daily times, and camera angles, which does not allow the precise association of the environmental variables related to the *Ae. aegypti* house biological cycle. In this study, the UAV photos were taken at 100 m, at approximately noon, and simultaneously with house surveys [12,51,52], making it possible to associate the vector's biological activity with landscape elements because it allows us to determine them precisely and according to the field data record. The dengue mosquito vector *Ae. aegypti* has expanded its distribution, as the mosquitoes have gradually adapted to climatic and environmental conditions, leading to a change in the epidemiology of viral diseases, facilitating their establishment in new ecozones inhabited by immunologically naive human populations [78]. Numerous research groups have generated evidence on the factors that facilitate vector mosquitoes' different life-cycle stages and that are associated with the increase in their populations [34] and disease transmission [79,80]. In addition to the scientific evidence, health programs must know the riches of other technologies to obtain high-resolution images that can help understand the vector-borne diseases (VBD) complex, dynamic, and multifactorial scenarios.

Although dengue and mosquito programs have indices and technological tools that can help understand the mosquito population dynamics, the role of UAVs as a newborn technology can help build a house index more robust that can permit the anticipation of the increase in the mosquito population in health programs.

The national dengue program needs a robust concept to determine an "aedic house" that can have characteristics for allowing *Ae. aegypti* life cycles. We approach the simultaneity of UAV high-resolution images with field surveillances; data can also be defined, taking into account different types of variables related not only to the dwelling but also population–socioeconomic and other characteristics related to the landscape, such as coverage, type of vegetation present, and topography. Due to the fact that most of the UAV studies have been made on simulation scenarios, the contributions of this study demonstrate the utility of UAVs in obtaining urban landscape data for building a strong, flexible, and robust risk index, the methodology of which can be adapted to different places on different socioeconomic neighbor levels in urban or semiurban areas.

Our model contemplates different types and natures of variables related to the immediate landscape of each house visited, including those made through specialized cartography by aerial images, such as the vegetation present through the coverage and height of trees. The vegetation indices that consider both the greenness and the plant density of the species and values of the digital surface model were included, representing all the elevations present on the surface, land, infrastructure, and vegetation. The plant species present, their height, and their foliage will affect the local temperature and wind speed due to the effect of shadows projected on the various surfaces. Indeed, evapotranspiration is the effect of evaporative cooling of the water that plants transpire. Another small contribution is due to soil moisture [81]. In addition, consider field survey data as the premise condition index [41], the modified house condition index [42], and the number of positive breeding sites for *Ae. aegypti*, as an indirect measure of the characteristics of the houses that allow them to offer conditions that facilitate the life cycle of mosquitoes and oviposition [82–84] and the overcrowding index, as a measure of the human population density can act, among other things, as one of the factors for the female mosquitoes to move in search of food sources [85]. The flight height is crucial to obtain the optimal resolution for identifying the landscape elements according to different scenarios and depends on the elements or object size needed to identify the landscape. It is specific to the research question. However, it is crucial to know the flight area—in urban environments, to avoid high obstacles such as telecommunications or wi-fi antennas, buildings, or tall infrastructure, and in rural and semirural environments, mainly to avoid tall trees. An evaluation was carried out to determine the best flight height for detecting and identifying specific plant communities. They found that the NDVI orthophotograph's best flight height to identify tolares of *Parastrephialepidophilla* was 25 m, followed by 50 m [86]. An optimum flight height for our study purposes was 100 m. For operational purposes, we recommend flights of 80 to 100 m. Moreover, we recommended following the national regulations for security flights. Another important flight issue is photo capture time, which should be at noon, to avoid shadows limiting the object or element detection and identification [12,52]. As dependent variables, all the immature stages of female and male *Ae. aegypti* mosquitoes that could explain the model were included and made simultaneous with the UAV flights.

Nowadays, there is an evident trend in the ML research community to use data from different sources in order to propose useful models for many phenomena of interest, and entomological surveillance and public health in general are not an exception. In addition to the variables coming from specialized cartography by aerial images explained before, we include several variables coming from field and household surveys (Section 3.2). Certainly, there could be many advantages to incorporating more information from many sources, because more information could enrich our knowledge of the phenomenon we are interested in, and arguably, could improve the model we used for solving some specific task. However, it is very important to pay attention to the way this information is used by our models, especially when different structures are present, giving us mixed-type variables. We think the latter is crucial, particularly for models in which a distance function among objects (houses, in our case) is involved, as is the case of ML-related models. This is because not all distance functions are capable of representing differences between objects with mixed-type data. Consider for example, the Euclidean distance, which is used by default in most of the supervised and nonsupervised ML algorithms, and is aimed at continuous (real-valued) variables; then, if we use a model based on Euclidean distance for categorical or mixed-type data, the results and conclusions we obtain with this model, may not be correct. In our proposed index, we take into account the different types of variables by using an efficient encoding procedure which gives us a shared metric space with a corresponding norm capable of representing distances between any two objects taking into account the two types of input variables we used, and we think this is one of our main contributions, given that similar works reported in the literature [38,39] leave unclear the way by which the mixed input data (categorical and continuous) are managed, or the distance function they used. Although in these works the model takes into account the characteristics (socioeconomic, climate change, dengue knowledge, and landscape factors) of a house as input and returns the prediction of the mosquito to abundance (high or low), our model takes into account the characteristics (environmental, demographic, and socioeconomic) of a house and its spatial relationship with other houses to predict the risk index for being classified as an "aedic" house. In addition, our implementation was carried out with open-source software.

Moreover, in our study, the El Vergel neighborhood presented conditions that determined, for most of the houses surveyed, a premise condition index with a medium risk level, which does not allow focusing prevention and control actions of vectors at the housing and block level in the area. Mexico's southern border has a wide diversity of ecotones, socioeconomic levels, climates, and a constant movement of migratory groups. The models must be adapted to the particular conditions of the endemic region under study. There is no ideal formula but a series of steps to determine which variables offer greater predictive value for a given area. New spatial analysis tools and mathematical models are fundamental methods in vector control activities [87], and even a multidisciplinary group for analyzing it. In addition, the mapping and spatial modeling technologies and methodologies need to increase affordability for use in public health [88], especially in endemic areas that require certainty, coverage, and opportunity. Predictive models can be used for decision making in preventing various arboviruses transmitted by *Aedes* in urban and semiurban endemic areas based on the evidence of overlapping *Aedes*-borne diseases (dengue, chikungunya, and zika) within geographic hotspots [89]. This methodology can adapt the model to the different regions of our country. This flexibility is allowed mainly by the availability of real-time information provided by drones, one of its most important contributions.

Our model can be improved in many ways, which we mention as development and future work. It is already advisable to have information at different times of the year to strengthen our model. Tapachula is an area where the dry and rainy seasons are strongly marked and determine a contrast in, at least, the availability and type of breeding sites where the vector develops. Likewise, it is necessary to validate the model in different neighborhoods of the municipality of Tapachula, where the conditions of the houses vary mainly depending on their socioeconomic level. Moreover, it is in our interest to explore some adaptations of PLS for count data, because it is not naturally designed to predict this kind of data, and compare with the log-counts approach. We advise exploring artificial intelligence methods, such as deep neural networks, to detect interesting objects (e.g., mosquito breeding sites) from the UAV images. Although there have been some attempts to use those methods, such as convolutional neural networks [90,91] for aerial surveillance of mosquitoes' oviposition [52], our aim is to use state-of-the-art, real-time object-detection methodologies, such as YOLO [92] and its different versions [93–95], and also, to explore methods based on geographic object-based image analysis (GEOBIA) [96,97] to develop tools for image segmentation and labeling to create training datasets, in order to expand the information of the model in real time, and to strengthen the community's perception of new technologies as is suggested by [98,99]. This helps to strengthen the timeliness and fidelity of entomological and epidemiological, sociodemographic, and environmental data.

6. Conclusions

This study shows that the use of UAVs can be incorporated into vector surveillance and control strategies, providing spatial and temporal data of the landscape components that allow dengue vectors to continue transmitting the disease in real time. On the other hand, the model based on ML and multivariate statistical techniques, proposed in this study to estimate the risk of dengue transmission, demonstrates high reliability in identifying high-risk areas within the study locality and managing control activities specifically.

Supplementary Materials: The following supporting information can be downloaded at: https://github.com/victorm0202/mosquito_borne_viral_disease_paper (accessed on 30 November 2022).

Author Contributions: Conceptualization, V.M.-S., K.M.V.-D. and R.D.-L.; methodology, V.M.-S., K.M.V.-D. and R.D.-L.; software, V.M.-S., F.J.H.-L. and D.A.M.-L.; validation, V.M.-S., K.M.V.-D. and R.D.-L.; formal analysis, V.M.-S. and F.J.H.-L.; investigation, V.M.-S., K.M.V.-D., F.J.H.-L., D.A.M.-L., R.D.-L. and G.G.-F.; resources, V.M.-S., K.M.V.-D., D.A.M.-L. and F.J.H.-L.; data curation, V.M.-S., K.M.V.-D., D.A.M.-L. and F.J.H.-L.; writing—original draft preparation, V.M.-S., K.M.V.-D., F.J.H.-L. and D.A.M.-L.; writing—review and editing, R.D.-L. and G.G.-F.; visualization, V.M.-S., K.M.V.-D., F.J.H.-L. and D.A.M.-L.; supervision, R.D.-L. and G.G.-F.; project administration, R.D.-L. and G.G.-F.; funding acquisition, R.D.-L. and G.G.-F. All authors have read and agreed to the published version of the manuscript.

Funding: This research was carried out with the support of four institutions of the Mexican government. (1) The vector control program of the health services of Chiapas for entomological activities in the field, (2) CENAPRED research group to carry out the flight of the drones and elaboration of cartography, (3) Group of CIMAT researchers for the elaboration of mathematical models and (4) CRISP/INSP technical personnel for the application of sociodemographic and Premise Condition Index surveys and researches for a biological point of view of results.

Institutional Review Board Statement: Ethics and Research Committee approval of Distrito Sanitario #7, Chiapas Health Services.

Informed Consent Statement: Informed consent was obtained from all subjects involved in the study.

Data Availability Statement: Not applicable.

Acknowledgments: We greatly appreciate the householders, authorities of El Vergel, Tapachula, Chiapas, and the authorities of Distrito Sanitario #7 of Chiapas Health Services for the facilities provided in this study. The authors also acknowledge the National Disaster Prevention Center of the Mexican Government for the UAVs flights, photographs, and cartography used in this study.

Conflicts of Interest: The authors declare no conflict of interest.

References

1. Nex, F.; Remondino, F. UAV for 3D mapping applications: A review. *Appl. Geomat.* **2014**, *6*, 1–15. [CrossRef]
2. Eskandari, R.; Mahdianpari, M.; Mohammadimanesh, F.; Salehi, B.; Brisco, B.; Homayouni, S. Meta-analysis of Unmanned Aerial Vehicle (UAV) Imagery for Agro-environmental Monitoring Using Machine Learning and Statistical Models. *Remote Sens.* **2020**, *12*, 3511. [CrossRef]
3. Rojas Viloria, D.; Solano Charris, E.L.; Muñoz Villamizar, A.; Montoya Torres, J.R. Unmanned aerial vehicles/drones in vehicle routing problems: A literature review. *Int. Trans. Oper. Res.* **2021**, *28*, 1626–1657. [CrossRef]
4. Bendig, J.; Bolten, A.; Bareth, G. Introducing a Low-Cost Mini-Uav for-and Multispectral-Imaging. *ISPRS Int. Arch. Photogramm. Remote Sens. Spat. Inf. Sci.* **2012**, *39B1*, 345–349. [CrossRef]
5. Landau, K.; Van Leeuwen, W. Fine Scale Spatial Urban Land Cover Factors Associated with Adult Mosquito Abundance and Risk in Tucson, Arizona. *J. Vector Ecol. J. Soc. Vector Ecol.* **2012**, *37*, 407–418. [CrossRef]
6. Ivosevic, B.; Han, Y.G.; Kwon, O. Calculating coniferous tree coverage using unmanned aerial vehicle photogrammetry. *J. Ecol. Environ.* **2017**, *41*, 1–8. [CrossRef]
7. Robb, C.; Hardy, A.; Doonan, J.H.; Brook, J. Semi-Automated Field Plot Segmentation From UAS Imagery for Experimental Agriculture. *Front. Plant Sci.* **2020**, *11*, 591886. [CrossRef]

8. González-Jorge, H.; Martínez-Sánchez, J.; Bueno, M.; Arias, P. Unmanned Aerial Systems for Civil Applications: A Review. *Drones* **2017**, *1*, 2. [CrossRef]
9. Hassanalian, M.; Abdelkefi, A. Classifications, applications, and design challenges of drones: A review. *Prog. Aerosp. Sci.* **2017**, *91*, 99–131. [CrossRef]
10. Fornace, K.M.; Drakeley, C.J.; William, T.; Espino, F.; Cox, J. Mapping infectious disease landscapes: Unmanned aerial vehicles and epidemiology. *Trends Parasitol.* **2014**, *30*, 514–519. [CrossRef]
11. Hardy, A.; Makame, M.; Cross, D.; Majambere, S.; Msellem, M. Using low-cost drones to map malaria vector habitats. *Parasites Vectors* **2017**, *10*, 1. [CrossRef] [PubMed]
12. Valdez-Delgado, K.M.; Moo-Llanes, D.A.; Danis-Lozano, R.; Cisneros-Vázquez, L.A.; Flores-Suarez, A.E.; Ponce-García, G.; Medina-De la Garza, C.E.; Díaz-González, E.E.; Fernández-Salas, I. Field effectiveness of drones to identify potential *Aedes aegypti* breeding sites in household environments from Tapachula, a dengue-endemic city in southern Mexico. *Insects* **2021**, *12*, 663. [CrossRef] [PubMed]
13. Haas-Stapleton, E.; Barretto, M.; Castillo, E.; Clausnitzer, R.; Ferdan, R. Assessing Mosquito Breeding Sites and Abundance Using An Unmanned Aircraft. *J. Am. Mosq. Control. Assoc.* **2019**, *35*, 228–232. [CrossRef] [PubMed]
14. Johnson, B.J.; Manby, R.; Devine, G.J. Performance of aerial Bacillus thuringiensis var. israelensis applications in mixed saltmarsh-mangrove systems and use of affordable unmanned aerial systems to identify problematic levels of canopy cover. *bioRxiv* **2020**. [CrossRef]
15. Lorenz, C.; Castro, M.C.; Trindade, P.M.P.; Nogueira, M.L.; de Oliveira Lage, M.; Quintanilha, J.A.; Parra, M.C.; Dibo, M.R.; Fávaro, E.A.; Guirado, M.M.; et al. Predicting *Aedes aegypti* infestation using landscape and thermal features. *Sci. Rep.* **2020**, *10*, 21688. [CrossRef]
16. Markwardt, R.; Sorosjinda-Nunthawarasilp, P. *Innovations in the Entomological Surveillance of Vector-Borne Diseases*; Cambridge Scholars Publisher: Cambridge, UK, 2021.
17. Wilke, A.B.B.; Vasquez, C.; Carvajal, A.; Moreno, M.; Fuller, D.O.; Cardenas, G.; Petrie, W.D.; Beier, J.C. Urbanization favors the proliferation of *Aedes aegypti* and *Culex quinquefasciatus* in urban areas of Miami-Dade County, Florida. *Sci. Rep.* **2021**, *11*, 22989. [CrossRef]
18. de Jesús Crespo, R.; Rogers, R.E. Habitat Segregation Patterns of Container Breeding Mosquitos: The Role of Urban Heat Islands, Vegetation Cover, and Income Disparity in Cemeteries of New Orleans. *Int. J. Environ. Res. Public Health* **2022**, *19*, 245. [CrossRef]
19. Tsuda, Y.; Suwonkerd, W.; Chawprom, S.; Prajakwong, S.; Takagi, M. Different spatial distribution of *Aedes aegypti* and *Aedes albopictus* along an urban-rural gradient and the relating environmental factors examined in three villages in northern Thailand. *J. Am. Mosq. Control. Assoc.* **2006**, *22*, 222–228. [CrossRef]
20. Vásquez-Trujillo, A.; Cardona, D.; Segura Cardona, A.; Camara, D.; Alves-Honório, N.; Parra-Henao, G. House-Level Risk Factors for *Aedes aegypti* Infestation in the Urban Center of Castilla la Nueva, Meta State, Colombia. *J. Trop. Med.* **2021**, *2021*, 12. [CrossRef]
21. Liu, M.; Wang, X.; Zhou, A.; Fu, X.; Ma, Y.; Piao, C. UAV-YOLO: Small Object Detection on Unmanned Aerial Vehicle Perspective. *Sensors* **2020**, *20*, 2238. [CrossRef]
22. DJI. Matrice 600®. Available online: https://www.dji.com/mx/downloads/products/matrice600 (accessed on 19 October 2019)
23. Thibbotuwawa, A.; Bocewicz, G.; Nielsen, P.; Banaszak, Z. Unmanned aerial vehicle routing problems: A literature review. *Appl. Sci.* **2020**, *10*, 4504. [CrossRef]
24. Diario Oficial de la Federación. NORMA Oficial Mexicana NOM-032-SSA2-2014, Para la Vigilancia Epidemiológica, Promoción, Prevención y Control de las Enfermedades Transmitidas por Vectores, 2015. Available online: https://www.dof.gob.mx/nota_detalle.php?codigo=5389045&fecha=16/04/2015 (accessed on 18 April 2019).
25. Hardy, A.; Proctor, M.; MacCallum, C.; Shawe, J.; Abdalla, S.; Ali, R.; Abdalla, S.; Oakes, G.; Rosu, L.; Worrall, E. Conditional trust: Community perceptions of drone use in malaria control in Zanzibar. *Technol. Soc.* **2022**, *68*, 101895. [CrossRef] [PubMed]
26. Carrillo-Larco, R.; Moscoso-Porras, M.; Taype-Rondan, A.; Ruiz-Alejos, A.; Bernabe-Ortiz, A. The use of unmanned aerial vehicles for health purposes: A systematic review of experimental studies. *Glob. Health Epidemiol. Genom.* **2018**, *3*, e13. [CrossRef] [PubMed]
27. Laksham, K.B. Unmanned aerial vehicle (drones) in public health: A SWOT analysis. *J. Fam. Med. Prim. Care* **2019**, *8*, 342–346. [CrossRef] [PubMed]
28. Williams, G.M.; Wang, Y.; Suman, D.S.; Unlu, I.; Gaugler, R. The development of autonomous unmanned aircraft systems for mosquito control. *PLoS ONE* **2020**, *15*, e0235548. [CrossRef]
29. Faraji, A.; Haas-Stapleton, E.; Sorensen, B.; Scholl, M.; Goodman, G.; Buettner, J.; Schon, S.; Lefkow, N.; Lewis, C.; Fritz, B.; et al. Toys or Tools? Utilization of Unmanned Aerial Systems in Mosquito and Vector Control Programs. *J. Econ. Entomol.* **2021**, *114*, 1896–1909. [CrossRef]
30. Bouyer, J.; Culbert, N.J.; Dicko, A.H.; Pacheco, M.G.; Virginio, J.; Pedrosa, M.C.; Garziera, L.; Pinto, A.T.M.; Klaptocz, A.; Germann, J.; et al. Field performance of sterile male mosquitoes released from an uncrewed aerial vehicle. *Sci. Robot.* **2020**, *5*, eaba6251. [CrossRef]
31. Marina, C.; Liedo, P.; Bond, G.; Osorio, A.; Valle Mora, J.; Angulo, R.; Gomez-Simuta, Y.; Fernandez Salas, I.; Dor, A.; Williams, T. Comparison of Ground Release and Drone-Mediated Aerial Release of *Aedes aegypti* Sterile Males in Southern Mexico: Efficacy and Challenges. *Insects* **2022**, *13*, 347. [CrossRef]

32. Carrasco-Escobar, G.; Manrique, E.; Ruiz-Cabrejos, J.; Saavedra, M.; Alava, F.; Bickersmith, S.; Prussing, C.; Vinetz, J.M.; Conn, J.E.; Moreno, M.; et al. High-accuracy detection of malaria vector larval habitats using drone-based multispectral imagery. *PLoS Neglected Trop. Dis.* **2019**, *13*, 1–24. [CrossRef]

33. AragÃ£o, F.V.; Cavicchioli Zola, F.; Nogueira Marinho, L.H.; de Genaro Chiroli, D.M.; Braghini Junior, A.; Colmenero, J.C. Choice of unmanned aerial vehicles for identification of mosquito breeding sites. *Geospatial Health* **2020**, *15*, 92–100. [CrossRef]

34. Sallam, M.F.; Fizer, C.; Pilant, A.N.; Whung, P.Y. Systematic Review: Land Cover, Meteorological, and Socioeconomic Determinants of *Aedes* Mosquito Habitat for Risk Mapping. *Int. J. Environ. Res. Public Health* **2017**, *14*, 5317. [CrossRef] [PubMed]

35. Estallo, E.; Sangermano, F.; Grech, M.; Ludueña-Almeida, F.; Frías-Cespedes, M.; Ainete, M.; Almirón, W.; Livdahl, T. Modelling the distribution of the vector *Aedes aegypti* in a central Argentine city: Modelling *Aedes aegypti* distribution. *Med. Vet. Entomol.* **2018**, *32*, 451–461. [CrossRef] [PubMed]

36. Obenauer, J.; Joyner, T.A.; Harris, J.B. The importance of human population characteristics in modeling *Aedes aegypti* distributions and assessing risk of mosquito-borne infectious diseases. *Trop. Med. Health* **2017**, *45*, 1–9. [CrossRef] [PubMed]

37. Cianci, D.; Hartemink, N.; Ibáñez-Justicia, A. Modelling the potential spatial distribution of mosquito species using three different techniques. *Int. J. Health Geogr.* **2015**, *14*, 10. [CrossRef] [PubMed]

38. Chen, S.; Whiteman, A.; Li, A.; Rapp, T.; Delmelle, E.; Chen, G.; Brown, C.L.; Robinson, P.; Coffman, M.J.; Janies, D.; et al. An operational machine learning approach to predict mosquito abundance based on socioeconomic and landscape patterns. *Landsc. Ecol.* **2019**, *34*, 1295–1311. [CrossRef]

39. Rahman, M.S.; Pientong, C.; Zafar, S.; Ekalaksananan, T.; Paul, R.E.; Haque, U.; Rocklöv, J.; Overgaard, H.J. Mapping the spatial distribution of the dengue vector *Aedes aegypti* and predicting its abundance in northeastern Thailand using machine-learning approach. *One Health* **2021**, *13*, 100358. [CrossRef]

40. INEGI. Censo de Población y Vivienda 2010. Website. 2010. Available online: https://www.inegi.org.mx/programas/ccpv/2010/ (accessed on 18 October 2022).

41. Tun-Lin, W.; Kay, B.H.; Barnes, A. The Premise Condition Index: A Tool for Streamlining Surveys of *Aedes aegypti*. *Am. J. Trop. Med. Hyg.* **1995**, *53*, 591–594. [CrossRef]

42. Moloney, J.; Skelly, C.; Weinstein, P.; Maguire, M.; Ritchie, S.R. Domestic *Aedes aegypti* breeding site surveillance: limitations of remote sensing as a predictive surveillance tool. *Am. J. Trop. Med. Hyg.* **1998**, *59*, 261–264. [CrossRef]

43. Silver, J. *Mosquito Ecology: Field Sampling Methods*; SpringerLink: Springer e-Books; Springer: Dordrecht, The Netherlands, 2008.

44. Darsie, R.; Ward, R.; Chang, C.; Litwak, T. *Identification and Geographical Distribution of the Mosquitoes of North America, North of Mexico*; University Press of Florida: Gainesville, FL, USA, 2016.

45. Arredondo-Jimenez, J.I.; Valdez-Delgado, K.M. *Aedes aegypti* pupal/demographic surveys in southern Mexico: Consistency and practicality. *Ann. Trop. Med. Parasitol.* **2006**, *100*, 17–32. [CrossRef]

46. DJI. Zenmuse X5®. Available online: https://www.dji.com/mx/zenmuse-x5/info#specs (accessed on 10 October 2019).

47. Mica Sense Inc. Red Edge®. 2019. Available online: https://support.micasense.com/hc/en-us/articles/115003537673-RedEdge-M-User-Manual-PDF- (accessed on 10 October 2019).

48. CENAPRED. Centro Nacional de Prevención de Desastres. Website. 2022. Available online: https://www.gob.mx/cenapred/ (accessed on 18 October 2022).

49. Diario Oficial de la Federación. NORMA Oficial Mexicana NOM-107-SCT3-2019, Que Establece los Requerimientos para Operar un Sistema de Aeronave Pilotada a Distancia (RPAS) en el Espacio aéreo Mexicano. 2019. Available online: http://www.dof.gob.mx/normasOficiales/8006/sct11_C/sct11_C.html (accessed on 15 April 2022).

50. Pix4D. Capture®, 2019. Available online: https://www.pix4d.com/es/producto/pix4dcapture (accessed on 15 October 2019).

51. Suduwella, C.; Amarasinghe, A.; Niroshan, L.; Elvitigala, C.; De Zoysa, K.; Keppetiyagama, C. Identifying Mosquito Breeding Sites via Drone Images. In Proceedings of the 3rd Workshop on Micro Aerial Vehicle Networks, Systems, and Applications (DroNet '17), Niagara Falls, NY, USA, 23 June 2017; Association for Computing Machinery: New York, NY, USA, 2017; pp. 27–30. [CrossRef]

52. Case, E.; Shragai, T.; Harrington, L.; Ren, Y.; Morreale, S.; Erickson, D. Evaluation of unmanned aerial vehicles and neural networks for integrated mosquito management of *Aedes albopictus* (Diptera: Culicidae). *J. Med. Entomol.* **2020**, *57*, 1588–1595. [CrossRef]

53. Pix4D. Mapper®. 2019. Available online: https://www.pix4d.com/es/centro-de-descarga (accessed on 2 December 2019).

54. L3HARRIS GEOSPATIAL. Getting Started with ENVI. Website. 2022. Available online: https://www.l3harrisgeospatial.com/docs/GettingStartedWithENVI.html (accessed on 20 October 2022).

55. Díaz, J. Estudios de índices de Vegetación a Partir de Imágenes Aéreas Tomadas Desde RPAS y Aplicaciones de estos a la Agricultura de Precisión. Master's Thesis, Universidad Complutense de Madrid, Madrid, Spain, 2015.

56. ArcGIS. Raster Calculator. Website. 2022. Available online: https://desktop.arcgis.com/en/arcmap/latest/tools/spatial-analyst-toolbox/raster-calculator.htm (accessed on 20 October 2022).

57. Jolliffe, I. *Principal Component Analysis*; Springer: Berlin, Germany, 1986.

58. Greenacre, M.; Blasius, J. *Multiple Correspondence Analysis and Related Methods*; Chapman & Hall/CRC Statistics in the Social and Behavioral Sciences; CRC Press: Boca Raton, FL, USA, 2006.

59. Izenman, A.J. *Modern Multivariate Statistical Techniques: Regression, Classification, and Manifold Learning*; Springer Publishing Company: Berlin, Germany, 2008.

60. Garg, A.; Tai, K. Comparison of regression analysis, artificial neural network and genetic programming in handling the multicollinearity problem. In Proceedings of the 2012 Proceedings of International Conference on Modelling, Identification and Control, Wuhan, China, 24–26 June 2012; pp. 353–358.
61. Borg, I.; Groenen, P. *Modern Multidimensional Scaling: Theory and Applications*; Springer: Berlin, Germany, 2005.
62. Gower, J.C. A General Coefficient of Similarity and Some of Its Properties. *Biometrics* **1971**, *27*, 857–871. [CrossRef]
63. Lopez-Arevalo, I.; Aldana-Bobadilla, E.; Molina-Villegas, A.; Galeana-Zapién, H.; Muñiz-Sanchez, V.; Gausin-Valle, S. A Memory-Efficient Encoding Method for Processing Mixed-Type Data on Machine Learning. *Entropy* **2020**, *22*, 1391. [CrossRef] [PubMed]
64. Pagès, J. *Multiple Factor Analysis by Example Using R*; Chapman & Hall/CRC The R Series; Taylor & Francis: Boca Raton, FL, USA, 2014.
65. Pagès, J. Analyse factorielle de données mixtes. *Rev. Stat. AppliquÉe* **2004**, *52*, 93–111.
66. Davidow, M.B.; Matteson, D. Factor Analysis of Mixed Data for Anomaly Detection. *arXiv* **2020**, arXiv:abs/2005.12129.
67. Murtagh, F. A Survey of Algorithms for Contiguity-constrained Clustering and Related Problems. *Comput. J.* **1985**, *28*, 82–88. [CrossRef]
68. Wagstaff, K.; Cardie, C.; Rogers, S.; Schrödl, S. Constrained K-Means Clustering with Background Knowledge. In Proceedings of the Eighteenth International Conference on Machine Learning (ICML '01), San Francisco, CA, USA, 28 June–1 July 2001; Morgan Kaufmann Publishers Inc.: San Francisco, CA, USA, 2001; pp. 577–584.
69. Pedregosa, F.; Varoquaux, G.; Gramfort, A.; Michel, V.; Thirion, B.; Grisel, O.; Blondel, M.; Prettenhofer, P.; Weiss, R.; Dubourg, V.; et al. Scikit-learn: Machine Learning in Python. *J. Mach. Learn. Res.* **2011**, *12*, 2825–2830.
70. Helland, I.S. Partial Least Squares Regression and Statistical Models. *Scand. J. Stat.* **1990**, *17*, 97–114.
71. Rosipal, R.; Krämer, N. Overview and Recent Advances in Partial Least Squares. In *Subspace, Latent Structure and Feature Selection*; Saunders, C., Grobelnik, M., Gunn, S., Shawe-Taylor, J., Eds.; Springer: Berlin, Germany, 2006; pp. 34–51.
72. Wegelin, J.A. *A Survey of Partial Least Squares (PLS) Methods, with Emphasis on the Two-Block Case*; Technical Report; Department of Statistics, University of Washington: Seattle, WA, USA, 2000.
73. Wold, H. Path Models with Latent Variables: The NIPALS Approach. In *Quantitative Sociology*; Blalock, H., Aganbegian, A., Borodkin, F., Boudon, R., Capecchi, V., Eds.; International Perspectives on Mathematical and Statistical Modeling; Academic Press: Cambridge, MA, USA, 1975; pp. 307–357.
74. Bureau, P.A.S. *Dengue and dengue hemorrhagic fever in the Americas: Guidelines for Prevention and Control*; Pan American Health Organization, Pan American Sanitary Bureau, Regional: Washington, DC, USA, 1994.
75. Garjito, T.A.; Hidajat, M.C.; Kinansi, R.R.; Setyaningsih, R.; Anggraeni, Y.M.; Mujiyanto; Trapsilowati, W.; Jastal; Ristiyanto; Satoto, T.B.T.; et al. Stegomyia Indices and Risk of Dengue Transmission: A Lack of Correlation. *Front. Public Health* **2020**, *8*, 328. [CrossRef]
76. Focks, D.A.; Chadee, D.D. Pupal Survey: An Epidemiologically Significant Surveillance Method for *Aedes aegypti*: An Example Using Data from Trinidad. *Am. J. Trop. Med. Hyg.* **1997**, *56*, 159–167. [CrossRef]
77. Tun-Lin, W.; Kay, B.H.; Barnes, A.; Forsyth, S. Critical Examination of *Aedes aegypti* Indices: Correlations with Abundance. *Am. J. Trop. Med. Hyg.* **1996**, *54*, 543–547. [CrossRef]
78. Näslund, J.; Ahlm, C.; Islam, K.; Evander, M.; Bucht, G.; Lwande, O.W. Emerging Mosquito-Borne viruses linked to *Aedes aegypti* and *Aedes albopictus*: Global status and preventive strategies. *Vector-Borne Zoonotic Dis.* **2021**, *21*, 731–746. [CrossRef] [PubMed]
79. Lee, S.A.; Jarvis, C.I.; Edmunds, W.J.; Economou, T.; Lowe, R. Spatial connectivity in mosquito-borne disease models: A systematic review of methods and assumptions. *J. R. Soc. Interface* **2021**, *18*, 20210096. [CrossRef] [PubMed]
80. Aswi, A.; Cramb, S.; Moraga, P.; Mengersen, K. Bayesian spatial and spatio-temporal approaches to modelling dengue fever: A systematic review. *Epidemiol. Infect.* **2019**, *147*, 1–14. [CrossRef] [PubMed]
81. Ochoa de la Torre, J. La Vegetación Como Instrumento Para el Control Microclimático. Ph.D. Thesis, UPC, Departament de Construccions Arquitectòniques I, Barcelona, Spain, 1999. Available online: http://hdl.handle.net/2117/93436 (accessed on 10 October 2019).
82. Powell, J.R.; Tabachnick, W.J. History of domestication and spread of *Aedes aegypti*—A review. *MemÓRias Inst. Oswaldo Cruz* **2013**, *108*, 11–17. [CrossRef]
83. Ibañez-Bernal, S.; Dantes, H.G. Los vectores del dengue en México: Una revisión crítica. *Salud Pública México* **1995**, *37*, 53-63.
84. Cheong, W. Preferred *Aedes aegypti* larval habitats in urban areas. *Bull. World Health Organ.* **1967**, *36*, 586. [PubMed]
85. Ritchie, S.; Gubler, D.; Ooi, E.; Vasudevan, S.; Farrar, J. Dengue vector bionomics: Why *Aedes aegypti* is such a good vector. In *Dengue and Dengue Hemorrhagic Fever*; CAB International: Oxfordshire UK, 2014; Chapter 24.
86. Estrada Zúñiga, A.C.; Ñaupari Vásquez, J. Detección e identificación de comunidades vegetales altoandinas, Bofedal y Tolar de Puna Seca mediante ortofotografías RGB y NDVI en drones "Sistemas Aéreos no Tripulados". *Sci. Agropecu.* **2021**, *12*, 291–301. [CrossRef]
87. Baak-Baak, C.M.; Cigarroa-Toledo, N.; Pinto-Castillo, J.F.; Cetina-Trejo, R.C.; Torres-Chable, O.; Blitvich, B.J.; Garcia-Rejon, J.E. Cluster Analysis of Dengue Morbidity and Mortality in Mexico from 2007 to 2020: Implications for the Probable Case Definition. *Am. J. Trop. Med. Hyg.* **2022**, 106, 1515-1521. [CrossRef]
88. Eisen, L.; Lozano-Fuentes, S. Use of mapping and spatial and space-time modeling approaches in operational control of *Aedes aegypti* and dengue. *PLoS Neglected Trop. Dis.* **2009**, *3*, e411. [CrossRef]

89. Dzul-Manzanilla, F.; Correa-Morales, F.; Che-Mendoza, A.; Palacio-Vargas, J.; Sánchez-Tejeda, G.; González-Roldan, J.F.; López-Gatell, H.; Flores-Suárez, A.E.; Gómez-Dantes, H.; Coelho, G.E.; et al. Identifying urban hotspots of dengue, chikungunya, and Zika transmission in Mexico to support risk stratification efforts: A spatial analysis. *Lancet Planet. Health* **2021**, *5*, e277–e285. [CrossRef]

90. LeCun, Y. Generalization and Network Design Strategies. In *Connectionism in Perspective*; Pfeifer, R., Schreter, Z., Fogelman, F., Steels, L., Eds.; Elsevier: Zurich, Switzerland, 1989.

91. LeCun, Y.; Bengio, Y. Convolutional networks for images, speech, and time series. *Handb. Brain Theory Neural Netw.* **1995**, *3361*, 1995.

92. Redmon, J.; Divvala, S.; Girshick, R.; Farhadi, A. You Only Look Once: Unified, Real-Time Object Detection. In Proceedings of the 2016 IEEE Conference on Computer Vision and Pattern Recognition (CVPR), Las Vegas, NV, USA, 27–30 June 2016; pp. 779–788. [CrossRef]

93. Redmon, J.; Farhadi, A. YOLO9000: Better, Faster, Stronger. In Proceedings of the 2017 IEEE Conference on Computer Vision and Pattern Recognition (CVPR), Honolulu, HI, USA, 21–26 July 2017; pp. 6517–6525.

94. Redmon, J.; Farhadi, A. YOLOv3: An Incremental Improvement. *arXiv* **2018**, arXiv:abs/1804.02767.

95. Bochkovskiy, A.; Wang, C.; Liao, H.M. YOLOv4: Optimal Speed and Accuracy of Object Detection. *arXiv* **2020**, arXiv:abs/2004.10934. Available online: http://xxx.lanl.gov/abs/2004.10934 (accessed on 18 October 2022).

96. Kucharczyk, M.; Hay, G.J.; Ghaffarian, S.; Hugenholtz, C.H. Geographic object-based image analysis: A primer and future directions. *Remote Sens.* **2020**, *12*, 2012. [CrossRef]

97. Stanton, M.C.; Kalonde, P.; Zembere, K.; Hoek Spaans, R.; Jones, C.M. The application of drones for mosquito larval habitat identification in rural environments: A practical approach for malaria control? *Malar. J.* **2021**, *20*, 1–17. [CrossRef] [PubMed]

98. Annan, E.; Guo, J.; Angulo-Molina, A.; Yaacob, W.F.W.; Aghamohammadi, N.; Yavaşoglu, S.; Bardosh, K.; Dom, N.; Zhao, B.; Lopez-Lemus, U.; et al. Community acceptability of dengue fever surveillance using unmanned aerial vehicles: A cross-sectional study in Malaysia, Mexico, and Turkey. *Travel Med. Infect. Dis.* **2022**, *49*, 102360–102360. [CrossRef]

99. Bartumeus, F.; Costa, G.B.; Eritja, R.; Kelly, A.H.; Finda, M.; Lezaun, J.; Okumu, F.; Quinlan, M.M.; Thizy, D.C.; Toé, L.P.; et al. Sustainable innovation in vector control requires strong partnerships with communities. *PLoS Neglected Trop. Dis.* **2019**, *13*, 1–5. [CrossRef]

Article

A Conceptual Framework for Economic Analysis of Different Law Enforcement Drones

Nikolaos Tsiamis [1,*], Loukia Efthymiou [1] and Konstantinos P. Tsagarakis [2,*]

1 Department of Environmental Engineering, Democritus University of Thrace, 67100 Xanthi, Greece; lefthymi@env.duth.gr
2 School of Production Engineering and Management, Technical University of Crete, 73100 Chania, Greece
* Correspondence: ntsiamis@env.duth.gr (N.T.); ktsagarakis@tuc.gr (K.P.T.)

Abstract: The widespread use of drones in various fields has initiated a discussion on their cost-effectiveness and economic impact. This article analyzes in detail a methodological evaluation framework for the levelized cost of drone services for law enforcement purposes. Based on the data availability, we compared two vehicles: Phantom 4 Pro and Thunder-B. Moreover, we calculated their levelized costs per surveillance time and trip distance. Our approach helps users calculate the real costs of their vehicles' services and produce equations for rapid estimations. We observed economies of scale for time and distance and showed differentiations per aircraft capacity. Furthermore, using the produced equations, we formulated a case study and compared the costs in a 4 km area constantly monitored by the two types of drones to support the best vehicle selection. We found that the Phantom 4 Pro costs less than the Thunder-B drone, for example. Thus, we demonstrate how, by applying this methodology beforehand, decision makers can select the most appropriate vehicle for their needs based on cost. Cost research estimations will improve UAV use and will help policymakers include UAV technology in crime prevention programs, especially when more data are available.

Keywords: cost-effectiveness; economics of drones; law enforcement; operating cost; unmanned aerial vehicles

Citation: Tsiamis, N.; Efthymiou, L.; Tsagarakis, K.P. A Conceptual Framework for Economic Analysis of Different Law Enforcement Drones. *Machines* **2023**, *11*, 983. https://doi.org/10.3390/machines11110983

Academic Editors: Octavio Garcia-Salazar, Anand Sanchez-Orta and Aldo Jonathan Muñoz-Vazquez

Received: 18 September 2023
Revised: 10 October 2023
Accepted: 13 October 2023
Published: 24 October 2023

1. Introduction

Drones, also known as unmanned aerial vehicles (UAVs), have received increasing interest in recent years for applications in various fields [1,2]. Drones are categorized by weight, flight range, purpose of flight, altitude capacity, etc. [3,4]. They have been employed for several purposes, including forest monitoring [5,6], fire prevention [7], and deforestation and illegal logging [8], while Balcerzak et al. [9] demonstrate that unmanned aerial vehicles could be helpful for international firefighting and crisis management missions. In addition, unmanned aerial vehicles are an effective solution in agriculture [10], as images taken from high altitudes can be easily obtained and help farmers with crop growth monitoring [11,12], irrigation management [13], and crop health [14,15]. Small drones have also been used to monitor seagrass in coastal waters, which are sensitive to environmental changes [16]. Also, drones can be used for search and rescue [17], disaster management [18], geographic mapping applications [19,20], geology applications [21], archeological site observations [22], and weather predictions [23]. Finally, drones have been used for health purposes, including protection against malaria [24] and during the COVID-19 pandemic [25].

Security forces have considered the advantages of using drones in line with the rapidly growing market and have actively involved them in law enforcement. In Italy, security forces have used drones for environmental monitoring [26], while in Africa they have been used for locating illegal poachers [27]. Boakye [28] found that aerial patrols can help detect crime and improve law enforcement effectiveness. Zhou et al. [29] also found that drones can be used for ship monitoring in terms of regulation violations.

The Police use drones for tracking missing people [30] and for traffic management as a viable solution against expensive manned helicopter searches [31]. In Europe, Frontex has used drones for border surveillance [32]. In the USA, drones are used for water rescues and disaster response, traffic trash response, investigation of suspects, crime scene analysis, surveillance and crowd monitoring [33]. In Poland, Police used drones to sample domestic chimney exhaust gasses to prevent air pollution due to inappropriate fuels or burning materials [34].

Furthermore, drones can be used to protect cybersecurity [35], so it is necessary to initiate changes in legislation [36] due to citizens' privacy. However, the benefits from their use outweigh any ethical or security issues if well-regulated [37].

As the extensive use of drones tends to replace conventional modes of monitoring and surveillance, it is essential to refer to their costs calculated by a standard methodology for estimating economic reference values. This paper aims to introduce an evaluation methodology and then compare the cost of using drones by law enforcement agencies. We analyzed two types of drones used by law enforcement officers, as well as by the Greek Police [38,39]. We finally presented a case study using the two drones and concluded which was the best choice based on cost estimates.

2. Literature Review

Authorities in various countries have incorporated their drone use into their legal systems, which may differ considerably [40]. Today, several applications can be identified in different fields of economic activity, which merits a discussion on the cost-effective use of drones. The literature is scarce, but some applications have been examined. For example, Sudbury and Hutchinson [41] calculated the cost based on the flight duration and conclude that Amazon's drone delivery of their packages is economically feasible [41]. Applications in health with the delivery of medical supplies are an essential need. Wright et al. [42] report that depending on geography and cargo characteristics, drone delivery could be a viable solution. Ochieng et al. [43] found that delivery with a motorcycle is more effective than drone delivery. However, as the delivery distance increases, drone systems become more effective than motorcycle delivery. Delivery with drones in hard-to-reach or inaccessible areas, like dense forests, may not yet be cost-effective. Still, technological improvements are expected to make this application cost-effective soon [44]. Sozzi et al. [45] compared the costs of fauna photos for vegetation indices taken by UAVs, airplanes, and satellites; they concluded that the cost depends on the analysis of the photos and the chosen platform. Finally, they concluded that drones have higher costs but could be more efficient than satellites or planes for taking high-resolution images in agriculture.

Yowtak et al. [46] calculated the delivery costs for grocery transport by comparing three different types of delivery (drones, engine vehicles, and battery vehicles). They concluded that each type of delivery has advantages and disadvantages. The economic cost comparison of the three types of delivery method found that UAVs still need to be more efficient. Christensen [47] employed a Monte Carlo simulation to estimate the cost and benefits of different scenarios of drone involvement in fire management. He concluded that there is a potential cost effectiveness of drone involvement in fire management compared to conventional control using helicopters. Zailani et al. [48] observed that using drones has high potential in blood product transportation. They found that drone transportation costs more than an ambulance, but they believe it is the best choice for developing nations. Borghetti et al. [49] compared vans, bicycles, scooters, and UAVs for last-mile delivery; they reported that UAVs could be the best choice for last-mile delivery if the package is small and light. However, UAVs experience some limitations, mainly derived from the restrictions posed by regulations. Finally, White et al. [50] estimated the cost and benefits of coastal surveying, comparing three alternatives: UAVs, manned aircraft, and walkovers. They concluded that the drone surveys were the most expensive method but faster than the other two, while walkover had the highest personnel cost.

Valerdi [51] provided cost metrics and a model based on weight and purchasing price for UAVs, developing a parametric cost model that relies on weight and endurance with the cost reference being pound/hour per thousand dollars. The authors of this study report limitations such as a lack of data availability. Malone et al. [52] estimated the ownership cost for UAV systems, including fixed and variable costs based, among other factors, on endurance, speed, altitude, payload, software design, and training of operators.

Banazadeh and Jafari [53] developed a framework to estimate the costs of aerospace systems. Then, they described a method as a case study scenario for estimating the cost of unmanned aerial vehicles. They used three indexes: "(1) acquisition cost; (2) acquisition cost divided by maximum takeoff weight; and (3) acquisition cost divided by empty weight" and concluded that this technique is better compared to others.

The above examples show that the literature on drone costs is scarce and not uniformly reported, as the technology is relatively new and expanding in scope. We see that drone technology is improving, which can reduce flight time costs. It is, therefore, essential to monitor parameters affecting costs related to UAV technology to give users an economic dimension for using this technology.

3. Materials and Methods

This section presents the methodological approach we followed for estimating the cost of drone use for monitoring and surveillance purposes. Due to data availability limitations, we selected two vehicles with access to some primary cost data. The first is the DJI Phantom 4 Pro, a slightly rotating wind semi-professional vehicle, while the second is the Thunder-B, a fixed-wing professional vehicle. These two types were selected because law enforcement in Greece has used them, so our estimations apply to real case applications and can benefit users. Apart from the manual reference data, we benefited from any information from personal contacts and press releases, but obtaining cost data from manufacturers was challenging. When necessary, we made assumptions and discussed them as follows: First, our cost analysis considered ideal weather conditions and did not include potential accidents or wearing equipment for longer assumed flights. Ideally, cost calculations will consider, among other factors, takeoff time, landing time, overlap time, hovering time, wind, precipitation, and other weather-related delays. Our analyses were based on hovering time, as the other times mentioned consist of a very small portion of a flight trip. Finally, we presented a case study in the Xanthi region. We selected a 4 km distance starting from the old town of Xanthi, crossing the Kosinthos River, a peri-urban grove of trees, and ending in the nearby village of Kimmeria. The peri-urban grove of trees and river constitute a small sample of flora and fauna. The surveillance time was assumed to last 2 hours (h) per day. The surveillance aimed to identify potential perpetrators of littering, setting fire to the grove, and polluting the Kosinthos water body via illegal discharges or waste dumping.

3.1. Vehicle Description and Characteristics

Phantom 4 Pro weighs 1375 gr using a 6000 mAh LiPo battery for operation and propellers for flight with 350 mm diagonal size. The battery flight capacity is about 30 min with 1.5 h of charging time. Its maximum flight time is 30 min, and its maximum wind speed resistance is 10 m/s, without rain. It has a maximum ascent speed of 5 m/s and a maximum descent speed of 4 m/s.

The purchase cost is EUR 1699. A battery replacement costs EUR 189, and a thermal camera costs about EUR 2149. The charger costs EUR 99, and the standard annual service from the mother company is EUR 169 excluding the cost of damaged spare parts [54], while the software is provided for free by the company. The lifetime of batteries varies from 300 to 500 cycles [55]. We will consider 400 charges as an average battery life in our estimations. Thunder-B is a "small tactical unmanned aerial vehicle (UAV) developed by Israeli company BlueBird Aero Systems," which is also used by the Greek police. It weighs 32 kilograms (kg), and its flight range is 150 kilometers (km). The maximum flight altitude is 16,000 feet (ft) (4870 m), while it can remain in the air for 24 h with one fuel tank or 12 h depending on the payload, speed, and weather conditions. It has a tank of 12 liters (L), and its maximum flight speed is 32–72 knots (kt) (16–37 m/s or 60–137 km/h). Also, it can fly with an airspeed of up to 45 knots (kt) and in rain of up to 10 millimeters/hour (mm/h). The Thunder-B requires a mobile system involving hardware and sensors at different costs, depending on extras, from USD 100,000 to USD 200,000. Based on press releases, the cost of purchasing a Thunder-B system (drone, software, and hardware) is EUR 200,000. However, obtaining written cost data has not been possible, despite our efforts, for this vehicle. There should be one annual service as reported by the manufacturer.

The Phantom 4 Pro can be operated by one operator, while the Thunder-B needs two operators. The operator's cost is calculated according to the average salary of police officers with 10 to 15 years in service, estimated at EUR 20,820 per year [56]. Table 1 summarizes the essential characteristics of the two studied vehicles.

Table 1. Characteristics of Phantom 4 Pro and Thunder-B drones.

UAV Characteristics	Phantom 4 Pro	Thunder-B
Wingspan	350 mm	4 m
Weight	1.375 kg	32 kg
Maximum speed	S-mode: 45 mph (72 kph) P-mode: 31 mph (50 kph)	137 kph Cruise speed 80 kph
Flight range	5 km	150 km
Endurance	30 min	up to 24 h/12 h with cargo capsules/vtol
Operating altitude		1820 m/6000 ft
Maximum altitude	19,685 ft/6000 m	4870 m/16,000 ft
Temperature range	0–40 °C	
Covert operation		Aprox. 500 m
Cost	EUR 1699	EUR 100,000–200,000
Fuel source	-	12 lt
Payload		up to 4 kg
Wind speed resistance	10 m/s	
Airspeed	10 m/s	60–137 kmh/32–72 knots
Battery	6000 mAh LiPo	-
Severe weather operation	Without rain and in winds of up to 10 m/s	In winds of up to 45 knots and rain of up to 10 mm/h

3.2. Cost Calculation Methodology

In this section, we detail how to estimate the cost of flight time of the two vehicles, based on the unit cost. These costs will be calculated based on the Total Annual Economic Cost (TAEC) formula:

$$\text{TAEC} = (C_c \times \text{CRF}) + C_a \tag{1}$$

The Capital Recovery Factor (CRF) is given by:

$$\text{CRF} = \frac{r(1+r)^t}{(1+r)^t - 1} \tag{2}$$

where C_c is the capital/purchasing cost
C_a is the annual operation and maintenance cost
t are the years of operation and
r the opportunity cost of capital (*OCC*).

By calculating the annual equivalent cost, we can refer to a unit for levelized incurred costs; that is, per kilometer or hour. Also, we considered the lifetime of a thermal camera and a charger to be 5 years, alongside the lifetime of batteries of 400 charge cycles. Each battery-charging kilowatt hour (kWh) cost is calculated according to the Greek electricity market based on March 2021 [57]. The price of fuels is EUR 1.575 based on prices in March

2021 when we ran the analysis. We consider that the speed of the drones is about the average speed reported in their manuals. Also, this study assumes that for Thunder-B, the service cost is 2.5% of the purchase cost, as no data were provided. Furthermore, we considered the number of operators that are necessary for surveillance flights. The law enforcement officer's wage was considered to be the same for the two drones regardless of the level of operating risk or the characteristics of each drone. Finally, the basic software cost was not calculated separately because the software for the Phantom 4 Pro is free (open), and that for the Thunder-B drone is included in the purchase cost. Thus, any software or future development cost cannot be separated from the purchase or maintenance costs we have considered.

3.3. Unit Cost Calculation

In this section, we describe the basic cost elements and any assumptions considered to compare the cost of flight time of the two vehicles based on the unit cost. These costs were calculated based on the TAEC provided by Formulas 1 and 2. We considered the lifetime of the drones to be 5 and 10 years for the Phantom 4 Pro and Thunder-B, respectively, with an OCC of 5%. We also considered the Phantom 4 Pro and Thunder-B drones' 5 km of surveillance for different flight times. This study assumes that the service cost for the Thunder-B drone is 2.5% of the purchase cost. We assume this drone can fly with one full tank for 16 h (out of the 24 h reference value). Thus, this drone needs 0.75 L of fuel per hour. The price of fuel is EUR 1.575. For the Phantom 4 Pro, one operator is necessary for half an hour of surveillance. Therefore, two operators with two vehicles are needed for surveillance flights from 1 to 8 h. The second operator will fly the second drone as soon as the first lands to change the battery, so the surveillance is uninterrupted, and vice versa. For a 10 to 16 h period, we need four operators and two vehicles, due to the second shift involved. On the other hand, a Thunder-B drone needs two operators for surveillance periods between 0.5 and 8 h and four operators (a second shift) for 8 to 16 h periods.

4. Results

This section presents analytical cost estimations for a specific 4 h trip, flight costs for different distance levels, and flight cost per hour. With this case scenario, we assume that the aggregated annual surveillance period is 1460 h if a 4h flight for each day of the year is necessary. We then calculated the cost per kilometer of flight, considering the vehicle's travel distance capacity per time. This is an analytical approach intended for budgeting law enforcement applications. Finally, we provide a case study example based on our findings.

4.1. Cost Estimations for 4 h Flight

Considering a typical flight surveillance period of 4 h, the flight cost, with the best possible details for each of the two drone types selected, was calculated. The different needs of the necessary units for achieving the planned time flights are presented in Appendix A. The required number of units for the 4-h case, we selected to present, are shown in Table 2.

4.1.1. Cost Estimations for Phantom 4 Pro

The purchase cost for two Phantom 4 Pro drones is EUR 3398 with an estimated economic life of 5 years. The necessary cost for equipping the vehicles with two thermal cameras is EUR 4298. The cost of the batteries is EUR 1512. We need eight batteries for 4 h flight surveillance per year. After this load, they will be worn out and will have to be replaced. The cost of three chargers is EUR 297. As discussed in Sections 3.2 and 3.3, all this equipment is necessary for 4 h uninterrupted operation. The equivalent operation and maintenance (O&M) costs of this equipment are EUR 338 for service, EUR 39.42 for energy, and EUR 41,640 for the two operators. Table 3 shows the calculations of annuitized capital and annual O&M costs, which result in a TAEC of EUR 45,451.20.

Table 2. Individual number (no) of units considered for the uninterrupted 4 h surveillance period, per vehicle.

Units	Phantom 4 Pro	Thunder-B
Drones (no)		
Camera (no)		
Personnel (no)		
Fuel (L)	N/A	
Batteries (no)		N/A
Charger (no)		N/A

Table 3. Total annual cost estimations for Phantom 4 Pro.

Costs	Units	Cost per Unit (EUR)	Cost (EUR)	*t* (yr)	CRF	Annual Cost (EUR)
Vehicle	2	1699	3398	5	0.231	784.85
Thermal camera	2	2149	4298	5	0.231	992.73
Battery	8	189	1512	1	1.050	1587.60
Charger	3	99	297	5	0.231	68.60
Sum of capital						3433.78
Basic service						338
Energy						39.42
Operator						41,640
Sum of O&M						42,017.42
TAEC						45,451.20

The DJI Phantom 4 Pro has a flight range of 5 km. So, the total covered kilometers are 7300 km [surveillance (1460 h), $*$ range of flight (5 km)]. Therefore, to calculate the total cost per hour, we divided the total cost by the total operating hours. The above calculation gives 45,451.20/1460 = EUR 31.13/h.

4.1.2. Cost Estimation for Thunder-B

We consider the Thunder-B drone system to have a purchase cost of EUR 200,000, with camera and software costs embedded. Based on our research methodology, we calculated the CRF and TAEC the same way as with the Phantom 4 Pro. The Thunder-B drone's lifetime is estimated to be ten years, with two operators needed per flight. The annual service cost is estimated to be EUR 5000. The total annual cost was calculated to be EUR 25,900.91. We consider that a fuel tank is sufficient for a 16 h flight. This type of drone needs 0.75 L of fuel per hour. So, the annual fuel cost is EUR 1724.63 for 4 h flights. The cost for the four operators is the same as for the operators of the Phantom 4 Pro, equaling EUR 41,640. The annual cost is EUR 74,265.54 (Table 4).

Table 4. Total annual cost estimations for Thunder-B.

Costs	Units	Cost per Unit (EUR)	Cost (EUR)	t (yr)	CRF	Annual Cost (EUR)
Vehicle	1	200,000	200,000	10	0.1295	25,900.91
SUM						25,900.91
O&M						
Basic service						5000
Fuels						1724.63
Operator						41,640
SUM						48,364.63
TAEC						74,265.54

The cost per hour and km are given if we divide the total annual cost by the total covered km and the total functional hours. With this case scenario, we concluded that the total monthly monitoring hours are 1460. The Thunder-B drone has a range of 50 km flights. So, the total covered kilometers are 73,000 km [surveillance (1460 h), * range of flight (50 km)]. To calculate the total cost for the user per hour, we divide the total cost by operating hours to calculate the unit cost per hour. This gives 74,265.54/1460 = EUR 50.87/h.

4.2. Cost Estimations per Flight Duration

Based on the analytical estimations of the previous section, we proceeded with further calculations per flight duration from 0.5 to 16 h of surveillance per day for the Thunder-B drone (Figure 1a) and from 0.5 to 8 h of surveillance per day for the Phantom 4 Pro drone (Figure 1b). Therefore, the total annual flight times equaled 182.5, 2920, and 5840 annual hours for 0.5, 8, and 16 surveillance hours per day, respectively. Economies of scale are shown graphically and from the estimated equations of Cost = $a \cdot x^b$, where x is flight time or distance covered. The "b" coefficient was estimated with a negative sign for both equations due to the aforementioned economies of scale. All calculations followed the fitted equations except for the 0.5 h surveillance period with the Phantom 4 Pro drone, as observed in Figure 1. We believe that a 0.5 h flight is a short period compared to the analysis we performed, and it is calculated separately; thus, it does not contribute to the regression of Figure 1. If we take, for example, the cost of surveillance for a 1 h trip, we see that the cost drops from about EUR 200/h to about EUR 100/h if the trip lasts 2 h for the Thunder-B.

Similarly, the cost of surveillance for a one-hour trip and two two-hour trips is about EUR 122 and EUR 62, respectively, for the Phantom 4 Pro drone. However, it is evident that the Phantom 4 Pro costs less than the Thunder-B for all calculated times. Figure 1 shows an excellent rapid cost estimation for different surveillance trips.

In Figure 2, we calculated the cost of surveillance per km based on the drones' typical average speeds. We calculated the cost per km for the Thunder-B drone when the trip lasts 50 km (Figure 2a). On the other hand, we calculated the cost per km for the Phantom 4 Pro drone when the trip lasts 5 km (Figure 2b). All points fit the equations Cost = $a \cdot x^b$, with similar evidence of economies of scale. In this case, the cost per kilometer for the Phantom 4 Pro drone is higher than that for the Thunder-B drone at all points. For example, 4 h of surveillance will cost EUR 1.02 per kilometer for a Thunder-B drone, while a Phantom 4 Pro drone will cost EUR 6.23 per kilometer. Although the two vehicles may be used for different applications, we present these values as an example of comparison and how this methodology can help estimate service costs.

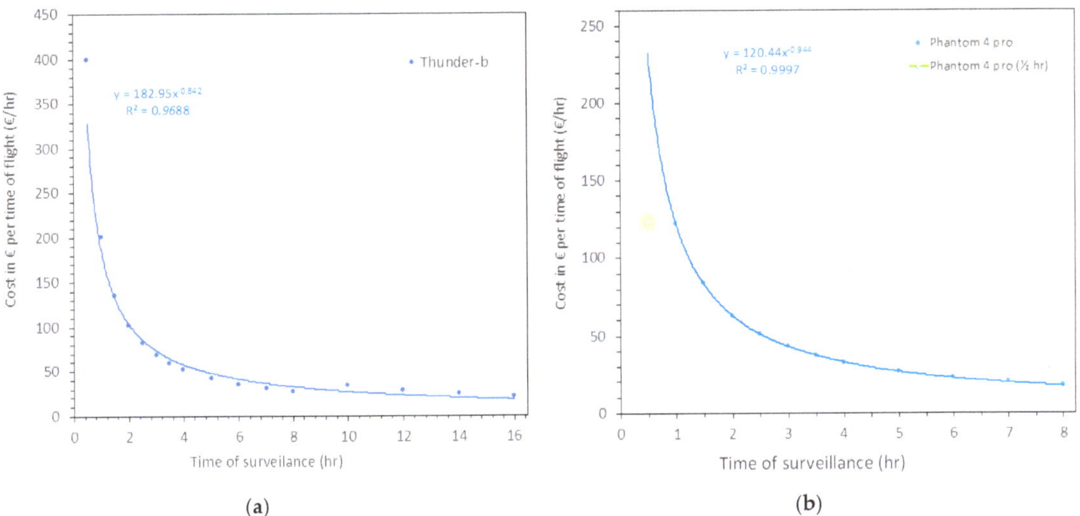

Figure 1. Cost per time-of-flight (**a**): Thunder-B drone and (**b**): Phantom 4 Pro drone.

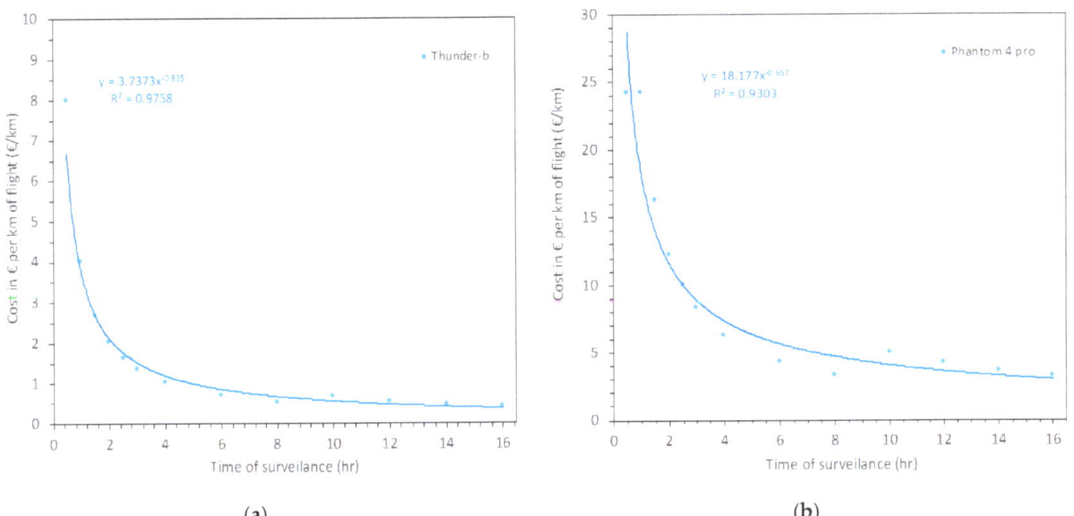

Figure 2. Cost per km of flight for (**a**): Thunder-B drone (50 km surveillance) and (**b**): Phantom 4 Pro drone (5 km surveillance).

4.3. Case Study

In our case study of surveillance by the enforcement authorities, we selected a small area in Xanthi, city, located in Greece. We used the two drones analyzed in the previous section. The 4 km surveillance area is shown in Figure 3, as an abstract from google maps. The line shows the distance, starting from the old town of Xanthi, crossing the Kosinthos River, a peri-urban grove of trees, and ending in the village of Kimmeria. The peri-urban grove of trees and river constitute a small sample of flora and fauna of the region.

Figure 3. Surveillance area, from the old town of Xanthi to Kimmeria village.

The surveillance time lasted for 2 h per day of scoping to ensure compliance with environmental legislation in this area. The surveillance aimed to provide identification of perpetrators of littering, setting fire to the grove, and polluting the Kosinthos water body.

In this case study, the distance was short, and we needed two Phantom 4 Pro drones with four batteries for two hours of uninterrupted surveillance. One battery lasts 30 min and needs charging for 1.5 h. The maximum speed of the Phantom 4 Pro is 50 km/h. If we fly the Phantom 4 Pro with an average speed of 25 km/h, it will cover the distance of 4 km in 9.6 min. So, the Phantom 4 Pro drone surveillance will be able to cover the distance two and a half times with one battery charge. The journey is short and the surveillance will be almost continuous. Therefore, for two hours of surveillance in this area, four Phantom 4 Pro batteries per day, two vehicles, two operators, two thermal cameras, and three chargers are necessary to retain the flying capacity. The summary of employing the Phantom 4 Pro drone for this case study is presented in Figure 4. We also state the limitations of our assumptions; that is, that there will be surveillance from a specified height, no significant winds or precipitations, and no other risks (i.e., operating risk), which would inevitably affect costs.

On the other hand, we needed one Thunder-B drone for two hours of surveillance. Also, a 1.5 L tank and two operators are necessary. The maximum speed of the Thunder-B drone is 137 km/h. In this case, the drone can fly at a recommended cruise speed of 80 km/h (note that 66 km/h speed is the minimum possible cruise speed). Under these circumstances, it will cover a distance of 4 km in 3.6 min. So, using the Thunder-B drone for will cover the distance faster than the Phantom 4 Pro, but the time period is also short, so the surveillance is continuous. The summary of employing the Thunder-B drone for this case study is presented in Figure 5.

Using the functions from Section 3.2, we assume the costs for one-year surveillance. In this case, the cost of the Phantom 4 Pro per hour is EUR 61.17 and per kilometer is EUR 15.29, while the cost of the Thunder-B drone per hour is EUR 100.55 and per kilometer is EUR 25.14.

This approach has some limitations. The flight will occur at the same altitude, assuming no operating or personal risk. We accepted that flights would occur in clear weather without rain and wind. The software cost was not calculated separately as discussed, and we assumed no critical accidents would occur.

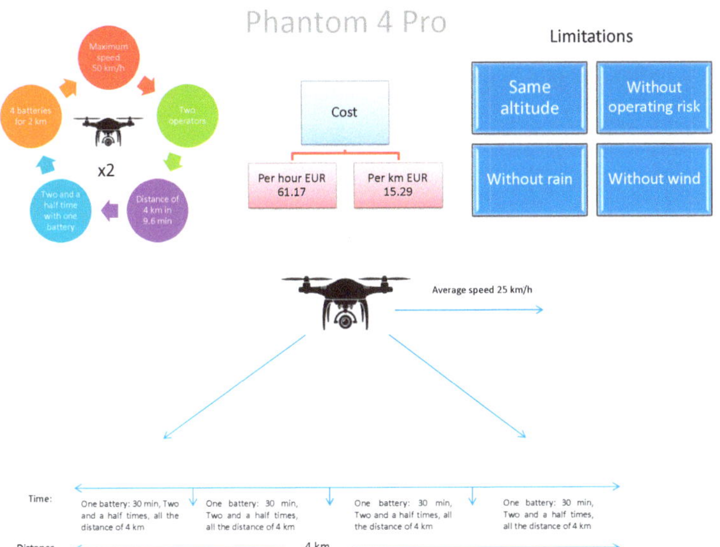

Figure 4. Surveillance with Phantom 4 Pro.

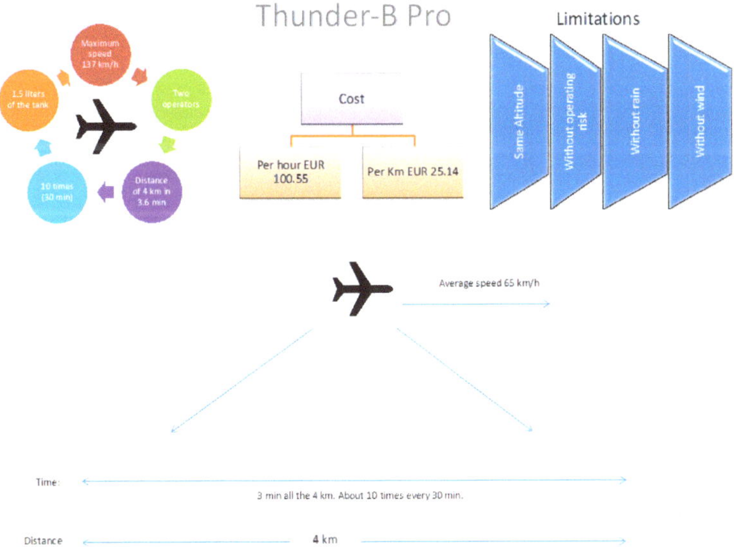

Figure 5. Surveillance with Thunder-B drone.

As noticed, the cost of the two drones is quite different. The cost of the Phantom 4 Pro is lower than that of the Thunder-B, while surveillance is continuous for both vehicles. We are thus able to select the first drone because it has a lower cost (Figure 6). The cost is reasonable compared to traditional methods of monitoring; for instance, with patrol cars. Furthermore, when observation is continuous, violation of the law can be minimized, and law enforcement officers using drones are more efficient. On the other hand, law enforcement officers can save time and operate in a safer working environment, and perpetrator identification is faster. Finally, in the case of a drone accident, the damage is limited to equipment and not to human lives.

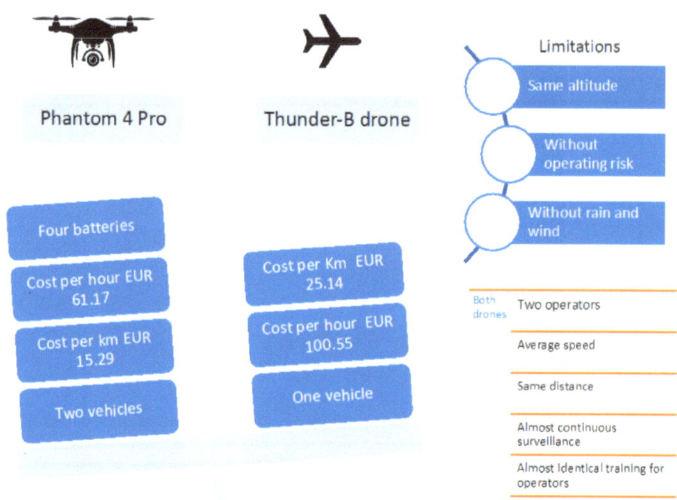

Figure 6. Comparison between Phantom 4 Pro and Thunder-B drones.

5. Discussion

In this work, we compared and levelized costs for two drones, the Phantom 4 Pro and Thunder-B, used for surveillance. The calculated cost for these drones concerns their use and not the total ownership. Valerdi [51] proposed that cost depends on weight and endurance calculated on pound/hour per thousand dollars. On the other hand, Malone et al. [52] described an estimation of total ownership cost for UAV systems. In this work, we calculated the unit cost of drones to be able to compare them.

We analyzed the operating cost for each drone's different time durations and distances. We observed that if the surveillance hours increased, the operating cost for these drones decreased due to economies of scale. The cost per kilometer constantly decreased if the km covered increased. On the other hand, we saw that the operating cost of the Phantom 4 Pro drone was lower than that of the Thunder-B drone. As shown in Figure 1, the cost of one or two hours of surveillance for a Phantom 4 Pro drone is lower than that for a Thunder-B drone. For Phantom 4 Pro, we conducted our analysis used batteries that last for half an hour and then should be replaced once the drone has landed.

Furthermore, there are additional restrictions on adverse weather conditions. The Phantom 4 Pro cannot fly in rain, snow, or wind. In contrast, Thunder-B drones can operate in rainy and windy conditions. Combining the covered distance with a fuel tank leads us to conclude that this drone covers more are than the smaller Phantom 4 Pro drone. As revealed from the cost analysis, the Thunder-B drone is cost-effective for large uninterrupted missions for up to 16 h of surveillance.

As revealed from the case study, for 2 h of constant surveillance in an area of 4 km distance with the examined drones, the cost with Phantom 4 Pro was EUR 61.17/h or EUR 15.29/km, while with the Thunder-B drone, the equivalent costs were EUR 100.55/h and EUR 25.14/km, respectively. The analysis we performed for this case study also has some limitations. We considered the same altitude, average speed, no operating risks or accidents, and ideal environmental conditions. The average speed of these drones differs based on the reference range. The cost is lower for the Phantom 4 Pro than for the Thunder-B, so we selected the first one based only on cost figures (Figure 6).

Employing drones for surveillance can add an extra cost for law enforcement agencies. Still, this case study shows that the cost is reasonable and can be more efficient than conventional methods. We assumed continuous surveillance without operational accidents or risks. On the other hand, law enforcement officers can save time via faster and more efficient drone monitoring in place of conventional patrolling. Real-time data collection is a significant aid for law enforcement officers in many sections, such as perpetrator identification, recognition of missing people, environmental monitoring in real time, aviation environmental violations, prevention of air pollution, or water rescues and disaster response. Drones are increasingly applied in various fields for law enforcement officers. So, their working efficiency is high in terms of safety and surveillance. The working environment is also safer because drone use decreases potential personnel risks. For example, if a drone crashes, the cost is lower than that of a human-crewed helicopter, where the pilot's life is the highest protected good in all societies. UAV technology helps to fight crime in a cost-effective manner. Law enforcement policymakers should include aerial surveillance in crime prevention programs. Indeed, there are also issues about the use of drones [40], especially concerns about the violation of human rights, which will not be analyzed here but only mentioned as there is an ongoing debate.

Data availability is a limitation of this work. We also recognize some further limitations on assumptions made on operating cost, altitude, average speed, environmental conditions, personal risk, and environmental impact, as the operating costs depend on these factors. Significantly, the cost will increase if the altitude of the drone flight is higher than calculated because the drone will need more energy. In the same way, if the speed is increased or the wind is strong, the drone requires more power for the same flight. Also, accidents will increase costs because if drones crash or land accidentally, further repair work and spare parts are necessary. Additionally, the cost will be increased if the environment is challenging (e.g., complex landscapes, rivers, or topography).

Moreover, the Phantom 4 Pro battery lifetime is considered to be 400 cycles. The basic software cost was not calculated separately because the software for Phantom 4 Pro is free (open) and the software for the Thunder-B drone is included in the purchase cost. Thus, any software or future development cost cannot be separated from the purchase or maintenance costs we have considered. Furthermore, there are no available real data for costs related to cyber security, navigation, software, and the impact of accidents; we excluded it to ensure our calculations were clear and feasible. Additionally, a law enforcement officer's wage was considered the same for the two drones regardless of the level of operating risk or the characteristics of each drone, which is the case in the region of our case study. Note that by this, we do not mean to oversimplify our approach, but to provide a practical and integrated way to obtain reasonable cost estimations and comparisons. Nevertheless, any assumptions are based on the available data and experience since no previous studies have analyzed the cost of using drones. Most researchers analyze the unit cost from the perspective of construction and ownership purchase rather than from their use. Cost values based on the data provided by the companies should be further validated with statistical and empirical findings following the extended use of the studied models. Also, the fast improvement of UAV technology itself may render the concluded costs outdated soon. Nevertheless, valid economic conclusions can be drawn by applying the same principles in this work.

6. Conclusions

The use of drones by law enforcement has been increasing in recent years. Some law enforcement applications need to be constantly observed, while others are partial. Economic analysis is essential for all purposes, as well as for security forces to be able to budget surveillance duties in advance. We describe a methodology that can be used for calculating levelized costs per flight time or distance covered and produce equations for further calculations. These equations show economies of scale in flight time and vehicle size. Economic data are necessary to integrate UAV technology into operational activities for entrepreneurial and security protection. Our case study shows that drones can be cost-effective for law enforcement monitoring. As an example, we showed how to estimate the monitoring cost. Last but not least, policymakers could include UAV technology in crime prevention programs.

To our knowledge, this study is the first economic analysis of drone use employing real data on a life-cycle cost basis. Suggestions for future works include keeping detailed records of all fixed and variable costs to produce more informative equations for more vehicles. Also, flight records under different weather conditions can add more precision to the estimations. Another suggestion that will pay off is training users and creating a repository of cost data from different vehicles. Finally, further research should involve the comparison of these findings with other conventional means of surveillance.

Author Contributions: N.T. conceived the idea; N.T. collected and decoded the data; N.T. and K.P.T. performed the analysis; N.T., L.E., and K.P.T. wrote the paper. All authors have read and agreed to the published version of the manuscript.

Funding: This research received no external funding.

Data Availability Statement: Available upon request.

Conflicts of Interest: The authors declare no conflict of interest.

Appendix A

This appendix provides the necessary unit components for uninterrupted surveillance time flights of 0.5–16 h duration for the Phantom 4 Pro (Table A1) and Thunder-B (Table A2).

Table A1. Necessary units for a continuous flight for Phantom 4 Pro.

Hours		0.5	1	1.5	2	2.5	3	3.5	4	5	6	7	8	10	12	14	16
Batteries per year		2	2	3	4	8	8	8	8	12	12	16	16	20	24	28	32
Drones		1	2	2	2	2	2	2	2	2	2	2	2	2	2	2	2
Personnel		1	2	2	2	2	2	2	2	2	2	2	2	4	4	4	4
Camera		1	2	2	2	2	2	2	2	2	2	2	2	2	2	2	2
Charger		1	1	2	3	3	3	3	3	3	3	3	3	3	3	3	3

Table A2. Necessary units for a continuous flight for Thunder-B.

		Hours											
Units		0.5	1	1.5	2	2.5	3	4	8	10	12	14	16
Fuel	= 1 L	0.375	0.75	1.125	1.5	1.875	2.25	3	6	7.5	9	10.5	16.2
Drone		1	1	1	1	1	1	1	1	1	1	1	1
Personnel		2	2	2	2	2	2	2	2	4	4	4	4

References

1. Merz, M.; Pedro, D.; Skliros, V.; Bergenhem, C.; Himanka, M.; Houge, T.; Matos-Carvalho, J.P.; Lundkvist, H.; Cürüklü, B.; Hamrén, R.; et al. Autonomous UAS-Based Agriculture Applications: General Overview and Relevant European Case Studies. *Drones* **2022**, *6*, 128. [CrossRef]
2. Mohsan, S.A.H.; Khan, M.A.; Noor, F.; Ullah, I.; Alsharif, M.H. Towards the Unmanned Aerial Vehicles (UAVs): A Comprehensive Review. *Drones* **2022**, *6*, 147. [CrossRef]
3. Saranya, C.; Pavithira, L.; Premsai, N.; Lavanya, H.; Govindarajan, R. Recent Trends of Drones in the Field of Defense. *Int. J. Electr. Appl.* **2015**, *1*, 47–55.
4. Klemas, V.V. Coastal and Environmental Remote Sensing from Unmanned Aerial Vehicles: An Overview. *J. Coast. Res.* **2015**, *31*, 1260–1267. [CrossRef]
5. Bossoukpe, M.; Faye, E.; Ndiaye, O.; Diatta, S.; Diatta, O.; Diouf, A.A.; Dendoncker, M.; Assouma, M.H.; Taugourdeau, S. Low-Cost Drones Help Measure Tree Characteristics in the Sahelian Savanna. *J. Arid Environ.* **2021**, *187*, 104449. [CrossRef]
6. Manfreda, S.; McCabe, M.F.; Miller, P.E.; Lucas, R.; Pajuelo Madrigal, V.; Mallinis, G.; Ben Dor, E.; Helman, D.; Estes, L.; Ciraolo, G. On the Use of Unmanned Aerial Systems for Environmental Monitoring. *Remote Sens.* **2018**, *10*, 641. [CrossRef]
7. Lloret, J.; Garcia, M.; Bri, D.; Sendra, S. A Wireless Sensor Network Deployment for Rural and Forest Fire Detection and Verification. *Sensors* **2009**, *9*, 8722–8747. [CrossRef]
8. Lawlor, K.; Olander, L.; Boyd, W.; Niles, J.; Madeira, E. *Addressing the Causes of Tropical Deforestation: Lessons Learned and the Implications for International Forest Carbon Policy*; International Forest Carbon and the Climate Change Challenge Series—Brief Number 5 June; Nicholas Institute for Environmental Policy Solutions, Duke University: Durham, NC, USA, 2009.
9. Balcerzak, A.T.; Jasiuk, B.E.; Fellner, C.A.; Feltynowski, D.M. The Polish Perspective of Using Unmanned Aerial Vehicle Systems in International Firefighting and Crisis Management Missions—Legal and Technological Analysis. In Proceedings of the 2021 International Conference on Unmanned Aircraft Systems (ICUAS), Athens, Greece, 15–18 June 2021; pp. 1478–1487.
10. Herwitz, S.R.; Johnson, L.F.; Dunagan, S.E.; Higgins, R.G.; Sullivan, D.V.; Zheng, J.; Lobitz, B.M.; Leung, J.G.; Gallmeyer, B.A.; Aoyagi, M.; et al. Imaging from an Unmanned Aerial Vehicle: Agricultural Surveillance and Decision Support. *Comput. Electron. Agric.* **2004**, *44*, 49–61. [CrossRef]
11. Han, L.; Yang, G.; Yang, H.; Xu, B.; Li, Z.; Yang, X. Clustering Field-Based Maize Phenotyping of Plant-Height Growth and Canopy Spectral Dynamics Using a UAV Remote-Sensing Approach. *Front. Plant Sci.* **2018**, *9*, 1638. [CrossRef]
12. Shafian, S.; Rajan, N.; Schnell, R.; Bagavathiannan, M.; Valasek, J.; Shi, Y.; Olsenholler, J. Unmanned Aerial Systems-Based Remote Sensing for Monitoring Sorghum Growth and Development. *PLoS ONE* **2018**, *13*, e0196605. [CrossRef]
13. Ronchetti, G.; Mayer, A.; Facchi, A.; Ortuani, B.; Sona, G. Crop Row Detection through UAV Surveys to Optimize On-Farm Irrigation Management. *Remote Sens.* **2020**, *12*, 1967. [CrossRef]
14. Mora, A.; Vemprala, S.; Carrio, A.; Saripalli, S. Flight Performance Assessment of Land Surveying Trajectories for Multiple UAV Platforms. In Proceedings of the 2015 Workshop on Research, Education and Development of Unmanned Aerial Systems (RED-UAS), Cancun, Mexico, 23–25 November 2015; IEEE: Piscataway, NJ, USA, 2015; pp. 1–7. [CrossRef]
15. Albetis, J.; Jacquin, A.; Goulard, M.; Poilvé, H.; Rousseau, J.; Clenet, H.; Dedieu, G.; Duthoit, S. On the Potentiality of UAV Multispectral Imagery to Detect Flavescence Dorée and Grapevine Trunk Diseases. *Remote Sens.* **2019**, *11*, 23. [CrossRef]
16. Riniatsih, I.; Ambariyanto, A.; Yudiati, E.; Redjeki, S.; Hartati, R. Monitoring the Seagrass Ecosystem Using the Unmanned Aerial Vehicle (UAV) in Coastal Water of Jepara. *Proc. IOP Conf. Ser. Earth Environ. Sci.* **2021**, *674*, 012075. [CrossRef]
17. Półka, M.; Ptak, S.; Kuziora, Ł. The Use of UAV's for Search and Rescue Operations. *Procedia Eng.* **2017**, *192*, 748–752. [CrossRef]
18. Griffin, G.F. The Use of Unmanned Aerial Vehicles for Disaster Management. *Geomatica* **2014**, *68*, 265–281. [CrossRef]
19. Zulkipli, M.A.; Tahar, K.N. Multirotor UAV-Based Photogrammetric Mapping for Road Design. *Int. J. Opt.* **2018**, *2018*, 1871058. [CrossRef]
20. Koeva, M.; Muneza, M.; Gevaert, C.; Gerke, M.; Nex, F. Using UAVs for Map Creation and Updating. A Case Study in Rwanda. *Surv. Rev.* **2018**, *50*, 312–325. [CrossRef]
21. Giordan, D.; Adams, M.S.; Aicardi, I.; Alicandro, M.; Allasia, P.; Baldo, M.; De Berardinis, P.; Dominici, D.; Godone, D.; Hobbs, P.; et al. The Use of Unmanned Aerial Vehicles (UAVs) for Engineering Geology Applications. *Bull. Eng. Geol. Environ.* **2020**, *79*, 3437–3481. [CrossRef]
22. Smith, S.L. Drones over the "Black Desert": The Advantages of Rotary-Wing UAVs for Complementing Archaeological Fieldwork in the Hard-to-Access Landscapes of Preservation of North-Eastern Jordan. *Geosciences* **2020**, *10*, 426. [CrossRef]
23. Leuenberger, D.; Haefele, A.; Omanovic, N.; Fengler, M.; Martucci, G.; Calpini, B.; Fuhrer, O.; Rossa, A. Improving High-Impact Numerical Weather Prediction with Lidar and Drone Observations. *Bull. Am. Meteorol. Soc.* **2020**, *101*, E1036–E1051. [CrossRef]
24. Stanton, M.C.; Kalonde, P.; Zembere, K.; Hoek Spaans, R.; Jones, C.M. The Application of Drones for Mosquito Larval Habitat Identification in Rural Environments: A Practical Approach for Malaria Control? *Malar. J.* **2021**, *20*, 244. [CrossRef] [PubMed]
25. Kunovjanek, M.; Wankmüller, C. Containing the COVID-19 Pandemic with Drones—Feasibility of a Drone Enabled Back-up Transport System. *Transp. Policy* **2021**, *106*, 141–152. [CrossRef] [PubMed]

26. Massarelli, C.; Muolo, M.R.; Uricchio, V.F.; Dongiovanni, N.; Palumbo, R. Improving Environmental Monitoring against the Risk from Uncontrolled Abandonment of Waste Containing Asbestos. The DroMEP Project. Geomatics Workbooks n° 12—"FOSS4G Europe Como 2015". Available online: https://www.researchgate.net/profile/Carmine-Massarelli/publication/281811563_Improving_environmental_monitoring_against_the_risk_from_uncontrolled_abandonment_of_waste_containing_asbestos_The_DroMEP_project/links/6124d74b0c2bfa282a6707c4/Improving-environmental-monitoring-against-the-risk-from-uncontrolled-abandonment-of-waste-containing-asbestos-The-DroMEP-project.pdf (accessed on 17 September 2023).
27. Cheteni, P. An Analysis of Anti-Poaching Techniques in Africa: A Case of Rhino Poaching. *Environ. Econ.* **2014**, *5*, 63–70.
28. Boakye, J. Enforcement of Logging Regulations in Ghana: Perspectives of Frontline Regulatory Officers. *For. Policy Econ.* **2020**, *115*, 102138. [CrossRef]
29. Zhou, F.; Pan, S.; Chen, W.; Ni, X.; An, B. Monitoring of Compliance with Fuel Sulfur Content Regulations through Unmanned Aerial Vehicle (UAV) Measurements of Ship Emissions. *Atmos. Meas. Tech.* **2019**, *12*, 6113–6124. [CrossRef]
30. Golcarenarenji, G.; Martinez-Alpiste, I.; Wang, Q.; Alcaraz-Calero, J.M. Efficient Real-Time Human Detection Using Unmanned Aerial Vehicles Optical Imagery. *Int. J. Remote Sens.* **2021**, *42*, 2440–2462. [CrossRef]
31. Rosenfeld, A. Are Drivers Ready for Traffic Enforcement Drones? *Accid. Anal. Prev.* **2019**, *122*, 199–206. [CrossRef]
32. Marin, L.; Krajčíková, K. Deploying Drones in Policing Southern European Borders: Constraints and Challenges for Data Protection and Human Rights. In *Drones and Unmanned Aerial Systems*; Springer: Berlin/Heidelberg, Germany, 2016; pp. 101–127.
33. Daniels, J. (Doug) Drones for Police, Fire & Other Emergency Responders Senior Law Enforcement Training Officer Remote Pilot-In-Command Ohio Peace Officer Training Academy Direct Line: (740) 845-6304. Available online: https://ceas.uc.edu/content/dam/aero/docs/fire/DRONES%20FOR%20POLICE%2C%20FIRE%20%26%20OTHER%20EMERGENCY.pdf (accessed on 30 May 2021).
34. Police Using Scentroid's Drone Environmental Monitoring to Combat Smog. Canada, 22 February 2018. Available online: https://scentroid.com/police-using-scentroid-dr1000-flying-lab-to-combat-smog/ (accessed on 4 February 2021).
35. Pyzynski, M. Cybersecurity of the Unmanned Aircraft System (UAS). In Proceedings of the 2020 International Conference on Unmanned Aircraft Systems (ICUAS), Athens, Greece, 1–4 September 2020; pp. 1265–1269.
36. Bassi, E. From Here to 2023: Civil Drones Operations and the Setting of New Legal Rules for the European Single Sky. *J. Intell. Robot. Syst.* **2020**, *100*, 493–503. [CrossRef]
37. Konert, A.; Dunin, T. A Harmonized European Drone Market?—New EU Rules on Unmanned Aircraft Systems. *Adv. Sci. Technol. Eng. Syst. J.* **2020**, *5*, 93–99. [CrossRef]
38. ΕΛΣΜΕ ΓΕΕΘΑ 20181127 1.2 Δημήτριος Γκριτζάπης «Επιχειρησιακή Δράση ΣμηΕΑ Στην ΕΛ.ΑΣ.». Available online: https://www.slideshare.net/helissme/20181127-12 (accessed on 4 February 2021).
39. Τα Ηλεκτρονικά "Γεράκια" Της ΕΛ.ΑΣ. Available online: https://www.zougla.gr/greece/article/i-enaeri-filakes-kata-tou-eglimatos (accessed on 5 February 2021).
40. Tsiamis, N.; Efthymiou, L.; Tsagarakis, K.P. A Comparative Analysis of the Legislation Evolution for Drone Use in OECD Countries. *Drones* **2019**, *3*, 75. [CrossRef]
41. Sudbury, A.W.; Hutchinson, E.B. A Cost Analysis of Amazon Prime Air (Drone Delivery). *J. Econ. Educ.* **2016**, *16*, 1–12.
42. Wright, C.; Rupani, S.; Nichols, K.; Yasmin, C.; Matiko, M. *What Should You Deliver by Unmanned Aerial Systems?* JSI Research & Training Institute, Inc.: Arlington, VA, USA, 2018.
43. Ochieng, W.O.; Ye, T.; Scheel, C.; Lor, A.; Saindon, J.; Yee, S.L.; Meltzer, M.I.; Kapil, V.; Karem, K. Uncrewed Aircraft Systems versus Motorcycles to Deliver Laboratory Samples in West Africa: A Comparative Economic Study. *Lancet Glob. Health* **2020**, *8*, e143–e151. [CrossRef] [PubMed]
44. Meier, P.; Bergelund, J. *Field-Testing the First Cargo Drone Deliveries in the Amazon Rainforest*; WeRobotics: Geneva, Switzerland, 2017.
45. Sozzi, M.; Kayad, A.; Gobbo, S.; Cogato, A.; Sartori, L.; Marinello, F. Economic Comparison of Satellite, Plane and UAV-Acquired NDVI Images for Site-Specific Nitrogen Application: Observations from Italy. *Agronomy* **2021**, *11*, 2098. [CrossRef]
46. Yowtak, K.; Imiola, J.; Andrews, M.; Cardillo, K.; Skerlos, S. Comparative Life Cycle Assessment of Unmanned Aerial Vehicles, Internal Combustion Engine Vehicles and Battery Electric Vehicles for Grocery Delivery. *Procedia CIRP* **2020**, *90*, 244–250. [CrossRef]
47. Christensen, B.R. Use of UAV or Remotely Piloted Aircraft and Forward-Looking Infrared in Forest, Rural and Wildland Fire Management: Evaluation Using Simple Economic Analysis. *N. Z. J. For. Sci.* **2015**, *45*, 16. [CrossRef]
48. Zailani, M.A.; Azma, R.Z.; Aniza, I.; Rahana, A.R.; Ismail, M.S.; Shahnaz, I.S.; Chan, K.S.; Jamaludin, M.; Mahdy, Z.A. Drone versus Ambulance for Blood Products Transportation: An Economic Evaluation Study. *BMC Health Serv. Res.* **2021**, *21*, 1308. [CrossRef]
49. Borghetti, F.; Caballini, C.; Carboni, A.; Grossato, G.; Maja, R.; Barabino, B. The Use of Drones for Last-Mile Delivery: A Numerical Case Study in Milan, Italy. *Sustainability* **2022**, *14*, 1766. [CrossRef]
50. White, S.M.; Schaefer, M.; Barfield, P.; Cantrell, R.; Watson, G.J. Cost Benefit Analysis of Survey Methods for Assessing Intertidal Sediment Disturbance: A Bait Collection Case Study. *J. Environ. Manag.* **2022**, *306*, 114386. [CrossRef]
51. Valerdi, R. Cost Metrics for Unmanned Aerial Vehicles. In *Infotech@ Aerospace*; Massachusetts Institute of Technology: Cambridge, MA, USA, 2005; p. 7102.
52. Malone, P.; Apgar, H.; Stukes, S.; Sterk, S. Unmanned Aerial Vehicles Unique Cost Estimating Requirements. In Proceedings of the 2013 IEEE Aerospace Conference, Big Sky, MT, USA, 2–9 March 2013; IEEE: Piscataway, NJ, USA, 2013; pp. 1–8.

53. Banazadeh, A.; Jafari, M.H. A Heuristic Complexity-Based Method for Cost Estimation of Aerospace Systems. *Proc. Inst. Mech. Eng. Part G J. Aerosp. Eng.* **2013**, *227*, 1685–1700. [CrossRef]
54. DJI Phantom 4 Pro. Available online: https://www.dji.com/gr/phantom-4-pro/info#specs (accessed on 20 February 2021).
55. What's the Best Battery? Available online: https://batteryuniversity.com/article/whats-the-best-battery (accessed on 20 March 2021).
56. Ανάλυση Του Μισθολόγιου Των Αστυνομικών. Available online: http://policenet.gr/article/%CE%B1%CE%BD%CE%AC%CE%BB%CF%85%CF%83%CE%B7-%CF%84%CE%BF%CF%85-%CE%BC%CE%B9%CF%83%CE%B8%CE%BF%CE%BB%CF%8C%CE%B3%CE%B9%CE%BF%CF%85-%CF%84%CF%89%CE%BD-%CE%B1%CF%83%CF%84%CF%85%CE%BD%CE%BF%CE%BC%CE%B9%CE%BA%CF%8E%CE%BD (accessed on 3 March 2021).
57. Υπολογισμός Κόστους Ηλεκτρικές Ενέργειας Για Κάθε Οικιακή Συσκευή. Available online: https://www.helppost.gr/dei/ypologismos-reuma-katanalosi/ (accessed on 25 March 2021).

MDPI AG
Grosspeteranlage 5
4052 Basel
Switzerland
Tel.: +41 61 683 77 34

Machines Editorial Office
E-mail: machines@mdpi.com
www.mdpi.com/journal/machines